Analytical Dynamics
of Discrete Systems

Reinhardt M. Rosenberg
University of California, Berkeley

Volume 4 in
Mathematical Concepts and Methods
in Science and Engineering

Analytical Dynamics of Discrete Systems, providing seniors and beginning graduate students in engineering and the natural sciences with an advanced textbook in dynamics, examines the development of classical particle mechanics from Newton to Lagrange. Every concept is clearly defined before it is used, every significant result is stated mathematically as well as verbally, and the domain of applicability of each is explicitly stated. Most of the chapters contain a large number of carefully worked out examples, as well as a set of suggested exercises.

The author adopts a geometrical approach to the study of dynamics. He provides an orderly transition from Newtonian to Lagrangean mechanics by demonstrating the need for a basically different classification of forces in these two theories and the necessity of replacing Newton's third law by d'Alembert's principle. In the first seven chapters, he includes detailed reviews of

- Newtonian mechanics, with attention paid to the historical setting in which it developed
- the representation of motion as a trajectory in configuration space, event space, and other spaces
- constraints
- rigid body kinematics and kinetics

The major portion of the book deals with the theory and application of Lagrangean mechanics, beginning with precise definitions of "virtual displacements," "virtual velocity," and "virtual work." The author then discusses the principles of Hamilton and of least action, the theory of contemporaneous and non-contemporaneous variations, the theory of generalized coordinates and forces, and derivations of Lagrange's equations. Special chapters on celestial problems, gyrodynamics, and impulsive motion are included as well.

Analytical Dynamics
of Discrete Systems

MATHEMATICAL CONCEPTS AND METHODS IN SCIENCE AND ENGINEERING

Series Editor: Angelo Miele
Mechanical Engineering and Mathematical Sciences
Rice University, Houston, Texas

Volume 1 **INTRODUCTION TO VECTORS AND TENSORS**
Volume 1: Linear and Multilinear Algebra
Ray M. Bowen and C.-C. Wang

Volume 2 **INTRODUCTION TO VECTORS AND TENSORS**
Volume 2: Vector and Tensor Analysis
Ray M. Bowen and C.-C. Wang

Volume 3 **MULTICRITERIA DECISION MAKING AND DIFFERENTIAL GAMES**
Edited by George Leitmann

Volume 4 **ANALYTICAL DYNAMICS OF DISCRETE SYSTEMS**
Reinhardt M. Rosenberg

Volume 5 **TOPOLOGY AND MAPS**
Taqdir Husain

Volume 6 **REAL AND FUNCTIONAL ANALYSIS**
A. Mukherjea and K. Pothoven

Volume 7 **PRINCIPLES OF OPTIMAL CONTROL THEORY**
R. V. Gamkrelidze

Volume 8 **INTRODUCTION TO THE LAPLACE TRANSFORM**
Peter K. F. Kuhfittig

Volume 9 **MATHEMATICAL LOGIC**
An Introduction to Model Theory
A. H. Lightstone

Analytical Dynamics
of Discrete Systems

Reinhardt M. Rosenberg
University of California, Berkeley

PLENUM PRESS · NEW YORK AND LONDON

Library of Congress Cataloging in Publication Data

Rosenberg, Reinhardt Mathias.
Analytical dynamics of discrete systems.

(Mathematical concepts and methods in science and engineering)
Bibliography: p.
Includes index.
1. Dynamics. I. Title.
QA845.R63 531'.11'01515 77-21894
ISBN 978-1-4684-8320-8 ISBN 978-1-4684-8318-5 (eBook)
DOI 10.1007/978-1-4684-8318-5

Softcover reprint of the hardcover 1st edition 1977

A Division of Plenum Publishing Corporation
227 West 17th Street, New York, N.Y. 10011

If ρ be the distance to O
Mr. Newton said he could show,
That the force of attraction
Behaves like the fraction
Of one over the square of rho.

R.M.R.

Preface

This book is to serve as a text for engineering students at the senior or beginning graduate level in a second course in dynamics. It grew out of many years experience in teaching such a course to senior students in mechanical engineering at the University of California, Berkeley. While temperamentally disinclined to engage in textbook writing, I nevertheless wrote the present volume for the usual reason—I was unable to find a satisfactory English-language text with the content covered in my intermediate course in dynamics.

Originally, I had intended to fit this text very closely to the content of my dynamics course for seniors. However, it soon became apparent that that course reflects too many of my personal idiosyncracies, and perhaps it also covers too little material to form a suitable basis for a general text. Moreover, as the manuscript grew, so did my interest in certain phases of the subject. As a result, this book contains more material than can be studied in one semester or quarter. My own course covers Chapters 1 to 5 (Chapters 1, 2, and 3 lightly) and Chapters 8 to 20 (Chapter 17 lightly). Insofar as the preparation of the student is concerned, the demands are satisfied by present-day methods of teaching mathematics, physics, and mechanics during the first three undergraduate years of an engineering curriculum. Students are expected to have studied kinematics and kinetics in a first course at the sophomore or junior level by the methods now current, and to be familiar with the fundamental principles of Newtonian mechanics and their applications in two and three dimensions. Their preparation in mathematics should include the elements of determinant and matrix theory, the calculus, and a first course in ordinary differential equations, and they must know how to manipulate, multiply, and differentiate vectors. It may be of some slight help to them to be familiar with set-theoretical symbols, but the demands in this respect are so modest that they can easily acquire this familiarity while studying its application.

In my opinion, a first course in dynamics should do more than only impart to the student the techniques needed to solve problems. Similarly, a second course in dynamics should do more than help the student learn new techniques more sophisticated than those he or she knows already; it should also deepen his or her understanding of the fundamentals. And so, a considerable portion of this text is devoted to a new, a longer, and a more penetrating look at a familiar subject—Newtonian mechanics. Not only does this seem to me to be one of the proper functions of a second course in dynamics, but it becomes altogether unavoidable when the transition is made from the Newtonian to the Lagrangean† point of view.

In the review of Newtonian mechanics some attention is paid to the foundations of that discipline, the problem of classical mechanics is defined with some precision, and much care is devoted to the theory of constraints. In all this I have stressed geometric interpretations not only because they appeal to me, but because I have found that they appeal to the student as well.

Rigid body mechanics has been touched lightly, as has motion relative to moving frames, because these subjects are usually discussed in a first course in dynamics. The theory of rotations has been treated as an illustration of orthogonal matrix transformations because, to my knowledge, that theory is almost never included in a first course in mechanics; Poinsot's representation is included for the same reason.

This book is intended for the student unfamiliar with Lagrangean mechanics; the theory and application of that theory forms the major portion new to him.

I regard Lagrangean mechanics not primarily as a mechanical process for producing equations of motion, but as a bold departure from Newtonian viewpoints, as the crowning touch to a development begun by Bernoulli and d'Alembert. Its formulation of the general theory of a constrained dynamical system is a subtle and aesthetically satisfying product. I have attempted to describe it that way.

Almost every chapter contains solved problems illustrating the theory in it. For one thing, I regard the application of theories as an important

† This spelling is phonetically equivalent to the more common "Lagrangian." It reflects my reluctance to mutilate Lagrange's name and was agreed to by the publisher to please me.

learning aid; for another, it is essential that knowledge of a way to solve a problem (or merely one's faith in the possession of this knowledge) not be confused with actually producing the solution.

Every author setting out to write a textbook must make certain decisions with respect to notation. Whatever they are, he is sure not to please everyone. In this respect, his position is perhaps not unlike that of the elected official, judged in a public-opinion poll; some readers will approve, some will disapprove, and some will have no opinion.

In general, I have followed conventional, and perhaps old-fashioned, notation. I have not used the double index summation convention even though it would have resulted in more compact formulas. It seems to me that the added burden placed on the student by its use should be reserved for fields in which most of the quantities dealt with are tensors, and in which tensor transformations form an essential part of the theory. Also, I have not used special symbols to differentiate between a function and the value of a function at a point. Thus, having defined a function f on some domain X, I say that the value of f at x is $f(x)$. On the rare occasions where the distinction is important I write: $f(x_i)$ is the value of f at $x_i \in X$.

Perhaps the only departure in my notation from that commonly used in elementary texts on dynamics is that I do not use bold print to denote a vector, and I use unit vectors sparingly. Thus, an n vector is usually written as $x = (x_1, x_2, \ldots, x_n)$. When unit vectors are useful, I use the symbol $\hat{e}_r$ for the unit vector in the r direction; thus,

$$x = \sum_{i=1}^{n} x_i \hat{e}_i .$$

When the space is $\mathscr{E}^3$ and Cartesian coordinates are used I write $x\hat{\imath}+y\hat{\jmath}+z\hat{k}$ without explanatory phrases.

On some occasions the same symbol is used in different places in the text to denote different quantities. Where this has been done intentionally, some explanatory text has been added pointing to this change. If it has also occurred unintentionally, I apologize in advance, and I would be grateful to readers who call my attention to it.

It is evident that a new text on an old and well-established subject can contain little that is new. This book is no exception; many of the problems treated here are classical, and much of the contents can also be found elsewhere. I hope, nevertheless, that some of the material will be novel to many readers.

In writing this book, I owe a great deal to others. Some of the excellent books which I found of great help are[†]: *An Introduction to the Use of Generalized Coordinates in Mechanics and Physics* by W. E. Byerly, *Classical Mechanics* by H. Goldstein, *Mechanics* by L. D. Landau and E. M. Lifshitz, *Analytical Mechanics* (in Russian) by A. I. Lur'e, *Classical Dynamics* by J. L. Synge, *The Dynamics of Particles* by A. G. Webster, and *A Treatise on the Analytical Mechanics of Particles and Rigid Bodies* by E. T. Whittaker. There are, however, two books without which the present volume could not have been written. One of these is a truly great and an altogether admirable book by L. A. Pars: *A Treatise on Analytical Dynamics*; the other is a very deep book (in German) by G. Hamel entitled *Theoretische Mechanik*.

Many of the examples in this book have originated with one of the sources listed above. Where this is the case, the source has been acknowledged. However, the treatment here is never a direct quotation and it is usually done differently as well as more extensively than in the quoted source.

Undoubtedly, this book owes more to Pars' treatise than to any other source. I have learned much from his careful and clear definitions and from his unambiguous treatment of concepts which emerge from most books as somewhat nebulous and indistinct shapes.

My love of dynamics came initially from a study of Hamel's inspiring book. It is from him that I first learned that the comparison arcs in Hamilton's principle are not, in general, possible paths, and that therefore Hamilton's is not, in general, a variational principle. Hamel had a very thorough understanding of dynamics and he had the rare ability of combining analytical skill with physical insight. Moreover, his book contains one of the most interesting collections of problems to be found anywhere. Many of the problems in this text are due to Hamel. His book is not easily read, his notation is often cumbersome and rarely conventional, and the organization of its content could be improved. Nevertheless, it makes rewarding reading for those who make the effort.

Unfortunately, neither Pars' nor Hamel's book is suitable as an undergraduate text. Pars' treatise is too comprehensive for this purpose (as he says, it gives "a reasonably complete account of the subject [the entire subject of dynamics] as it now stands") and it lacks the problem collection expected in such texts. Hamel's book not only has the disadvantage (to English-speaking students) of being written in German, but it is perhaps

† Books which are frequently referred to are listed in the bibliography. Where only rare references are made to a source, it is given in a footnote.

more suitable for the student who already has an acquaintance with the subject matter.

It gives me great pleasure to acknowledge my debt to many. First and foremost, I want to thank the students who have attended my second course of dynamics; I have learned much from them.

My special thanks go to my colleagues Professors C. S. Hsu and G. Leitmann, and to my teaching assistants Messrs. Wen-Fan Lin and James Casey; they have read the manuscript critically and have made many suggestions which have materially improved it. In particular, Messrs. Lin and Casey have checked all formulas, equations, and examples. Without their devoted effort, this book would contain a myriad of errors which are not now in it.

Finally, I wish to acknowledge with gratitude aid from the National Science Foundation, which has supported my work in the geometry of dynamics, some of which is published here for the first time.

Reinhardt M. Rosenberg

Contents

1

Introduction

In looking back on the history of the development of particle mechanics to its present state of near perfection one can hardly fail to notice that, rather than having evolved in a steady progression, bursts of exciting activity were preceded by more or less barren periods. For instance, few important discoveries in mechanics were made between the lifetimes of Aristotle and Archimedes on the one hand, and those of Kepler and Galileo on the other. However, Galileo's epochal recognition of the importance of acceleration initiated one of the most fruitful periods of discovery, brought to a temporary conclusion in Newton's *Philosophiae naturalis principia mathematica* (1687). The theory of particle mechanics which was, in a sense, concluded by Newton is what we call today *Newtonian particle mechanics.* It comprises that method of dynamical analysis in which the fundamental problem of particle mechanics is formulated by means of Newton's second law, i.e., by the Newtonian equations of motion.

The next epoch of important discoveries is intimately linked with the names of Johann Bernoulli, Euler, d'Alembert, and Lagrange. It was summarized a hundred years after the appearance of Newton's *Principia* in Lagrange's monumental *Mécanique Analytique* (1788), and its subject matter is called today *Lagrangean mechanics.* Here, the formulation of the fundamental problem of particle mechanics requires the notion of the virtual displacement, and it is formulated by Lagrange's equations of motion.

Subsequent developments in the analytical methods[†] of classical

[†] The term "analytical" is not taken to mean "nongeometrical," rather it has here its original meaning of the structure of an entire science, based on a few fundamental principles.

mechanics are in large measure due (among others) to Poisson, Hamilton, Jacobi, and Gauss. While no contemporary treatise summarizing these new ideas was ever written by any of their discoverers, they are nevertheless associated with a single name—that of Hamilton—because, here, the fundamental problem of particle mechanics is formulated by the so-called Hamiltonian canonical equations, and this branch of mechanics is called today *Hamiltonian mechanics.*

It is often said (and in a restricted sense it is true) that one may formulate the problems of particle mechanics by the fundamental postulates of any one of these three theories. Nevertheless, it is inconceivable that the sequence of their discovery could have been different from that of their historical emergence. The reason for this opinion is that the only basic postulate of classical mechanics which can be tested experimentally, i.e., which is accessible to verification in the real world of observable events, is Newton's second law. It is this law on which Newtonian particle mechanics relies. The other two theories require the abstract notion of virtual displacements, i.e., imagined configuration changes not involving time. Clearly, such principles, resting on a thought experiment as they do, cannot be tested experimentally. Moreover, the mathematical apparatus required to deal with these theories increases in sophistication as we ascend the historical sequence of developments of particle mechanics. Many of the required mathematical tools emerged only as they were needed in the solution of dynamics problems. Thus, these mathematical discoveries were often motivated, not merely utilized, by the requirements of dynamical analysis.

In the last hundred and fifty years, nothing has appeared in the development of *classical* particle mechanics that can be compared in fundamental importance to the enunciation of Newton's second law. Nevertheless, these were not idle years. Not only were refinements added to the existing theory but, more important, the emerging trend toward generalization in mathematics permitted much deeper insight into what one may call the "inner connections between the functional relations of dynamical analysis." In consequence, classical particle dynamics is one of the aesthetically most satisfying scientific structures modeling observable events of the real world, and at the same time it is one of the most completely evolved theories. Today, there exists a very complete and internally consistent science of particle mechanics (and of rigid bodies, which are special systems of particles); the practitioner (i.e., the physical scientist concerned with solving specific problems of particle mechanics) has a formidable array of methods and tools available, whether he or she is interested in problems of space flight, variable-mass systems, or problems of impact.

It is natural that a science which is in such a state of near perfection should have stimulated activity toward its axiomatic foundation. The axiomatic method consists in setting forth certain basic statements or axioms about the concepts to be studied, and to deduce from them theorems by the methods of logical inference and deduction.

The famous mathematician David Hilbert once said, "I think that everything that can be an object of scientific thought at all, as soon as it is ripe for the formation of a theory, falls into the lap of the axiomatic method."† If Hilbert's opinion is justified, it is clear that classical particle mechanics is a suitable object for the application of the axiomatic method.

Here, Hilbert echoed Newton's viewpoint on this question, because the structure of Newton's *Principia* was modeled on Euclid's axiomatic *Elements of Geometry*. Both are divided into "books," both begin with "definitions" followed by "axioms," and these are followed in both works by "propositions," which are classified in both either as "theorems" or as "problems." Newton began with eight definitions of such concepts as matter, quantity of motion (i.e., momentum), force, etc., and with the assertion that words like time, space, place, and motion are "well known to all" and are only in need of certain refinements, not of definition. Then follow the three famous *axiomata sive leges motus* (laws of motion); the major portion of the *Principia* is taken up by 193 propositions, of which 111 are theorems and the remaining ones problems.

While Newton attempted an application of the axiomatic method in his presentation of the science of mechanics, one must conclude that by modern standards of logic this attempt was not successful. For instance, it is clear that the second law contains the first. The second law states:

> *The change of motion (in the modern idiom: of momentum) is proportional to the impressed force...*

Thus, it implies that, if no impressed force is present, the motion remains unchanged. But this last statement is the same as that contained in the first law (the law of inertia). Hence, the first law is, in fact, not an axiom but rather a theorem. Moreover, Newton's definition (IV) states:

> *An impressed force is an action exerted upon a body in order to change its state, either of rest or of uniform motion in a straight line.*

Thus, the second law is anticipated by the definition. One sees, then, that

† Quoted from Richard von Mises, *Positivism: A Study in Human Understanding*, Harvard University Press, Cambridge, Massachusetts, 1951.

Newton's axioms are not independent of each other, nor of his definitions. This fault in Newton's axiomatic presentation was basically due to the ideas of logic which existed in his and earlier times and which were not resolved until it was realized that the fundamental concepts are defined *by* the axioms, not independently of them.

With the clarification of the fundamentals of logic (brought about largely by Mach and Hilbert), many post-Newtonian studies of the axiomatic foundations of classical particle mechanics have been made. Those which appeared around the turn of the century are largely connected with the names of Mach, Kirchhoff, Poincaré, Hertz, and, somewhat later, Hamel. More recent contributions to the axiomatization of classical particle mechanics were made by Pendse, Simon, McKinsey, Sugar, Suppes, Bunge, and others.

The science of dynamics is the science of "motion," and motion can be "represented" in different ways. For instance, we might ask: What is the position of every particle of a system when the position of one of them is given? or: What is the velocity of every particle when the position of all of them is known? There are other questions, similar to the above two, that might be asked. The answer to each leads to a different "representation" of the motion. We are, thus, led naturally to the consideration of spaces in which the motion is representable by a point, or by a curve (called a trajectory) traced by a point. The spaces which will be considered here are: the configuration space, the event space, the state space, and the state–time space. It is surprising how much information about the general character of the trajectories in these spaces is deducible from the axioms alone, and without doing any computations.

Perhaps the most useful application of these geometrical representations is to "constraints." These can always be interpreted *by means of* and, in most cases, *as* surfaces or surface elements in these spaces. The notion of constraints must be introduced before the dynamics problem can be formulated. Therefore, we shall classify and examine constraints before carefully formulating the two problems of particle mechanics most frequently encountered.

There is a basic difference between Newtonian problems in which unbounded forces occur and in which they are excluded. It turns out that unbounded forces can be made to fit into the framework of Newtonian mechanics provided their impulse is bounded. Most of this book is devoted to a study of dynamics when the forces are bounded. However, in the last chapter we consider systems in which unbounded forces of bounded impulse act.

This book is aimed at a treatment of Lagrangean mechanics and that branch of mechanics could not exist without the central concept of the "virtual displacement" and the "virtual velocity." Moreover, the transition from the Newtonian to the Lagrangean points of view could not be made without the much maligned and often misunderstood principle of d'Alembert, nor without a careful and unambiguous classification of forces as internal or external, and as given or constraint forces.

There are basic differences between Newtonian and Lagrangean mechanics. Not only is Lagrangean mechanics formulated in terms of generalized coordinates while the Newtonian formulation requires that the coordinates be defined at the outset, but it turns out that there are two quite different types of mechanics problems (even when all forces are bounded): the holonomic and the nonholonomic, and their basically different natures cannot be fully appreciated without the structure of Lagrangean mechanics.

No matter which fundamental principles of mechanics are utilized as the points of departure, the formulation of the dynamics problem either results in, or can be reduced to, a set of differential equations. The desired solution is usually found by integrating these equations. Thus, it is only natural that certain aspects of the theory of integration should be discussed here.

Two applications of dynamical theory are of particular interest. One is the problem of planetary motion, the other is the problem of gyroscopics. The former has done more to advance the science of particle mechanics than any other, and it has a fascinating history besides. The latter has not only furnished a rational explanation of phenomena that seem to the uninitiated to border on the miraculous, but it is one of the more sophisticated applications of dynamical theory, and renewed interest in it has been shown in recent years because of the applications of gyroscopics to space-flight problems. It is for all these reasons that both these topics have been included in special chapters in this second course in dynamics.

2

Dynamical Systems

2.1. Particles

Classical particle mechanics deals with particle motion resulting from forces acting on particles. It is satisfying to have an axiomatic foundation for this discipline. However, the axiomatization of a science is certainly not essential for its comprehension, and so we do not treat this subject in the book.

Here, we want to give some fairly precise descriptions of particles and systems of particles, of Newton's second law in the form

$$\text{Impulse} = \text{Momentum change},$$

and of the quantities occurring in that law. We also want to expose the difference between that law and the more familiar form

$$\text{Force} = \text{Mass} \times \text{Acceleration}.$$

Therefore, we will be looking at a familiar subject, but not in a familiar way.

Because of the economy of space which they afford, we will use some set-theoretical symbols in these descriptions, but their use is not really essential; it is merely convenient. These symbols are defined where they are first introduced.

A particle P_r is a point in three-dimensional Euclidean space $\mathscr{E}^3$ such that each P_r carries with it permanently a "label" m_r called the *mass* of P_r. For this reason, the term "point mass" is frequently used in place of "particle." Every m_r $(r = 1, 2, \ldots, s)$ is a positive, real constant for all

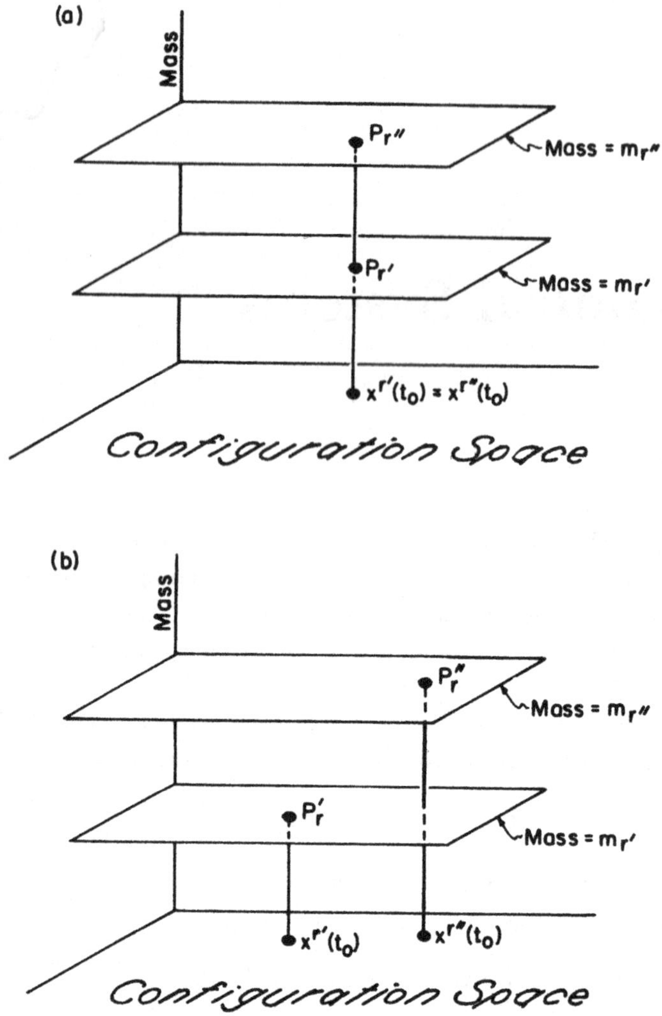

Fig. 2.1.1. Schematic representation of the impenetrability property.

time† $t \in \mathscr{T}$, where $\mathscr{T}$ is defined on the set of real numbers by‡

$$\mathscr{T} = \{t\colon -\infty < t < \infty\}. \tag{2.1.1}$$

The position vector $x^r(t)$ of P_r with respect to some fixed point in $\mathscr{E}^3$ is, for all $t \in \mathscr{T}$ and for every $r = 1, 2, \ldots, s$, a single-valued, continuous vector function of t. This statement is equivalent to saying that every particle

† $t \in \mathscr{T}$ is read: t belongs to the set $\mathscr{T}$.

‡ $\{t\colon -\infty < t < \infty\}$ is read: the set of all t such that t lies in the interval $-\infty < t < \infty$.

occupies one and only one position in space at any given instant of time. In addition, all P_r have the so-called "impenetrability property," which we state as follows: If there exists a single $t_0 \in \mathscr{T}$ at which any two particles $P_{r'}$ and $P_{r''}$ have the same position, e.g., $x^{r'}(t_0) = x^{r''}(t_0)$, then their positions coincide permanently, i.e., $x^{r'}(t) \equiv x^{r''}(t)$ for all $t \in \mathscr{T}$. If there exists a single $t_0 \in \mathscr{T}$ at which any two points $P_{r'}$ and $P_{r''}$ have different positions, e.g., $x^{r'}(t_0) \neq x^{r''}(t_0)$, then their positions never coincide, i.e., $x^{r'}(t) \neq x^{r''}(t)$ for all $t \in \mathscr{T}$. In the first case, the two points are indistinguishable from, and therefore equivalent to, a single particle of mass $m_r = m_{r'} + m_{r''}$. Hence, the impenetrability property is equivalent to the observation that one and only one particle can occupy a given position at any given instant of time. The above definition of impenetrability is illustrated in Figs. 2.1.1(a) and (b).

While the position vectors $x^r(t)$ are everywhere continuous, this is not necessarily true of the velocities. The velocities $dx^r/dt = \dot{x}^r$ $(r = 1, 2, \ldots, s)$ may have (at most) isolated, finite discontinuities at instants $t_j \in \mathscr{T}_j$ where†

$$\mathscr{T}_j = \{t_j : \dot{x}^r(t_j) \text{ not defined for one or more } r \text{ in } r = 1, 2, \ldots, s;\ j = 1, 2, \ldots\}. \tag{2.1.2}$$

2.2. Systems of Particles

To define the meaning of the phrase "system of particles" we proceed basically as follows: We specify a closed surface $\mathscr{S}$ in $\mathscr{E}^3$ called the "boundary." The points within and on the boundary are the "living space." Then the particles in the living space are called the "particles of the system."

Let the *boundary* $\mathscr{S}$ be the spherical surface‡

$$\mathscr{S} = \left\{x : x \in \mathscr{E}^3;\ \sum_{i=1}^{3} x_i^2 = R^2\right\}, \tag{2.2.1}$$

where R is a real constant.

The *living space* $\mathscr{L}(R)$ is the set of points

$$\mathscr{L}(R) = \left\{x : x \in \mathscr{E}^3;\ \sum_{i=1}^{3} x_i^2 \leq R^2\right\}. \tag{2.2.2}$$

† The symbol ; in (2.1.2) is read: moreover.

‡ In more general cases, $\mathscr{S}$ is a closed surface (or several closed surfaces) bounding a finite domain of $\mathscr{E}^3$. The surface $\mathscr{S}$ has been chosen here as spherical to illustrate the idea of the boundary in a simple way.

The set of particles whose position belongs to $\mathscr{L}(R)$ is defined by†

$$\Pi_{\mathscr{L}} = \{P_r[(x^r(t)), m_r]: x^r \in \mathscr{L}(R); r = 1, 2, \ldots, n(t) < \infty\}. \quad (2.2.3)$$

The totality of the members of this set is called the *population* of the living space, or *the particles of the system*, or *the system of particles*. We assume that $\Pi_{\mathscr{L}}$ is not empty, i.e., there is at least one particle in the living space.

The number of particles of the system $n(t)$ may change in time. If it does, particles must necessarily be acquired or lost by the system by passing through the boundary. Dynamical systems (within the compass of Newtonian particle mechanics) which acquire or lose particles by passage through the boundary are called "variable-mass" dynamical systems.

2.3. Forces and Laws of Motion

We postulate the existence of a *Galilean* or *inertial* reference frame $g = (\hat{e}_1, \hat{e}_2, \hat{e}_3)$, where the $\hat{e}_i$ are linearly independent unit vectors‡ immersed in $\mathscr{E}^3$, and the existence of a "universal" or absolute time $\mathscr{T}^*$ such that the velocity vectors $\dot{x}^r(t) = dx^r/dt$ with reference to that frame and time satisfy for all $t \in \mathscr{T}^*$ and for all $x^r(t) = x_1{}^r(t)\hat{e}_1 + x_2{}^r(t)\hat{e}_2 + x_3{}^r(t)\hat{e}_3$ the system of integral equations

$$\dot{x}^r(t) = \frac{1}{m_r}\int_{t_0}^{t} F^r(x^1, x^2, \ldots, x^n; \dot{x}^1, \dot{x}^2, \ldots, \dot{x}^n; \tau)\, d\tau + \dot{x}^r(t_0) \quad (2.3.1)$$

for all $r = 1, 2, \ldots, n$. The $\dot{x}^r(t_0)$ are arbitrary constant vectors of bounded magnitude; they are called *initial velocities*. The function F^r is called the *resultant* of the forces acting on P_r; its time integral over $[t_0, t] = t_0 \leq \tau \leq t$ is called the *impulse* of F^r in that time interval. It follows from the properties of $\dot{x}^r$ that the resultant forces must be integrable with respect to time over any time interval, i.e., over any connected subset of $\mathscr{T}^*$.

Equation (2.3.1) states that the change of linear momentum $m_r[\dot{x}^r(t) - \dot{x}^r(t_0)]$ of every particle P_r during any time interval $[t_0, t]$ is equal to the impulse of the resultant force over the same time interval. One sees that (2.3.1) is not Newton's second law, but rather the integrated form of that

† $P_r[(x^r(t)), m_r]$ is read: The points P_r having position vector $x^r(t)$ and mass m_r.

‡ The vectors $\hat{e}_i$ $(i = 1, 2, \ldots, n)$ are said to be linearly independent if one cannot find n scalar constants λ_i, not all zero, such that $\sum_{i=1}^{n}\lambda_i\hat{e}_i = 0$.

law. We prefer "*impulse equals momentum change*" to "*force equals the product of mass and acceleration*" because we noted earlier that velocities may have isolated, finite discontinuities. When the velocity has a discontinuity, the acceleration is not defined or, at such instants, Newton's second law is not an equation; however, (2.3.1) is valid at these instants.

Particle velocities can be discontinuous only at instants t_j when "impulsive forces" act. We define the set of these instants as[†]

$$\mathscr{T}_j = \left\{ t_j \in \mathscr{T}^* : \exists \lim_{t \to t_j} \int_{t_j}^{t} F^r \, d\tau \neq 0;\ j = 1, 2, \ldots \right\}. \tag{2.3.2}$$

A force F is called "*impulsive*" if the bounded quantity P satisfies

$$P = \lim_{t \to t_j} \int_{t_j}^{t} F(\tau) \, d\tau \neq 0. \tag{2.3.3}$$

But that equation implies that $|F(t_j)| = \infty$, i.e., impulsive forces have infinite magnitude. Such forces are frequently treated in dynamical problems, for instance, when an object is struck a blow, or when it impinges on a rigid surface.

The complement of $\mathscr{T}_j$ in $\mathscr{T}^*$ is denoted by

$$\mathscr{T}^{c*} = \mathscr{T}^* - \mathscr{T}_j. \tag{2.3.4}$$

Therefore, $\mathscr{T}^{c*}$ is the set of instants when the velocities are continuous.

From (2.3.1) one finds directly (by differentiation)

Newton's second law of motion: $m_r \ddot{x}^r(t) = F^r(x^1, x^2, \ldots, x^n; \dot{x}^1, \dot{x}^2, \ldots, \dot{x}^n; t)$ *holds for all* $t \in \mathscr{T}^{c*}$, *for all* $x^r \in \mathscr{E}^3$, *and for all* $r = 1, 2, \ldots, n(t)$. (2.3.5)

When a property holds always except on a set of times of measure zero, it is said, to hold "almost always." Therefore, the law (2.3.5) states that Newton's second law holds almost always.

One also has[‡]:

If and only if $\mathscr{T}_j = \varnothing$, *Newton's second law holds for all* $t \in \mathscr{T}^*$, *for all* $x^r \in \mathscr{E}^3$, *and for all* $r = 1, 2, \ldots, n(t)$. (2.3.6)

[†] $\exists \lim_{t \to t_j}$ is read: The limit $t \to t_j$ exists.

[‡] $\varnothing$ is the empty set.

Expressed differently, when no impulsive forces act, Newton's second law holds always. Because of these theorems, we use the following terminology:

When $\mathscr{E}_j = \varnothing$, the system of particles is called "strictly Newtonian," and it is denoted by (SN). When $\mathscr{E}_j \neq \varnothing$, the system is called "Newtonian" (N). (2.3.7)

2.4. Galilean Transformations

We postulated the existence of a Galilean frame $g = (\hat{e}_1, \hat{e}_2, \hat{e}_3)$ immersed in $\mathscr{E}^3$ in which the axiom of bounded momentum (2.3.1) holds. The set of all reference frames in which (2.3.1) is valid is denoted by

$$\mathscr{G} = \{g\}. \tag{2.4.1}$$

In other terms, we admit the existence of more than one Galilean reference frame, and the totality of these is the set $\mathscr{G}$.

If a transformation $(x^r, t) \to (x^{r\prime}, t')$ from one set of space–time coordinates to another leaves the left-hand side of the axiom of bounded momentum (2.3.1), or of Newton's second law (2.3.5), form-invariant, it is called a *Galilean transformation*. The most general form of such a transformation is

$$\begin{aligned} x^{r\prime} &= at + b + \Phi x^r, \\ t' &= \alpha t + \beta, \end{aligned} \tag{2.4.2}$$

where a and b are arbitrary constant vectors, Φ is an arbitrary constant rotation matrix,[†] and α and β are arbitrary constant scalars.

From the first of equations (2.4.2) one finds by direct differentiation

$$\begin{aligned} \dot{x}^{r\prime}(t) &= a + \Phi \dot{x}^r(t), \\ \ddot{x}^{r\prime}(t) &= \Phi \ddot{x}^r(t). \end{aligned} \tag{2.4.3}$$

It follows from the first of equations (2.4.3) that, when the time is left unchanged ($t = t'$) in a Galilean transformation, the velocity measurements involve only a shift in origin and a scale change. The physical interpretation of such a transformation is that the frame $g' = (\hat{e}_1', \hat{e}_2', \hat{e}_3')$, in which the velocity is $\dot{x}^{r\prime}$, moves with uniform constant velocity (it does not accelerate)

† The theory of rotations is treated in Section 10.3–10.8, where the rotation matrix is denoted by (b_{ij}), and where it is shown that this matrix is nonsingular.

with respect to the frame $g = (\hat{e}_1, \hat{e}_2, \hat{e}_3)$, in which the velocity is $\dot{x}^r$. In addition, a different scale is used in measuring these velocities.

When the transformation involves also a time change in accordance with the second of Eqs. (2.4.2), one finds from that equation

$$\frac{d}{dt'} = \frac{1}{\alpha}\frac{d}{dt},$$

or

$$\begin{aligned} \frac{dx^{r'}}{dt'} &= \frac{a}{\alpha} + \frac{\Phi}{\alpha}\frac{dx^r}{dt}, \\ \frac{d^2x^{r'}}{dt'^2} &= \frac{\Phi}{\alpha^2}\frac{d^2x^r}{dt^2} \end{aligned} \tag{2.4.4}$$

These are identical in form to (2.4.3) because a, α, and Φ are all constant and arbitrary. One concludes that Galilean transformations consist in introducing a new frame of reference and a new time. The measuring scales in the new quantities may differ from those in the old, and the new frame moves with uniform, constant velocity with respect to the old. The new time differs from the old by a change in origin, and a scale change.

A Galilean transformation (2.4.2) does not necessarily leave the equation of bounded momentum or Newton's second law form-invariant. We have already seen how it transforms the acceleration terms of this law. However, it also transforms the remaining terms. In order to leave Newton's second law form-invariant, the transformation must be such that one has after the transformation

$$m_r\ddot{x}^{r'} = F^r(x^{1'}, x^{2'}, \ldots, x^{n'}; \dot{x}^{1'}, \dot{x}^{2'}, \ldots, \dot{x}^{n'}; t')$$

(where dots denote d/dt') if one had originally

$$m_r\ddot{x}^r = F^r(x^1, x^2, \ldots, x^n; \dot{x}^1, \dot{x}^2, \ldots, \dot{x}^n; t)$$

(where dots denote d/dt, and the F^r are the same functions of their respective arguments). A similar set of requirements could have been written for the law of bounded momentum.

When the equations of motion remain form-invariant under a Galilean transformation, the forces are said to possess the property of "*objectivity*." Hence we ask: What must be the form of the forces and/or their arguments such that the forces possess the objectivity property? One way of answering the question is to say that the forces must depend only on "relative displacements" and "relative velocities" and on time. As an example consider a particle in rectilinear motion; assume it is immersed in a fluid and its velocity relative to this fluid produces a force on the particle which is proportional to that relative velocity. Let an observer note the velocity of the particle (relative to himself) to be $\dot{x}$, and that of the fluid to be u. Then the equation of motion is

$$m\ddot{x} = k(\dot{x} - u).$$

If we introduce, for example, the Galilean transformation in which Φ is the identity matrix, $\alpha = 1$, and $\beta = 0$, we find

$$x' = at + b + x,$$
$$t' = t.$$

From the first of these

$$\dot{x}' = a + \dot{x};$$

hence

$$u' = a + u,$$

where u' is the fluid velocity relative to the observer in the new Galilean frame,

$$\ddot{x}' = \ddot{x},$$

and from the second

$$d/dt' = d/dt.$$

Therefore, the transformed equation of motion is

$$\begin{aligned} m\ddot{x}' &= k(a + \dot{x}' - a - u') \\ &= k(\dot{x}' - u'), \end{aligned}$$

and hence is form-invariant.

Another way of illustrating the required properties which ensure objectivity is suggested by Synge (p. 9). He considers a system of four particles (for instance) forming a tetrahedron, as shown in Fig. 2.4.1. Each has a force F_i acting on it, and each has the velocity v_i relative to a a Galilean frame. This configuration at some instant t_0 is shown in Fig. 2.4.1. If the displacement dependence of the force is completely determined by the tetrahedron configuration only, it involves relative positions only. If, moreover, the force and velocity vectors are rigidly attached to the

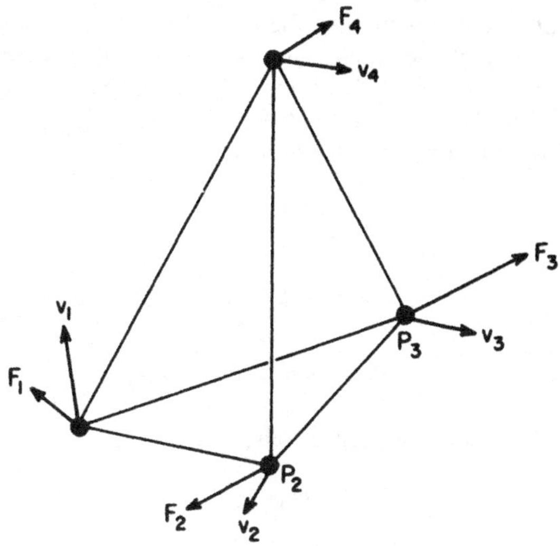

Fig. 2.4.1. Synge's schematic illustrating the objectivity property.

tetrahedron itself, i.e., they translate and rotate with the tetrahedron, then the forces have the objectivity property, or the laws of motion are form-invariant under a Galilean transformation.

2.5. Arguments of the Forces

The forces F^r in the laws of motion (2.3.1), (2.3.3), and (2.3.5) have been considered to depend on relative positions, relative velocities, and time, but *not* on accelerations. In the dynamics of *rigid bodies*, it is sometimes convenient to regard the resultant force acting on the body as a function of the acceleration of that body. This situation occurs for instance when the motion of a ship in water is considered. As the ship moves, it also moves some of the surrounding water, and the force exerted by that water on the hull is the product of its mass and acceleration, the latter being that of the body. The total mass of the ship and of the water carried with it is called by naval architects the "apparent mass." It is, therefore, of interest to inquire whether the resultant force acting on a *particle* can also be a function of the particle's acceleration. Pars (p. 11) has shown by a very simple argument that this cannot be the case in Newtonian particle mechanics if the initial position and velocity and the force acting on a particle determine its future position uniquely for all time.

Example 2.5.1. Consider the rectilinear motion of a single particle on which three experiments are conducted. In the first, the particle is subjected to a force which is a function φ of the acceleration; in the second it is subjected to a force which is a function ψ of the acceleration, and ψ is independent of φ; finally, in the third experiment the particle is subjected to a force of the form $\varphi + \psi$. Thus, in the first case, the force may be written as

$$F_1 = m\varphi(\ddot{x}_1), \tag{a}$$

in the second as

$$F_2 = m\psi(\ddot{x}_2) \tag{b}$$

and in the third as

$$F_3 = m[\varphi(\ddot{x}_3) + \psi(\ddot{x}_3)]. \tag{c}$$

The equations of motion are, respectively,

$$\ddot{x}_1 = \varphi(\ddot{x}_1) \tag{d}$$

$$\ddot{x}_2 = \psi(\ddot{x}_2) \tag{e}$$

$$\ddot{x}_3 = \varphi(\ddot{x}_3) + \psi(\ddot{x}_3). \tag{f}$$

But, in Newtonian mechanics, when two forces act simultaneously on a particle, their effect is the same as that of the action of their vector sum. Thus, we must have

$$\ddot{x}_3 = \ddot{x}_1 + \ddot{x}_2, \tag{g}$$

and the substitution of (g) in (f) gives

$$\ddot{x}_1 + \ddot{x}_2 = \varphi(\ddot{x}_1 + \ddot{x}_2) + \psi(\ddot{x}_1 + \ddot{x}_2). \tag{h}$$

One sees readily that (h) is not consistent with (d) and (e) because the sum of (d) and (e) is

$$\ddot{x}_1 + \ddot{x}_2 = \varphi(\ddot{x}_1) + \psi(\ddot{x}_2). \tag{i}$$

Since φ and ψ are independent, a comparison between (h) and (i) shows that (d) and (e) would be consistent with (h) if and only if one had

$$\begin{aligned} \varphi(\ddot{x}_1) &= \varphi(\ddot{x}_1 + \ddot{x}_2), \\ \psi(\ddot{x}_2) &= \psi(\ddot{x}_1 + \ddot{x}_2). \end{aligned} \tag{j}$$

Thus, the two first experiments are not independent of each other, which contradicts the statement that they are independent.

We conclude that:

In Newtonian particle mechanics, the forces acting on the particles may not, in general, depend on the particle accelerations.

The resultant force acting on the particle P_r is written as

$$F^r = \sum_{\substack{\alpha=1 \\ \alpha\neq r}}^{n} \Theta_\alpha^{\ r} + \sum_{\beta=n+1}^{s} \Phi_\beta^{\ r} + \sum_{j=1}^{A} \Psi_j^{\ r} \qquad (r = 1, 2, \ldots, n). \tag{2.5.1}$$

The forces $\Theta_\alpha^{\ r}$ are of the form

$$\Theta_\alpha^{\ r} = \Theta_\alpha^{\ r}(x^r, x^\alpha) = K(|\, x^r - x^\alpha \,|)\frac{x^r - x^\alpha}{|\, x^r - x^\alpha \,|} \neq 0, \tag{2.5.2}$$

where K is a scalar function of the distance $|\, x^r - x^\alpha \,| \neq 0$ between P_r and P_α; it depends on no other arguments. From (2.5.2) one has

$$\Theta_\alpha^{\ r}(x^r, x^\alpha) = -\Theta_r^{\ \alpha}(x^\alpha, x^r). \tag{2.5.3}$$

The $\Theta_\alpha^{\ r}$ are the interaction forces between the particles of the system, and they cancel in pairs. They are called the *internal forces*. In consequence of (2.5.2) and (2.5.3), they satisfy

$$\sum_{r=1}^{n} \sum_{\substack{\alpha=1 \\ \alpha\neq r}}^{n} \Theta_\alpha^{\ r} = 0. \tag{2.5.4}$$

Equation (2.5.3) is, in fact, Newton's third law.

The forces $\Phi_\beta^{\ r}$ are of the same form as the $\Theta_\alpha^{\ r}$, i.e., they also satisfy (2.5.2) and (2.5.3) when α is replaced by β. Therefore, they are the interactions which exists between the particles that belong to the system and those outside it.

The forces $\Psi_j^{\ r}$ are forces other than interaction forces between particles; they will not be further specified until later on. We call $\sum_\beta \Phi_\beta^{\ r} + \sum_j \Psi_j^{\ r}$ the *external forces*.

2.6. The Problems of Particle Mechanics

While it is difficult to distinguish between, or to classify, all problems of particle mechanics, two problems occur frequently. These may be stated loosely as follows:

> **Problem I.** *It is desired to have the particles of a system move in a specified manner. What forces are required to achieve this motion?*

and

> **Problem II.** *Completely specified forces act on the particles of a system. How will the particles move?*

The first problem includes the problems of statics because the specified motion of the statics problem is rest with respect to an inertial frame. In this way, the statics problem is regarded as one of finding forces such that no motion takes place when they act.

The second problem is, in general, the more interesting of the two. However, to formulate it completely and with precision, we need additional concepts, which will be introduced in the next sections. Here, we only observe that the second problem is regarded as completely solved when the number of particles in the system at every instant of time and the position of each particle at each instant of time are known.

3

Representations of the Motion

3.1. The Configuration Space

In a system having n particles P_r ($r = 1, 2, \ldots, n$), where the position vector of the rth particle at the time t is

$$x^r(t) = \big(x_1{}^r(t), x_2{}^r(t), x_3{}^r(t)\big), \tag{3.1.1}$$

the $3n = N$ numbers $x_1{}^r(t), x_2{}^r(t), x_3{}^r(t)$ ($r = 1, 2, \ldots, n$) specify uniquely the positions of all n particles at the time t. The set of numbers

$$C = \{x_1{}^r(t), x_2{}^r(t), x_3{}^r(t) \colon x^r \in \mathscr{E}^3;\, t \in \mathscr{T};\, r = 1, 2, \ldots, n\} \tag{3.1.2}$$

is called the *configuration* of the system; in (3.1.2)

$$\mathscr{T} = \{t \colon -\infty < t < \infty\} \tag{3.1.3}$$

is the time space.

If we construct a $3n$-dimensional Euclidean space with orthogonal basis such that the $x_i{}^r$ ($r = 1, 2, \ldots, n$; $i = 1, 2, 3$) are scalars giving the components of the positions of the particles P_r, then a given configuration of the system corresponds to a unique point in this space and, conversely, a given point in this space corresponds to a unique configuration. This space is called the *configuration space.*

The configuration space is homogeneous and isotropic.†

† If the distance between two points in a space is defined in terms of their coordinates, the space is said to be *homogeneous* if the distance is invariant under a coordinate translation, and *isotropic* if it is invariant under a coordinate rotation. If the distance between two points $x = (x_1, x_2, x_3)$ and $y = (y_1, y_2, y_3)$ in Euclidean 3-space $\mathscr{E}^3$ is defined by $d(x, y) = |x - y|$, it is very easy to show $\mathscr{E}^3$ is homogeneous and isotropic.

This follows from the properties of $\mathscr{E}^3$.

It is evident that, in the configuration space, all coordinates become equivalent, i.e., the distinction between their being components along the x_1^r, x_2^r, or x_3^r axes is submerged. Therefore, it is convenient to denote these coordinates by u_i ($i = 1, 2, \ldots, 3n = N$) in order to avoid unnecessary super- and subscript notation. One could, for instance, establish the following correspondence between the x_j^r ($j = 1, 2, 3$) and u_i coordinates ($i = 1, 2, \ldots, N$):

$$x_1^1 = u_1, x_2^1 = u_2, x_3^1 = u_3, x_1^2 = u_4, \ldots, x_3^n = u_N.$$

Then, the configuration space is $\mathscr{E}^N$, and

$$u = (u_1, u_2, \ldots, u_N) \in \mathscr{E}^N. \tag{3.1.4}$$

When the number $n(t) \geq 1$ of particles remains constant in time (variable-mass problems of Newtonian mechanics are excluded) the configuration space is useful for the representation of the motion of the entire system of particles.

The configuration at any instant t is given by a point in $\mathscr{E}^N$, and the sequence of configurations, *as time changes*, results in a curve in the configuration space, traced by this point. We shall call this point the *representative point* in $\mathscr{E}^N$, and the curve which it traces is called a *C trajectory*.

Below we list some general properties of C trajectories:

(i) *C trajectories are continuous.* This is obvious because every $u_i(t)$ is one of the $x_i^r(t)$, and these are all continuous.

(ii) *C trajectories may have multiple points.* A multiple point is one which is crossed more than once by a given trajectory. Thus, a multiple point corresponds to a configuration which is attained by the motion more than once. This is certainly possible as, for instance, in periodic motion.

(iii) *C trajectories may have corners.* The *direction* of a trajectory at a point, i.e., at a value of t when the configuration is $u(t)$, is given by its unit tangent vector at that point:

$$d = \frac{(\dot{u}_1, \dot{u}_2, \ldots, \dot{u}_N)}{\left[\sum_{i=1}^{N} (\dot{u}_i)^2\right]^{1/2}} = \frac{\sum_{i=1}^{N} \dot{u}_i \hat{e}_i}{\left[\sum_{i=1}^{N} (\dot{u}_i)^2\right]^{1/2}}.$$

A C trajectory can have a *corner* only where the direction is not defined.

(a) When there exists a $t = t^*$ such that $\dot{u}_i(t^*) = 0$ for all $i = 1, 2, \ldots, N$, the direction is not defined. An instant t^* for which all $\dot{u}_i$ vanish is called an *instant of rest*, and the corresponding configuration $u(t^*)$ is called a *rest point*.

(b) When there exists a $t = t^{**}$ such that some velocity component $\dot{u}_k$ is discontinuous at t^{**}, then

$$d(t^{**} - 0) \neq d(t^{**} + 0),$$

where

$$d(t^{**} \pm 0) = \lim_{\varepsilon \to 0} d(t^{**} \pm \varepsilon),$$

and again, the direction is not defined. This may occur when an impulsive force acts at the time t^{**}, and only then. We call this a *true corner*.

3.2. The Event Space

The combination of a given configuration and the time at which it is attained is called an *event*. Hence, an event is the set

$$E = \{x_1{}^r, x_2{}^r, x_3{}^r, t\colon r = 1, 2, \ldots, n;\ t \in \mathscr{T}\}. \tag{3.2.1}$$

If we construct the $(3n + 1)$-dimensional Euclidean space with orthogonal basis such that the $x_j{}^r$ ($r = 1, 2, \ldots, n$; $j = 1, 2, 3$) and t are scalars giving the positions of the particles P_r and the time at which they occupy these positions, then a given event corresponds to a unique point in this space and, conversely, a given point in this space corresponds to a unique event. This space is called the *event space*. Following the same procedure of replacing the $x_j{}^r$ by the u_i as was done in connection with (3.1.4), the event space is $\mathscr{E}^{N+1}$ and

$$(u, t) = (u_1, u_2, \ldots, u_N, t) \in \mathscr{E}^{N+1}; \qquad t \in \mathscr{T}. \tag{3.2.2}$$

The event space is homogeneous, but not isotropic.

This is easily seen from the following example.

Example 3.2.1. Consider the event space of a planar problem. If this three-dimensional (x, y, t) space were isotropic, one would have under a rotation of the x, y, t system

$$\begin{aligned} x &= l_1x' + l_2y' + l_3t', \\ y &= m_1x' + m_2y' + m_3t', \\ t &= n_1x' + n_2y' + n_3t', \end{aligned} \tag{a}$$

where the l_i, m_i, and n_i are constant. But, then, a simple calculation shows that, for instance,

$$\frac{dx}{dt} = \frac{l_1 \dfrac{dx'}{dt'} + l_2 \dfrac{dy'}{dt'} + l_3}{n_1 \dfrac{dx'}{dt'} + n_2 \dfrac{dy'}{dt'} + n_3}. \tag{b}$$

This does not have the form of (2.4.3) and, hence, does not leave Newton's second law form-invariant.

This example shows that the acceleration would not satisfy the second of equations (2.4.3) under a rotation of the t axis, and, in consequence, Newton's second law would not hold. Thus, all linear transformations of the event space are permitted provided the t axis remains parallel to itself.

When the number $n(t)$ of particles of the system remains constant (variable-mass problems are excluded) the event space is useful for the representation of the motion of a system of particles. (Note that when $N = 1$, the event space is the familiar time-displacement plane.)

We shall denote as the *representative point* in $\mathscr{E}^{N+1}$ the point defining an event of a given motion; the sequence of events is a curve traced by the representative point and is called an *E trajectory*.

Below we list some general properties of E trajectories. The demonstration of these properties is frequently identical to that made for C trajectories; hence, not all proofs will be given.

(i) *E trajectories are continuous.*

(ii) *E trajectories cannot have multiple points.*

This second property is actually a consequence of the following property:

(iii) *E trajectories are strictly monotone in time.* If a scalar function (or a curve) is *monotone* in its scalar argument, it either never decreases as the argument increases, or else it never increases as the argument increases. If it is *strictly monotone*, it either always increases with increasing argument, or else it always decreases with increasing argument. Property (iii) states that the latter is the case. We prove it first for the case (a) that the E trajectory is everywhere smooth. Next, we shall assume (b) that it has isolated corners.

(a) The direction (if it exists) of an E trajectory is the unit tangent vector

$$d = \frac{\sum_{i=1}^{N} \dot{u}_i \hat{e}_i + \hat{e}_t}{\left[\sum_{i=1}^{N} (\dot{u}_i)^2 + 1\right]^{1/2}}, \tag{3.2.3}$$

where the $\hat{e}_i$ $(i = 1, 2, \ldots, N)$ are the unit vectors along the u_i axes, and $\hat{e}_t$ is the unit vector along the t axis. We shall show that this unit tangent vector can never lie in a plane normal to the t axis. This is equivalent to the observation that smooth E trajectories are always strictly monotone in t. For, if the trajectory first rises along t and then falls (or if it first falls along t and then rises) the t component of the tangent vector in E space must pass through zero and, hence, is not strictly monotone. A unit vector in E space, in a plane normal to the t axis, is necessarily of the form

$$\bar{d} = \frac{\sum_{i=1}^{N} a_i \hat{e}_i}{\left[\sum_{i=1}^{N} (a_i)^2\right]^{1/2}}, \tag{3.2.4}$$

where the a_i are scalar quantities. But, by (3.2.3), the component of $\bar{d}$ along the t axis is

$$\left[\sum_{i=1}^{N} (\dot{u}_i)^2 + 1\right]^{-1/2}$$

and *not* zero. Thus, (3.2.4) can never be of the form (3.2.3).

(b) Assume that at t^{**}, the E trajectory has an isolated corner. [Obviously, this is the only type of corner than can occur (why?).] In fact, it may occur when an impulsive force acts at t^{**}. Now, if the E trajectory is not strictly monotone in t, then the corner is such that the E trajectory reverses its direction along the t axis. But, in that case, there exists an instant (in fact, a time interval) near t^{**} when the system has two different configurations at the same instant, as il-

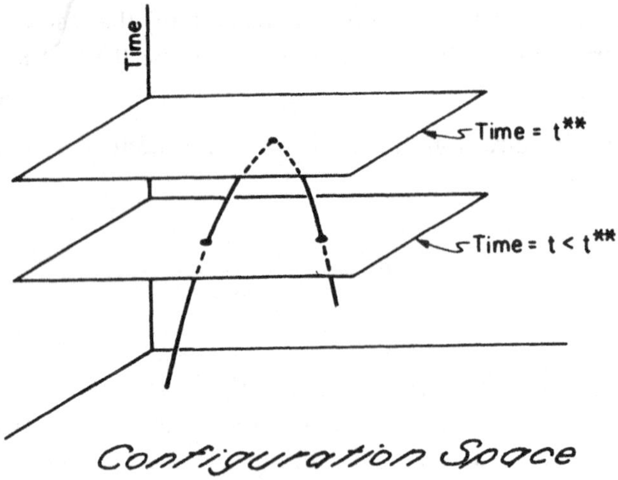

Fig. 3.2.1. E trajectories are strictly monotone in time.

lustrated in Fig. 3.2.1.[†] This contradicts the fundamental property that every $x_j^r(t)$ [and hence every $u_i(t)$] is a single-valued function of time. Note that property (iii) implies property (ii) because, if an E trajectory is to have multiple points, it must return down (or up) the t axis, and we have shown that this cannot be the case.

(iv) *E trajectories can have a direction parallel to the t axis.* The proof is left as an exercise. Note that instants at which E trajectories have directions parallel to the t axis are instants of rest.

(v) *E trajectories may have corners.* Note that the existence of a true corner on a C trajectory requires that the E trajectory have a corner since the C trajectory is simply the projection of an E trajectory on the configuration space.

3.3. The State Space

The combination of a given configuration $x^r(t) = (x_1^r(t), x_2^r(t), x_3^r(t))$ $(r = 1, 2, \ldots, N)$ at a time t and of the velocities

$$\dot{x}^r(t) = (\dot{x}_1^r(t), \dot{x}_2^r(t), \dot{x}_3^r(t))$$

† The exceptional case where the representative point stops at t^{**} and then retraces the E trajectory does not invalidate our argument because, in that case, trajectories like the one in Fig. 3.2.1 exist in the neighborhood of the exceptional case because of the continuity of E trajectories in initial conditions and parameter values.

at the same value of t defines the so-called *state* of a system of particles at t. Hence, the state is the set of numbers

$$S = \{x_1^r, x_2^r, x_3^r; \dot{x}_1^r, \dot{x}_2^r, \dot{x}_3^r: r = 1, 2, \ldots, n\}. \tag{3.3.1}$$

If we construct the $6n$-dimensional Euclidean space with orthogonal basis such that the x_j^r and the $\dot{x}_j^r$ ($r = 1, 2, \ldots, n$; $j = 1, 2, 3$) are scalars giving the positions and velocities of the particles P_r, then a given state of the system corresponds to a unique point in that space and, conversely, a given point in that space corresponds to a unique state of the system. This space is called the *state space.*

The state space is homogeneous and isotropic.

Following the same procedure of replacing the x_j^r by the u_i and writing $\dot{x}_j^r = \dot{u}_i$, the state space is $\mathscr{E}^{2N}$ and

$$(u, \dot{u}) = (u_1, u_2, \ldots, u_N, \dot{u}_1, \dot{u}_2, \ldots, \dot{u}_N) \in \mathscr{E}^{2N}. \tag{3.3.2}$$

When the number $n(t)$ of particles of the system is constant, the state space is useful for the representation of the motion of the system.

The sequence of states through which the system passes in time defines a curve in the state space traced by a representative point in $\mathscr{E}^{2N}$; it is called an *S trajectory*. Hence, the S trajectory is the locus of states of the system for a given motion.

Below, some general properties of S trajectories are given. Their demonstration is only made in nonobvious instances, and where it does not follow immediately from previous demonstrations of the properties of C and E trajectories.

(i) *S trajectories are continuous for all $t \in \mathscr{T}^c$ where $\mathscr{T}^c = \mathscr{T} - \mathscr{T}_j$ and* $\mathscr{T} = \{t: -\infty < t < \infty\}, \mathscr{T}_j = \{t_j$: one or more $\dot{u}_i(t_j)$ not defined$\}$. *S trajectories are piecewise continuous and piecewise smooth for all $t \in \mathscr{T}$.*

(ii) *S trajectories may have multiple points.*

(iii) *No two or more distinct S trajectories of a given system may cross or touch at the same instant.* If they did, the same initial conditions would result in distinct motions.

3.4. The State–Time Space

The combination of a given state

$$S = \big(x_1{}^r(t), x_2{}^r(t), x_3{}^r(t); \dot{x}_1{}^r(t), \dot{x}_2{}^r(t), \dot{x}_3{}^r(t)\big) \qquad (r = 1, 2, \ldots, n)$$

and the time t at which the state is attained is the set of numbers

$$T = \{x_1{}^r, x_2{}^r, x_3{}^r; \dot{x}_1{}^r, \dot{x}_2{}^r, \dot{x}_3{}^r; t: t \in \mathscr{T}; r = 1, 2, \ldots, n\}. \qquad (3.4.1)$$

If we construct the $(6n + 1)$-dimensional Euclidean space with orthogonal basis such that the $x_j{}^r$, the $\dot{x}_j{}^r$ $(r = 1, 2, \ldots, n;\ j = 1, 2, 3)$, and t are scalars giving the positions and velocities of the particles at the time t, then a given point in that space defines uniquely a given state and time, and a given state and time defines a unique point in that space. This space is called the *state–time space*.

Replacing again the $x_j{}^r$ by the u_i, with $\dot{x}_j{}^r = \dot{u}_i$, the state–time space is $\mathscr{E}^{2N+1}$ and

$$(u, \dot{u}, t) = (u_1, u_2, \ldots, u_N, \dot{u}_1, \dot{u}_2, \ldots, \dot{u}_N, t) \in \mathscr{E}^{2N+1}. \qquad (3.4.2)$$

Again, *when variable-mass problems of Newtonian mechanics are excluded*, the state–time space is useful for the representation of the motion of a system.

The state–time space is homogeneous but not isotropic.

The sequence of states and corresponding times defines a curve traced by a representative point in $\mathscr{E}^{2N+1}$; it is the locus of states and corresponding times of the system in a given motion and is called a *T trajectory*.

Some of the general properties of T trajectories are listed below:

(i) *T trajectories are piecewise continuous and piecewise smooth.*

(ii) *T trajectories cannot have multiple points.*

(iii) *T trajectories are strictly monotone in time.*

The demonstration of these properties either follows directly from earlier results, or is done in a similar manner.

3.5. Notions on the Concept of Stability

The idea of "stability" or "instability" of a motion is closely connected with the representation of motions as trajectories in the spaces just discussed.

Consider a given motion of some system as the T trajectory T^* in the state–time space. Let the same system also have another motion, whose trajectory is T^{**}; such a motion could be produced by initial conditions which are different from those of T^*. Let us also suppose that at some arbitrary point along T^*, the trajectory T^{**} lies near T^*. Then, if T^{**}, having been once near T^*, remains everywhere near T^*, the motion T^* is said to be *stable in the sense of Liapunov.*

Consider next the motion of some system as the S trajectory S^* in state space. Then, if we denote by S^{**} the S trajectory of another motion of the same system, the motion S^* is said to be *stable in the sense of Poincaré* if S^{**}, having been once near S^*, remains everywhere near it.

These two definitions of stability are not equivalent, as is shown in Fig. 3.5.1. In it, we show the trajectory T^* inside a small tube of radius ε. The projection of T^* along the t axis is the S^* trajectory. Of course, it lies inside the projection of the ε tube. In (a) we show a motion which is both Liapunov and Poincaré-stable, in (b) the motion is Liapunov-unstable but Poincaré-stable, and in (c), the motion is unstable in both senses.

A case which is frequently of interest in dynamics is motion in the neighborhood of an equilibrium position. Equilibrium is the special "motion" in which the configuration remains the same for all times. Hence, one may represent equilibrium as an E trajectory which is a straight line in the event space parallel to the t axis. Stable motion in the neighborhood of an equilibrium position is an E trajectory which remains for all t in an ε tube parallel to the t axis that is centered in an equilibrium position;

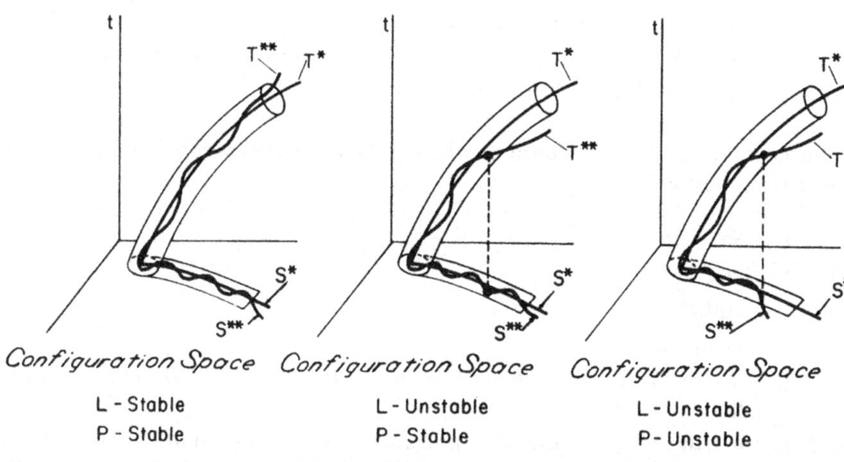

Fig. 3.5.1. Schematic representation of Liapunov and Poincaré stability in event space.

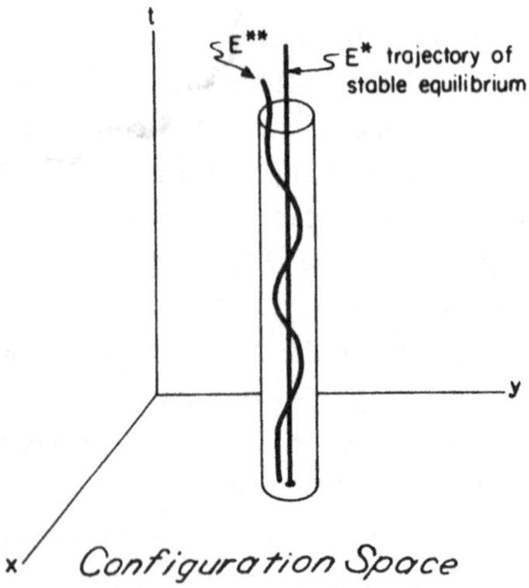

Fig. 3.5.2. The notion of stable equilibrium in event space.

this is illustrated in Fig. 3.5.2. We shall return to this problem later on (see Chapter 17).

3.6. Problems

In the next six problems, study the curves in the x, y plane and examine them for possible corners and/or multiple points. The quantities a, A, and b are real constants; t is the time. Can these curves be C trajectories? Give reasons. If your answer is conditional, state the conditions.

3.1. $x = t^{1/2}$; $(y^2 + t)^2 = a^2(t - y^2)$; $0 < t_0 \le t \le T < \infty$.

3.2. $x = A \cos^{1/2} bt$; $(y^2 + A^2 \cos bt)^2 = a^2(A^2 \cos bt - y^2)$.

3.3. $x = A \sin bt$; $(y^2 + ay + A^2 \sin^2 bt)^2 = a^2(A^2 \sin^2 bt + y^2)$; $a > 0$.

3.4. $x = A \sin bt$; $(y^2 + ay + A^2 \sin^2 bt)^2 = 4a^2(A^2 \sin^2 bt + y^2)$; $a > 0$.

3.5. $x = A \sin bt$; $(y^2 + 2ay + A^2 \sin^2 bt)^2 = a^2(A^2 \sin^2 bt + y^2)$; $a > 0$.

3.6. $x = t$; $y = t^{2/3}(t - 2)^2$; $-1 \le t \le 3$.

In the next four problems, determine whether the given curves can be E trajectories; b is a real constant.

3.7. $x = \sin^{-1} bt$.

3.8. $x = \cos^{-1} bt$.

3.9. $x = \sinh^{-1} bt$.

3.10. $x = \cosh^{-1} bt$.

3.11. Discuss the T trajectory of $x = A \sin \alpha t + B \cos \alpha t$; A, B, and α are real constants.

3.12. Discuss the C, E, S, and T trajectories of $x = e^{\beta t}(A \sin \alpha t + B \cos \alpha t)$ for the cases $\beta < 0$, $\beta = 0$, $\beta > 0$; A, B, α, and β are real constants.

4

Constraints

4.1. General Observations

In most problems of particle mechanics, the motion of the particles is "constrained" in some way. This is the term used to denote the condition that some motions or configurations are not admitted. One has, in fact, the rarely verbalized theorem:

In a system of two or more particles, unconstrained motion does not exist.

Proof. Particle motion has the property of impenetrability, i.e., the motion is constrained so that no more than one particle can occupy any one position in 3-space at any given instant of time.

While this constraint does not seem to be a strong limitation on the motion, it may in fact limit it severely.

Example 4.1.1. Consider two particles moving along a straight line, and let their positions at any instant of time be given by their distance from some fixed point on that line (see Fig. 4.1.1). At the instant shown, the particles do not coincide. Hence, by impenetrability, they can never coincide, i.e., the line $x_1 = x_2$ in configuration space is a forbidden line, as shown in Fig. 4.1.2.

It follows that half of the configuration space is here inaccessible to C trajectories because they must not touch or pierce the forbidden line.

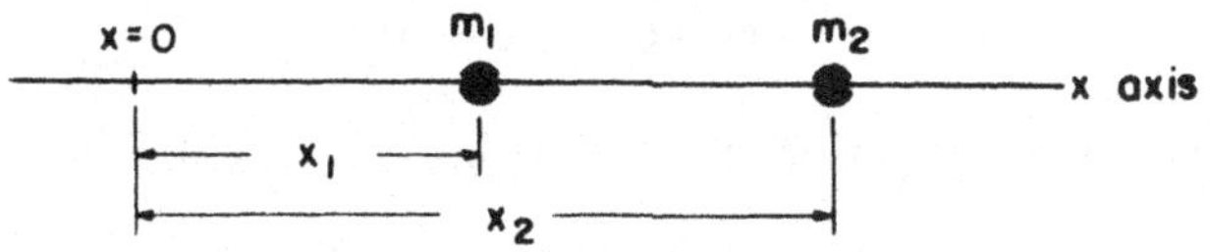

Fig. 4.1.1. Two particles moving on a straight line.

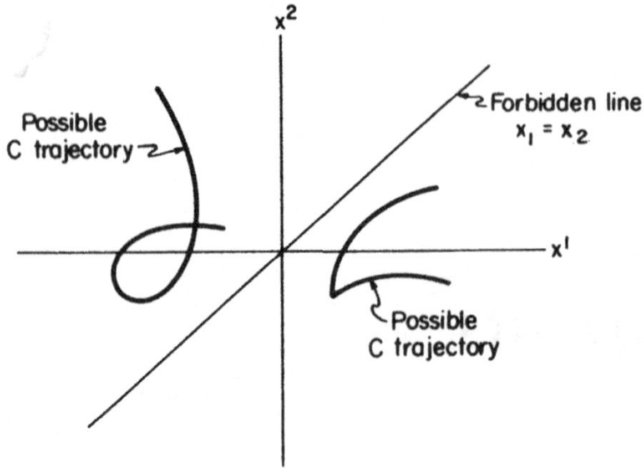

Fig. 4.1.2. Some possible C trajectories of two particles moving on a straight line.

Exercise. Determine the portion of the configuration space accessible to the C trajectory when three particles move on a line. (Ans. 1/6).

Besides the impenetrability constraint, there are many instances of constrained motion where the constraint is not usually introduced explicitly, but merely mentioned verbally. For instance, when it is mentioned that a problem of rectilinear motion is to be treated, the problem is usually formulated by the equation of motion

$$m\ddot{x} = X(x, \dot{x}, t),$$

where $X(x, \dot{x}, t)$ is a force acting in the direction of motion x. However, to formulate the problem fully, one must write

$$m\ddot{r} = F(r, \dot{r}, t)$$

where, for instance,

$$r = x\hat{\imath} + y\hat{\jmath} + z\hat{k},$$
$$F = X\hat{\imath} + Y\hat{\jmath} + Z\hat{k},$$

and the equations of constraint are

$$y \equiv 0, \qquad z \equiv 0.$$

Writing the vector equation of motion in component form, one has

$$m\ddot{x} = X(x, y, z; \dot{x}, \dot{y}, \dot{z}; t),$$
$$m\ddot{y} = Y(x, y, z; \dot{x}, \dot{y}, \dot{z}; t),$$
$$m\ddot{z} = Z(x, y, z; \dot{x}, \dot{y}, \dot{z}; t).$$

The constraints $y = z \equiv 0$ imply $\dot{y} = \ddot{y} = \dot{z} = \ddot{z} \equiv 0$, and these imply, in turn, $Y = Z \equiv 0$. Thus the formulation becomes simply

$$m\ddot{x} = X(x, 0, 0; \dot{x}, 0, 0; t),$$

identical with the original one.

Constraints like $y = 0$ are special cases of a constraint of the form

$$f(x, y, z) = 0. \tag{4.1.1}$$

Equations of constraint may be explicit functions of time. Consider, for instance, a particle which is constrained to move on a plane that moves parallel to itself in a prescribed way. The constraint on this particle may be put in the form

$$z - g(t) = 0,$$

where $g(t)$ is a given function. But this is a special case of a constraint of the form

$$f(x, y, z, t) = 0. \tag{4.1.2}$$

4.2. Holonomic Constraints

Equation (4.1.1) is a special case of an equation of the form

$$f(x_1^1, x_2^1, x_3^1, x_1^2, \ldots, x_1^n, x_2^n, x_3^n) = 0. \tag{4.2.1}$$

If we use the notation introduced in the section on configuration space, (4.2.1) is written as

$$f(u_1, u_2, \ldots, u_N) = 0 \tag{4.2.2}$$

and, if we generalize (4.1.2) in a similar manner, we have

$$f(u_1, u_2, \ldots, u_N, t) = 0. \tag{4.2.3}$$

A constraint of the form (4.2.3), *or reducible to that form, is called a* holonomic *constraint. Every constraint not of the form* (4.2.3), *or not reducible to it, is called* nonholonomic.

Among holonomic constraints we distinguish between those that depend explicitly on time and those that do not:

A holonomic constraint of the form (4.2.2), *or reducible to it, is called* scleronomic. *Every holonomic constraint not of the form* (4.2.2), *or not reducible to it, is called* rheonomic.

These names come from the Greek. The word "holonomic" means "altogether lawful," the word "scleronomic" means "rigid," and the word "rheonomic" means "flowing." The reasons for this terminology will become apparent shortly.

To lend meaning to the concept of constraints let us begin by considering (4.1.1) written as

$$f(u_1, u_2, u_3) = 0. \tag{4.2.4}$$

This equation defines a surface in the u_1, u_2, u_3 space; imposing the constraint (4.2.4) on the motion of a particle is equivalent to saying that the particle whose position coordinates are $u_1(t)$, $u_2(t)$, $u_3(t)$ must for all time move in the surface $f = 0$ defined by (4.2.4). If we should say instead, the u_1, u_2, u_3 space is the configuration space (of the single particle), then the constraint (4.2.4) defines a surface in the configuration space, and the C trajectory must lie in that surface for every admissible motion. With this terminology, we can now readily interpret (4.2.2). That equation defines an N–1-dimensional surface in the configuration space $\mathscr{E}^N$ given in (3.1.4), and the C trajectory of every motion of that system must lie in this surface.

The surface defined by (4.2.2) is a *rigid* surface in $\mathscr{E}^N$, i.e., it does not change, warp, or deform with time, hence, the term *scleronomic*.

With a similar interpretation, we may regard (4.2.3) as a surface in $\mathscr{E}^N$ which changes or deforms in time; hence, the term *rheonomic*.

It is worthwhile to note that a *changing* surface in $\mathscr{E}^N$ defines a *rigid* surface in the event space $\mathscr{E}^{N+1}$ given in (3.2.2), but the changing surface in $\mathscr{E}^N$ has N–1 dimensions while the corresponding surface in $\mathscr{E}^{N+1}$ has N dimensions. It is also clear that a holonomic scleronomic constraint defines a surface in $\mathscr{E}^{N+1}$ which is *cylindrical*, i.e., its projection on the configuration space $\mathscr{E}^N$ is the same for every value of t.

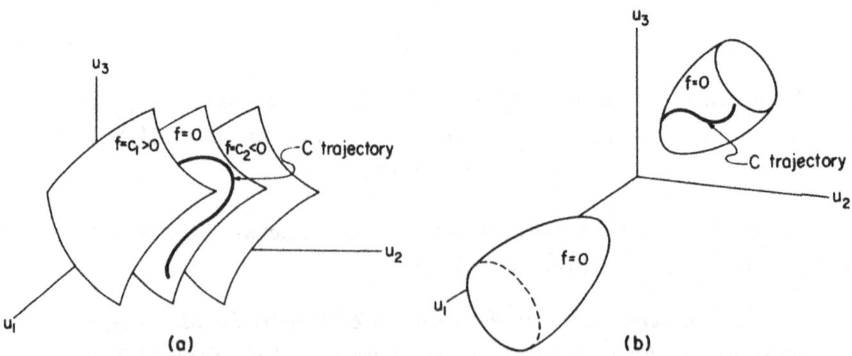

Fig. 4.2.1. Examples of a holonomic scleronomic constraint.

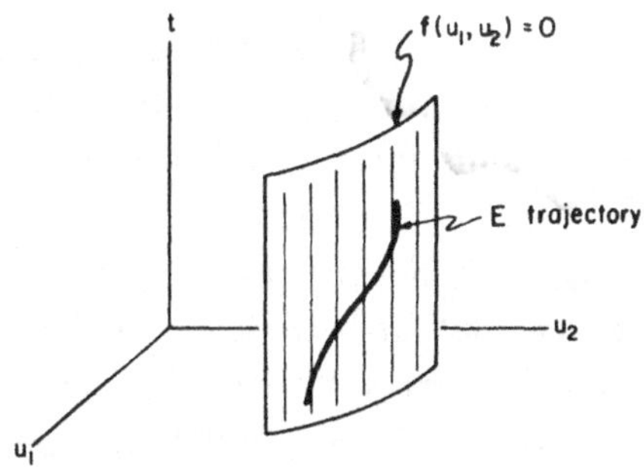

Fig. 4.2.2. Holonomic scleronomic constraint in event space is a cylindrical surface.

These interpretations of holonomic constraints are illustrated in Figs. 4.2.1 and 4.2.2. Figure 4.2.1 illustrates (holonomic) scleronomic constraints in configuration space: In (a) the surface defined by the constraint is connected; in (b) it is not. In Fig. 4.2.2, a (holonomic) scleronomic constraint is shown in the event space, and in Fig. 4.2.3, a (holonomic) rheonomic constraint in the event space is illustrated.

Equations like (4.2.2) and (4.2.3) impose conditions on *finite* displacements, i.e., every $u_i(t)$ ($i = 1, 2, \ldots, N$) must be such as to satisfy one of these equations. Now, it is interesting to inquire what conditions are imposed on *infinitesimal* displacements by these equations of constraint.

This question is difficult to answer in all generality. However, if the surface is smooth (i.e., all first partial derivatives of the function f with respect to all arguments exist and are continuous everywhere) one finds from (4.2.2) by differentiation

$$\sum_{s=1}^{N} \frac{\partial f}{\partial u_s} du_s = 0, \tag{4.2.5}$$

and from (4.2.3)

$$\sum_{s=1}^{N} \frac{\partial f}{\partial u_s} du_s + \frac{\partial f}{\partial t} dt = 0. \tag{4.2.6}$$

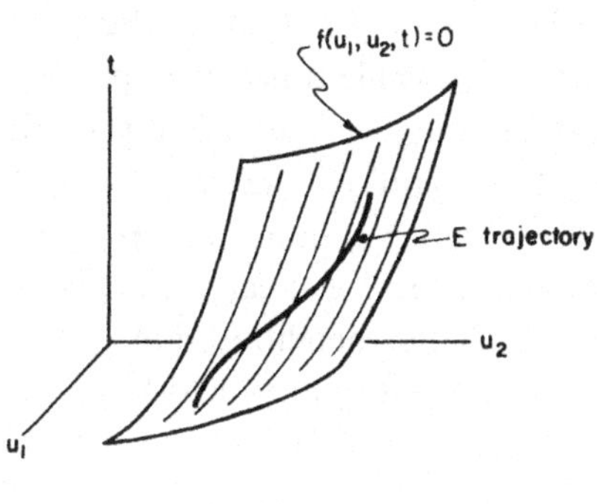

Fig. 4.2.3. Holonomic rheonomic constraint in event space is not a cylindrical surface.

Hence, infinitesimal displacements du_s ($s = 1, 2, \ldots, N$) of (holonomic) scleronomic constraints must satisfy (4.2.5), and those of rheonomic constraints must satisfy (4.2.6).

Every C trajectory lying in the surface of constraint is a *possible* trajectory, but it is evident that not every one of them corresponds to an *actual* motion because actual motions must also satisfy dynamical laws, and these have not been invoked.

It may be worthwhile to note that the phrase "or reducible to that form" was put into the definitions of holonomic constraints because (4.2.5) and (4.2.6) are not of the forms (4.2.2) or (4.2.3) but are reducible to them. For instance, (4.2.5) can be integrated to give

$$f(u_1, u_2, \ldots, u_N) = c,$$

where c is a constant, and this may be written as

$$g(u_1, u_2, \ldots, u_N) = 0,$$

where $g = f - c$; the latter is of the form (4.2.2).

It is not difficult to give (4.2.5) and (4.2.6) a geometrical interpretation. Consider, for instance, the surface (4.2.2) at the point $u^* = (u_1^*, u_2^*, \ldots, u_N^*)$. Let us use the notation

$$\left.\frac{\partial f}{\partial u_s}\right|_{u=u^*} = \left(\frac{\partial f}{\partial u_s}\right)^*.$$

In other terms, $(\partial f/\partial u_s)^*$ is the derivative $\partial f/\partial u_s$ evaluated at the point u^*. Then, the equation

$$\sum_{s=1}^{N} \left(\frac{\partial f}{\partial u_s}\right)^* du_s = 0$$

defines all points in the tangent plane of the surface $f = 0$ at u^*. But this last equation is an equation of constraint on the possible trajectories passing sufficiently near the point u^*. Expressed differently, in a sufficiently small neighborhood of u^*, the constraint equation requires that all possible trajectories must lie, not in the surface of constraint, but in its *tangent plane*, because a smooth surface and its tangent plane are identical within first-order terms in infinitesimals in a sufficiently small neighborhood. Thus, a smooth surface is then modeled by its tangent plane, as shown in Fig. 4.2.4.

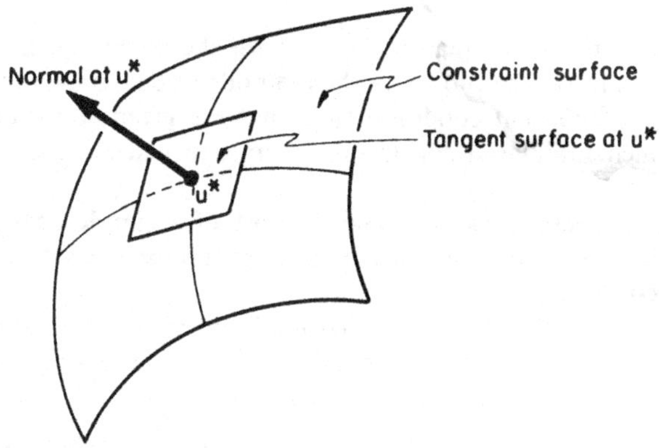

Fig. 4.2.4. A smooth surface in the neighborhood of a point is modeled by its tangent plane at that point.

Example 4.2.1. A spherical pendulum is a particle which moves under the gravitational force in a frictionless spherical surface. What are the constraints on finite and on infinitesimal displacements?

Let $R = \text{const}$ be the radius of the sphere, and let the origin of the x, y, z triad coincide with the center of the sphere. The configuration (x, y, z) must satisfy

$$x^2 + y^2 + z^2 - R^2 = 0,$$

and infinitesimal configuration changes must satisfy

$$x\,dx + y\,dy + z\,dz = 0.$$

Example 4.2.2. (Hamel, p. 618). A gutter is created by a parabola which remains parallel to itself while its apex translates and descends in a prescribed manner. Find the finite and infinitesimal constraints.

Let the parabola remain in a plane parallel to the y, z plane with the parabola open in the direction of positive z. Let the curve along which the apex moves be given by $z_0(x)$. The finite constraints are

$$z - z_0(x) - \tfrac{1}{2}ay^2 = 0.$$

The infinitesimal constraints are

$$-z_0'(x)\,dx - ay\,dy + dz = 0 \qquad (' = d/dx).$$

Example 4.2.3. Let a constraint on infinitesimal displacements be given by

$$dy - g(z)\,dx = 0$$

where $g(z) \neq \text{const}$ is a given function of z. Find the constraint on the finite displacements implied by the given constraint.

None exists. If one were to assume that finite and infinitesimal constraints satisfy the same constraint, one would have

$$y - g(z)x = \text{const}.$$

But that equation implies

$$dy - g(z)\,dx = xg'(z)\,dz,$$

and this last equation agrees with the given constraint only if $g(z) \equiv$ const. [It can in fact be shown by the methods of Section 4.5 that $g'(z) \equiv 0$ is the necessary and sufficient condition such that the given constraint on infinitesimal displacements implies a constraint on finite displacements.]

Example 4.2.4. The support of a simple pendulum is moved horizontally in a prescribed manner as time progresses. What are the finite and infinitesimal constraints?

Let the x axis be horizontal and let $f(t)$ be the distance of the point of suspension from the origin of the x, y plane. Moreover, let l be the length of the pendulum. The finite constraint is

$$[x - f(t)]^2 + y^2 - l^2 = 0.$$

The infinitesimal constraint is

$$[x - f(t)]\, dx + y\, dy - [x - f(t)]\dot{f}\, dt = 0.$$

Note that this is an example of a rheonomic constraint.

We examine now the question of the possible existence of more than one constraint. Suppose we deal with the problem of a particle constrained to move on a curve in 3-space. This curve may be regarded as the intersection of two surfaces

$$f_1(u_1, u_2, u_3) = 0, \qquad f_2(u_1, u_2, u_3) = 0 \tag{4.2.7}$$

such that their normals do not coincide anywhere along the curve. Therefore, two holonomic constraints in 3-space may be used to define a curve. Suppose, now, that we have three equations of constraint for the motion of a particle, i.e., we impose on the configuration the conditions

$$f_1(u_1, u_2, u_3) = 0, \qquad f_2(u_1, u_2, u_3) = 0, \qquad f_3(u_1, u_2, u_3) = 0,$$

such that the first two define a curve, and the third is a surface intersected by this curve. But this intersection defines a single point, thus, these three constraints impose the condition of no motion whatever with respect to the u_1, u_2, u_3 triad. Two intersecting constraint surfaces will always define a curve, and three a point, if the functions defining the surfaces are linearly independent. [The functions $f_r(u_1, u_2, \ldots, u_N, t)$, $r = 1, 2, \ldots, L$, are said to be linearly independent if one cannot find L quantities λ_r not all zero such that

$$\sum_{r=1}^{L} \lambda_r f_r(u_1, u_2, \ldots, u_N, t) = 0.$$

See also the definition of linearly independent vectors, Section 2.4.]

If the three constraints had been rheonomic instead, they would have defined a point which moves in 3-space in a manner prescribed by the constraints, but it does so independently of dynamical considerations. Thus, we see that the number of independent constraints must be less than the dimension of the configuration space if the motion of the particles of the system is to be a function of the forces acting on them. We use then definition

The number $N - L > 0$ is called the number of degrees of freedom *of a system of particles, where L is the number of independent equations of constraint.*

Then a dynamical system consisting of $N/3$ particles may have the holonomic constraints

$$f_r(u_1, u_2, \ldots, u_N, t) = 0 \qquad (r = 1, 2, \ldots, L < N) \tag{4.2.8}$$

and, if these constraints are all scleronomic, they are of the form

$$f_r(u_1, u_2, \ldots, u_N) = 0 \qquad (r = 1, 2, \ldots, L < N). \tag{4.2.9}$$

It should be noted that it may occur easily that some constraints are scleronomic while others are not.

The holonomic rheonomic constraints imposed on infinitesimal displacement are of the form

$$\sum_{s=1}^{N} \frac{\partial f_r}{\partial u_s} du_s + \frac{\partial f_r}{\partial t} dt = 0 \qquad (r = 1, 2, \ldots, L). \tag{4.2.10}$$

When they are scleronomic, they are

$$\sum_{s=1}^{N} \frac{\partial f_r}{\partial u_s} du_s = 0 \qquad (r = 1, 2, \ldots, L). \tag{4.2.11}$$

Both are linear forms of differentials. However, these constraints may be written equally well as linear forms of derivatives. Thus, if the u_s and t are considered as functions of some parameter α, we could write (with $' = d/d\alpha$)

$$\left.\begin{aligned} \sum_{s=1}^{N} \frac{\partial f_r}{\partial u_s} u_s' + \frac{\partial f_r}{\partial t} t' = 0, \\ \sum_{s=1}^{N} \frac{\partial f_r}{\partial u_s} u_s' = 0, \end{aligned}\right\} \qquad (r = 1, 2, \ldots, L).$$

The most frequently met parameter is t itself, in which case we find, in place of (4.2.10),

$$\sum_{s=1}^{N} \frac{\partial f_r}{\partial u_s} \dot{u}_s + \frac{\partial f_r}{\partial t} = 0 \qquad (r = 1, 2, \ldots, L) \tag{4.2.12}$$

and, instead of (4.2.11),

$$\sum_{s=1}^{N} \frac{\partial f_r}{\partial u_s} \dot{u}_s = 0 \qquad (r = 1, 2, \ldots, L). \tag{4.2.13}$$

Constraints in the form (4.2.10) and (4.2.11) involving differentials rather than derivatives are said to be in the *Pfaffian* form. One sees from the non-Pfaffian forms (4.2.12) and (4.2.13) that holonomic constraints imply conditions on the *velocities* as well as on the displacements.

Having introduced the possible existence of more than one equation of constraint, we now examine the question whether or not constraints on *finite* displacements always imply the same constraints on *infinitesimal* displacements.

Consider the scleronomic constraints (4.2.11). It is assumed that these equations are *linearly independent* of each other, i.e., none is implied by any of the others. Then, if we write, for short, $\partial f_r/\partial u_s = A_{rs}$, these equations may be written in the matrix form

$$\begin{bmatrix} A_{11} & A_{12} & \cdots & A_{1L} & \cdots & A_{1N} \\ A_{21} & A_{22} & \cdots & A_{2L} & \cdots & A_{2N} \\ \vdots & \vdots & & \vdots & & \vdots \\ A_{L1} & A_{L2} & \cdots & A_{LL} & \cdots & A_{LN} \end{bmatrix} \begin{bmatrix} du_1 \\ du_2 \\ \vdots \\ du_L \\ \vdots \\ du_N \end{bmatrix} = 0 \tag{4.2.14}$$

and, in virtue of the assumption of independence of the equations of constraint, the $L \times N$ matrix in (4.2.14) has maximum rank. This condition implies the nonvanishing of at least one of the following determinants:

$$\begin{vmatrix} A_{1,p+1} & A_{1,p+2} & \cdots & A_{1,p+L} \\ A_{2,p+1} & A_{2,p+2} & \cdots & A_{2,p+L} \\ \vdots & \vdots & & \vdots \\ A_{L,p+1} & A_{L,p+2} & \cdots & A_{L,p+L} \end{vmatrix} \qquad (p = 0, 1, 2, \ldots, N - L).$$

This is the condition necessary and sufficient for the infinitesimal displacements to be constrained by

$$\sum_{s=1}^{N} \frac{\partial f_s}{\partial u} du_s = 0 \qquad (r = 1, 2, \ldots, L) \tag{4.2.15}$$

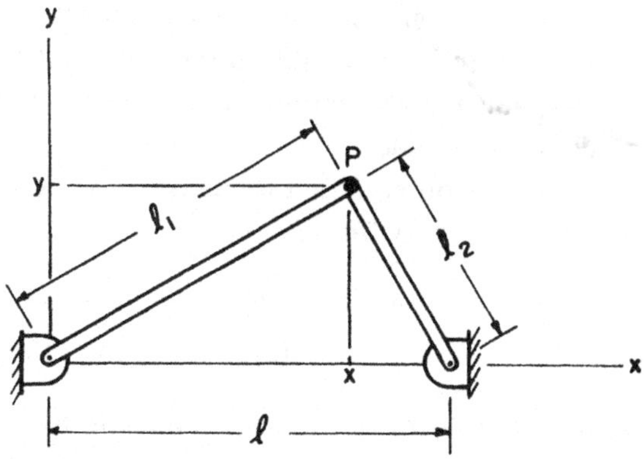

Fig. 4.2.5. The truss of Example 4.2.5.

if the finite displacements are constrained by

$$f_r(u_1, u_2, \ldots, u_N) = 0 \qquad (r = 1, 2, \ldots, L). \tag{4.2.16}$$

Example 4.2.5. (Hamel p. 86). Consider the linkage shown in Fig. 4.2.5. We examine the position of the point P. The equations constraining if from moving are

$$x^2 + y^2 - l_1^2 = 0,$$
$$(x - l)^2 + y^2 - l_2^2 = 0,$$

and, for the links to meet, we must have

$$l_1 + l_2 \geq l.$$

The infinitesimal constraints are

$$x\,dx + y\,dy = 0,$$
$$(x - l)\,dx + y\,dy = 0.$$

The value of the determinant is given by

$$\begin{vmatrix} x & y \\ x - l & y \end{vmatrix} = yl.$$

Hence, if $y \neq 0$, the only solutions are $dx = dy = 0$, i.e., finite as well as infinitesimal displacements of P are prevented by the constraints.

However, the special case $l_1 + l_2 = l$ corresponds to collinear links l_1 and l_2 and implies $y = 0$. Thus, finite displacements of P are still impossible, but the infinitesimal displacement dy is now possible.

4.3. Nonholonomic Constraints

As stated, every constraint that is not holonomic is nonholonomic. One will readily understand that it is not possible to give a general discussion of nonholonomic constraints such as can be done for holonomic ones because the latter is a narrowly circumscribed class while the former is not. (Thus, bananas are readily discussed, while nonbananas are not.) Nevertheless, some classification of frequently encountered nonholonomic constraints is possible.

First we observe that there are nonholonomic constraints which are of either of the two forms

$$\begin{aligned} f(u_1, u_2, \ldots, u_N) &\leq 0, \\ f(u_1, u_2, \ldots, u_N, t) &\leq 0, \end{aligned} \tag{4.3.1}$$

or reducible to them, but not reducible to (4.2.2) or (4.2.3).

Note that

$$f(u_1, u_2, \ldots, u_N) = c,$$

where c is a nonzero, real constant easily reduced to the form of (4.2.9); hence it is holonomic. However, the constraint

$$f(u_1, u_2, \ldots, u_N) \leq c \tag{4.3.2}$$

is not reducible to that form. An example of such a constraint is the admissible motion of a lion in a circular cage of radius R; the lion's position satisfies for all time

$$x^2 + y^2 - R^2 \leq 0.$$

Similarly, if the radius of the cage changes in a prescribed manner with time one has

$$x^2 + y^2 - R(t)^2 \leq 0.$$

If an object may rest on a table or rise above it, but cannot penetrate it, the constraint can be put into the form

$$z \geq 0.$$

The constraint imposed by the impenetrability property is nonholonomic because it is an inequality. Hence, we might have put the earlier theorem into the startling form

Every dynamical system of two or more particles is subject to at least one nonholonomic constraint.

The examples of nonholonomic constraints mentioned here have in common that they are finite relations (not differential) involving only the coordinates of the *configuration space* (and possibly t), and they are *inequalities*.

Consider the inequality constraints

$$\begin{aligned} f(u_1, u_2, \ldots, u_N, t) &< 0, \\ f(u_1, u_2, \ldots, u_N, t) &\leq 0. \end{aligned} \tag{4.3.3}$$

Every nonholonomic constraint of the forms (4.3.3) *or reducible to them is called a* configuration *constraint.*

Again, we distinguish between those that depend explicitly on time, and those that do not.

Every nonholonomic configuration constraint that does not depend explicitly on time t is called scleronomic. Every nonholonomic configuration constraint that is not scleronomic is rheonomic.

Evidently, the constraint

$$f(u_1, u_2, \ldots, u_N, t) \geq c,$$

where c is a real constant, is easily brought into the form (4.3.3) if one subtracts c from both sides, and then multiplies by -1.

The geometrical interpretation of these constraints is simple. In general, the surface

$$f(u_1, u_2, \ldots, u_N, t) = 0$$

divides the configuration space into at least two open domains; in one $f > 0$, and in the other $f < 0$. This surface changes with time if t occurs explicitly in f, otherwise it is rigid and fixed. Then, the first of equations (4.3.3) states that the C trajectory must remain in that domain of the configuration space where f is negative. The second admits C trajectories that may stay on the side $f < 0$, or they may touch, but not pierce, the surface $f = 0$; these situations are illustrated in Fig. 4.3.1.

Quite another type of nonholonomic constraint arises when one generalizes the Pfaffian form of holonomic constraints; the nonholonomic constraints generated in this way are *equality* constraints.

Let us write L independent holonomic constraints

$$f_r(u_1, u_2, \ldots, u_N, t) = c_r \qquad (r = 1, 2, \ldots, L), \tag{4.3.4}$$

where the c_r are real constants, in their differential form

$$\sum_{s=1}^{N} \frac{\partial f_r}{\partial u_s} du_s + \frac{\partial f}{\partial t} dt = 0 \qquad (r = 1, 2, \ldots, L). \tag{4.3.5}$$

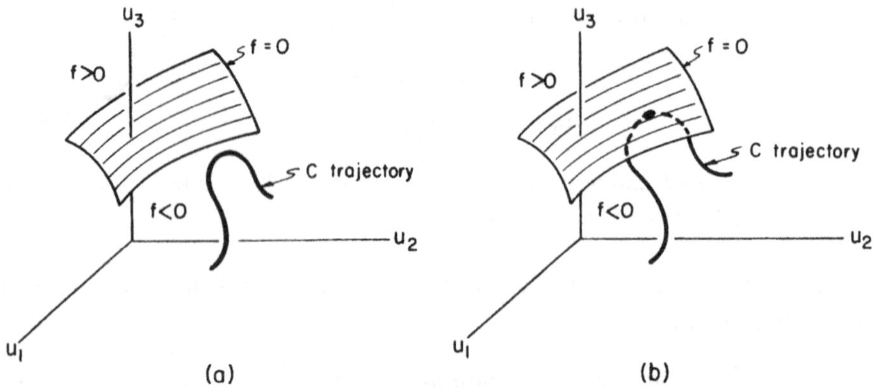

Fig. 4.3.1. a) C trajectory satisfying the constraint $f(u_1, u_2, u_3) < 0$; b) C trajectory satisfying the constraint $f(u_1, u_2, u_3) \leq 0$.

Then, these last equations are, in fact, a set of L first-order differential equations, and (4.3.4) are their integrals. For this relation to exist between L first-order differential equations and their integrals, it is not necessary that the differential equations be exact differentials. For instance, we may multiply each of (4.3.5) by an arbitrary function

$$g_r = g_r(u_1, u_2, \ldots, u_N, t) \qquad (r = 1, 2, \ldots, L)$$

and then form L independent linear combinations of the resulting equations. They will have the general form

$$\sum_{s=1}^{N} A_{rs}\, du_s + A_r\, dt = 0 \qquad (r = 1, 2, \ldots, L) \tag{4.3.6}$$

where the A_{rs} and A_r are functions of $u_1, u_2, \ldots, u_N, t$, and these equations will still have (4.3.4) as their integrals because (4.3.5) and (4.3.6) are completely equivalent, the only difference between them being that (4.3.5) are exact differentials while (4.3.6) are not.

The generalization of (4.3.6) to *nonholonomic* constraints consists in considering constraint equations of the form (4.3.6) which are, however, not integrable. Expressed more precisely:

> *If a system of L independent equations of constraint of the form* (4.3.6) *does not possess L integrals of the form* (4.3.4), *the system of constraints is* nonholonomic.

We have already encountered a nonholonomic constraint of this type in Example 4.2.3.

4.4. The Pfaffian Forms

The general form of equality constraints considered in classical mechanics is

$$\sum_{s=1}^{N} A_{rs}\, du_s + A_r\, dt = 0 \qquad (r = 1, 2, \ldots, L), \tag{4.4.1}$$

in which the A_{rs} and A_r are (at least once piecewise differentiable) functions of the u_s ($s = 1, 2, \ldots, N$) and of t. They are L linear differential forms in more variables than there are equations, i.e., $N + 1 > L$. If the number of equations and variables were equal, the set (4.4.1) would always be integrable under very broad conditions on the A_{rs} and A_r, i.e., this system would then be holonomic, in general. The constraint equations are said to be in *Pfaffian* form because the German mathematician Pfaff was the first to explain the meaning of nonintegrable differential forms that are fewer in number than that of the variables involved. However, the term "Pfaffian form" is used whether the system (4.4.1) is holonomic or not.

Obviously, the exact differentials (4.2.10) and (4.2.11) are Pfaffian forms. The first was obtained from a rheonomic set, the second from a scleronomic set. It may, therefore, seem entirely reasonable to retain the terms "rheonomic" and "scleronomic" for the Pfaffian forms (4.2.10) and (4.2.11), respectively. This is, in fact, done by many authors. They call a general Pfaffian form (4.4.1) scleronomic when t does not occur explicitly in it, and rheonomic otherwise, and they apply this nomenclature whether the system is holonomic, or not.

However, for reasons which will become evident later, it is more useful to distinguish between the two cases: $A_r \equiv 0$ for all $r = 1, 2, \ldots, L$, and $A_r \not\equiv 0$ for one or more than one r, rather than between scleronomic and rheonomic Pfaffian forms. We shall say:

> *A constraint equation of the form* (4.4.1) *is called* catastatic *when* $A_r \equiv 0$; *otherwise it is* acatastatic. *When every* $A_r \equiv 0$ ($r = 1, 2, \ldots, L$) *the* system *is catastatic, otherwise it is acatastatic.*

The word "catastatic" comes from the Greek and means "orderly."

The geometric interpretation of catastatic and acatastatic constraints is quite interesting. It is best illustrated by means of an example with two spatial coordinates only. Consider the catastatic constraint

$$a(x, y, t)\, dx + b(x, y, t)\, dy = 0 \tag{4.4.2}$$

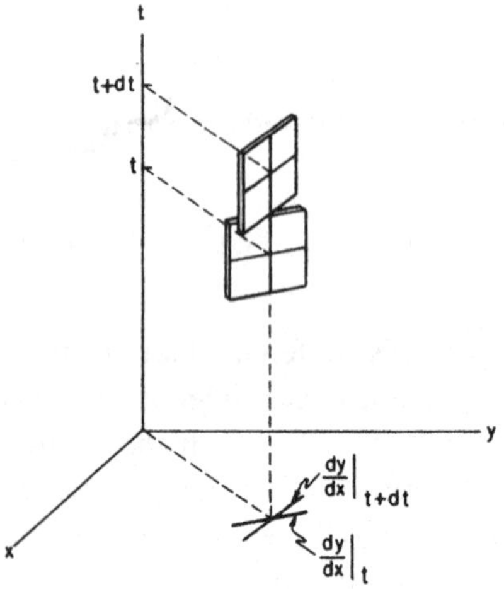

Fig. 4.4.1. Schematic representation of a catastatic constraint.

and the acatastatic constraint

$$a(x, y, t)\,dx + b(x, y, t)\,dy + c(x, y, t)\,dt = 0, \tag{4.4.3}$$

where $c \not\equiv 0$.

From (4.4.2), one has

$$\frac{dy}{dx} = -\frac{a(x, y, t)}{b(x, y, t)} \tag{4.4.4}$$

so that for fixed x, y, the slope dy/dx changes with time (t is treated as a parameter). In the x, y, t space the tangent plane at (x, y, t) defined by the constraint changes with time; this is shown in Fig. 4.4.1. It is the essential

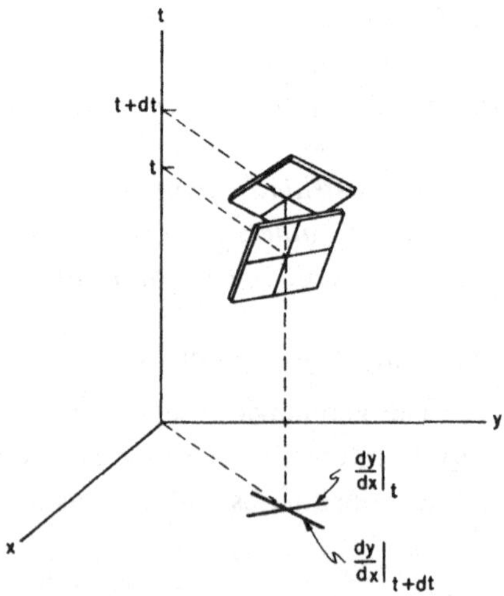

Fig. 4.4.2. Schematic representation of an acatastatic constraint.

property of the catastatic constraint that the tangent plane turns about a line parallel to the t axis for all t.

The tangent plane in the x, y, t space represented by an acatastatic constraint is shown in Fig. 4.4.2. The essential property distinguishing it from catastatic constraints is that, for fixed x, y, the tangent plane no longer rotates about a line parallel to the t axis as t is varied.

4.5. When is a System of Constraints Holonomic?

In holonomically constrained dynamical systems, the constraints are usually given in the finite form

$$f_r(u_1, u_2, \ldots, u_N, t) = c_r \qquad (r = 1, 2, \ldots, L < N), \tag{4.5.1}$$

where the c_r are constants. However, when they are given in the form

$$\sum_{s=1}^{N} A_{rs}\dot{u}_s + A_r = 0 \qquad (r = 1, 2, \ldots, L), \tag{4.5.2}$$

or in the Pfaffian form

$$\sum_{s=1}^{N} A_{rs}\, du_s + A_r\, dt = 0 \qquad (r = 1, 2, \ldots, L), \tag{4.5.3}$$

it may be very difficult, in general, to decide whether or not these constraints are holonomic or, expressed differently, whether or not there exist integrals of the form (4.5.1) of the differential equations (4.5.2).

Let us suppose that we are given a single constraint in the Pfaffian form

$$A\, dx + B\, dy + C\, dz = 0, \tag{4.5.4}$$

where A, B, and C are functions of x, y, and z. If (4.5.4) is an exact differential, there must exist a function f such that

$$A = \partial f/\partial x, \qquad B = \partial f/\partial y, \qquad C = \partial f/\partial z, \tag{4.5.5}$$

and the necessary and sufficient conditions for this to be true is that the first partial derivatives of A, B, and C with respect to x, y, and z exist, and that

$$\frac{\partial A}{\partial y} = \frac{\partial B}{\partial x}, \qquad \frac{\partial A}{\partial z} = \frac{\partial C}{\partial x}, \qquad \frac{\partial B}{\partial z} = \frac{\partial C}{\partial y}. \tag{4.5.6}$$

These last conditions are readily demonstrated by utilizing (4.5.5) and forming the indicated derivatives.

However, while (4.5.6) is necessary for the constraint to be an exact differential, it is not necessary for that constraint to be *integrable*. In fact, it is shown in elementary calculus books that the integrability condition for (4.5.4) is that the equation

$$A\left(\frac{\partial B}{\partial z}-\frac{\partial C}{\partial y}\right)+B\left(\frac{\partial C}{\partial x}-\frac{\partial A}{\partial z}\right)+C\left(\frac{\partial A}{\partial y}-\frac{\partial B}{\partial x}\right)=0 \quad (4.5.7)$$

be satisfied identically. If one defines a vector

$$V=(A, B, C)$$

equation (4.5.7) takes on the easily remembered form

$$V\cdot \operatorname{curl} V=0.$$

It is obvious that (4.5.7) is satisfied when the conditions (4.5.6) hold, i.e., when the constraint equation is an exact differential. However, the following example shows that the converse is not necessarily true.

Example 4.5.1. Show that the constraint

$$yz(y+z)\,dx+zx(z+x)\,dy+xy(x+y)\,dz=0$$

is holonomic.

Equating coefficients of the differentials between the given constraint and (4.5.4) and forming the first partial derivatives one finds

$$\frac{\partial A}{\partial y}=z(y+z)+yz, \qquad \frac{\partial A}{\partial z}=y(y+z)+yz,$$

$$\frac{\partial B}{\partial x}=z(z+x)+zx, \qquad \frac{\partial B}{\partial z}=x(z+x)+zx,$$

$$\frac{\partial C}{\partial x}=y(x+y)+xy, \qquad \frac{\partial C}{\partial y}=x(x+y)+xy.$$

Obviously, (4.5.6) are not satisfied because, for instance, $\partial A/\partial y$ contains y but not x, and $\partial B/\partial x$ contains x, but not y. However, if one substitutes in (4.5.7), one finds

$$\begin{aligned}yz(y+z)&\{x(z+x)+zx-[x(x+y)+xy]\}\\&+zx(z+x)\{y(x+y)+xy-[y(y+z)+yz]\}\\&+xy(x+y)\{z(y+z)+yz-[z(z+x)+zx]\}\equiv 0,\end{aligned}$$

as one sees readily by expanding the above expression.

When there exists a single constraint equation in N variables (one of which may be the time t), i.e., when the equation of constraint is of the form

$$\sum_{s=1}^{N} A_s(u_1, u_2, \ldots, u_N)\, du_s = 0 \tag{4.5.8}$$

it can be shown[†] that the necessary and sufficient condition for the existence of an integral of (4.5.8) of the form

$$f(u_1, u_2, \ldots, u_N) = \text{const}$$

is that the set of equations

$$\begin{aligned} A_\gamma\left(\frac{\partial A_\beta}{\partial u_\alpha} - \frac{\partial A_\alpha}{\partial u_\beta}\right) &+ A_\beta\left(\frac{\partial A_\alpha}{\partial u_\gamma} - \frac{\partial A_\gamma}{\partial u_\alpha}\right) \\ &+ A_\alpha\left(\frac{\partial A_\gamma}{\partial u_\beta} - \frac{\partial A_\beta}{\partial u_\gamma}\right) = 0 \qquad (\alpha, \beta, \gamma = 1, 2, \ldots, N) \end{aligned} \tag{4.5.9}$$

be simultaneously and identically satisfied. There are $N(N-1)(N-2)/6$ such equations, of which $(N-1)(N-2)/2$ are independent. If (4.5.8) is an exact differential one has

$$A_s = \partial f/\partial u_s \qquad (s = 1, 2, \ldots, N)$$

and these require that

$$\partial A_\alpha/\partial u_\beta = \partial A_\beta/\partial u_\alpha \qquad (\alpha, \beta = 1, 2, \ldots, N).$$

In that case, the equations (4.5.9) are seen to be trivially satisfied.

Finally, when there are L independent constraint equations of the form

$$\sum_{s=1}^{N} A_{rs}(u_1, u_2, \ldots, u_N)\, du_s = 0 \qquad (r = 1, 2, \ldots, L < N), \tag{4.5.10}$$

where one of the u_s may be the time t, it was shown by Frobenius[‡] that the necessary and sufficient condition for the existence of L independent integrals of (4.5.10) of the form

$$f_r(u_1, u_2, \ldots, u_N) = c_r \qquad (r = 1, 2, \ldots, L)$$

[†] See Ince, E. L., *Ordinary Differential Equations*, Dover Publications, Inc., New York (1953), p. 54.

[‡] Frobenius, F. G., *Gesammelte Abhandlungen*, Springer, Göttingen (1968), pp. 249–334.

with the c_r constants, is that the bilinear forms

$$\sum_{\alpha=1}^{N}\sum_{\beta=1}^{N}\left(\frac{\partial A_{r\beta}}{\partial u_\alpha}-\frac{\partial A_{r\alpha}}{\partial u_\beta}\right)x_\alpha y_\beta=0 \qquad (r=1,2,\ldots,L) \tag{4.5.11}$$

be satisfied simultaneously and identically, where the x_α and y_β are any two sets of solutions of the algebraic equations

$$\sum_{s=1}^{N} A_{rs}x_s=0 \qquad (r=1,2,\ldots,L). \tag{4.5.12}$$

As there are fewer equations (4.5.12) then there are variables, (4.5.12) will in general have more than one set of solutions. The Frobenius condition is difficult to apply, in general, because the test for integrability must be preceded by the computationally difficult task of finding two sets of solutions of (4.5.12); however, we do utilize it in Section 9.5.

Again, if each of the equations (4.5.10) is an exact differential, one must have

$$A_{rs}=\partial f_r/\partial u_s$$

and this requires that

$$\partial A_{r\beta}/\partial u_\alpha=\partial A_{r\alpha}/\partial u_\beta \qquad (\alpha,\beta=1,2,\ldots,N;\ r=1,2,\ldots,L).$$

When this is the case, the Frobenius conditions are trivially satisfied.

4.6. Accessibility (of the Configuration Space)

In Chapter 3, the motion of a system is represented as a sequence of configurations or in terms of a C trajectory in configuration space. In Chapter 4, holonomic constraints are interpreted as surfaces, and the imposition of these constraints is interpreted to mean that the C trajectory must lie in the intersection of the surfaces defined by the holonomic constraints. Hence, only the configurations lying in this intersection of surfaces are "accessible." In fact, it is evident that when $L < N$ *independent holonomic* constraints are imposed on a system whose configuration space has N dimensions, the dimensionality of the space of accessible configurations is $N - L$.

We have also examined the effect of *nonholonomic* configuration constraints, which are, in general, inequality constraints, and we noted that the dimensionality of the space of accessible configurations is the same as that of the configuration space. However, certain *domains* of the configuration space are not accessible, as illustrated for example in Fig. 4.3.1.

It remains to study the effect on the accessibility of configurations produced by *nonholonomic equality* constraints. We state the result:

> *All configurations accessible in the* absence *of nonholonomic equality constraints of the form* (4.4.1) *are also accessible in their* presence.

Pfaff has shown that the general nonintegrable Pfaffian equation of the form (4.4.1) can be reduced to a system of equations of the form†

$$dy - z\,dx = 0. \tag{4.6.1}$$

Let it be required to reach the arbitrary configuration $x = x_1$, $y = y_1$, $z = z_1$ from the origin of the configuration space. This can evidently be done by following the path

$$y = f(x), \qquad z = df/dx, \tag{4.6.2}$$

where $f(x)$ is any once-differentiable function that satisfies

$$f(0) = f'(0) = 0, \qquad f(x_1) = y_1, \qquad f'(x_1) = z_1, \tag{4.6.3}$$

with $' = d/dx$. Substitution of (4.6.2) in (4.6.1) shows that the constraint is identically satisfied, and of (4.6.3) in (4.6.2) that, when $x = x_1$, the configuration (x_1, y_1, z_1) is indeed reached; this is illustrated in Fig. 4.6.1.

The result quoted here is intuitively appealing by considering the classical nonholonomic problem of the skate, or the knife edge. The constraint on the skate is that it should always be directed tangent to its path, or

$$\cos\theta\,dy - \sin\theta\,dx = 0,$$

where (x, y) is the point of contact of the skate with the ice, and θ is the angle between the direction of the skate and the x axis. Thus, the constraint on the skate is of the form (4.6.1). Now, it is evident to any one who has skated that one may skate to any prescribed point (x, y) on the ice. Then, one merely need rotate the skate about the point of contact until it has any

† Quoted from Pars, p. 17.

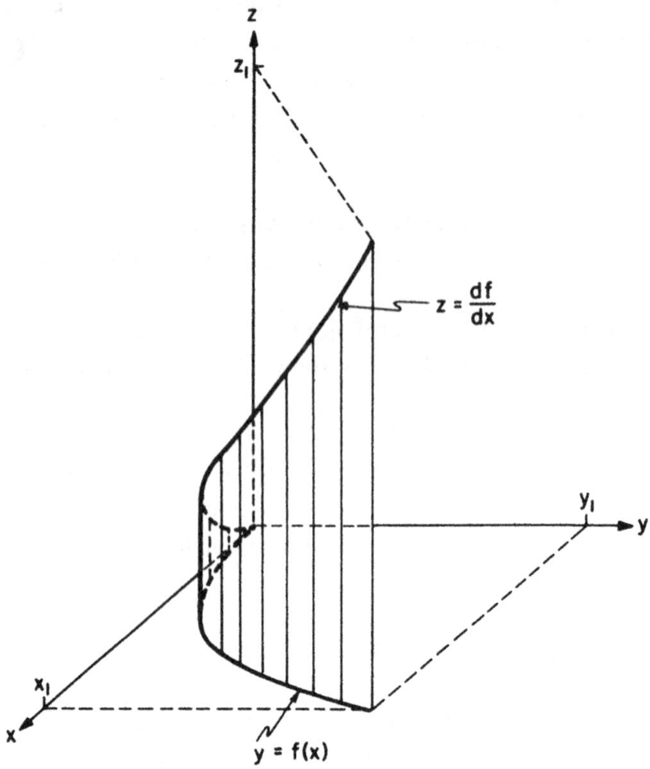

Fig. 4.6.1. Demonstration that under nonholonomic equality constraints, every configuration is accessible.

prescribed direction θ; in this way, any arbitrary configuration (x, y, θ) may be attained even though the problem is subject to a nonholonomic equality constraint.

We emphasize here one of the characteristic and important differences between holonomic and nonholonomic constraints:

The dimension of the space of accessible configurations is reduced by holonomic constraints, but not by nonholonomic constraints.

It is reasonable, therefore, to suppose that holonomic constraints permit a reduction of the number of coordinates needed to formulate a given problem, but nonholonomic constraints do not. This will, indeed, be found to be the case; this least number of coordinates is called a set of generalized coordinates (see Chapter 11).

4.7. Problems

4.1. Two particles having Cartesian coordinates (x_1, y_1, z_1) and (x_2, y_2, z_2), respectively, are attached to the extremities of a bar whose length $l(t)$ changes with time in a prescribed fashion. Give the equations of constraint on the finite and infinitesimal displacements of the Cartesian coordinates.

4.2. What are the equations of constraint on the finite and infinitesimal coordinates (x_1, y_1, z_1) and (x_2, y_2, z_2) of the bobs of a double spherical pendulum of lengths l_1 and l_2, respectively?

4.3. A thin bar of length $l < 2r$ can move in a plane in such a way that its endpoints are always in contact with a circle of radius r. If the Cartesian coordinates of its endpoints are (x_1, y_1) and (x_2, y_2), respectively, what constraints on finite and infinitesimal displacements must these coordinates satisfy?

4.4. Answer the same questions as in Problem 4.3 if the circle is replaced by an ellipse having major axis $2a$ and minor axis $2b$, and $l < 2b$. Are these constraints holonomic?

4.5. Discuss the changes in the answer to Problem 4.4 if $2b < l < 2a$. Is this constraint holonomic?

4.6. The motion of an otherwise unconstrained particle is subject to the conditions $\dot{z} = \dot{x}\dot{y}$. Discuss the constraint on the infinitesimal and finite displacements.

4.7. A disk of radius r is permanently in the vertical plane and rolls without sliding on a given profile $g(x)$, as shown. A rod of length l remains in the same vertical plane as the disk. One of its extremities is attached to, and can rotate about, the disk center; the other extremity is moved on a straight line parallel to the x axis in a prescribed fashion $f(t)$. Express the wheel angular velocity as a function of l, r, $g(x)$, and $f(t)$. Classify this constraint equation.

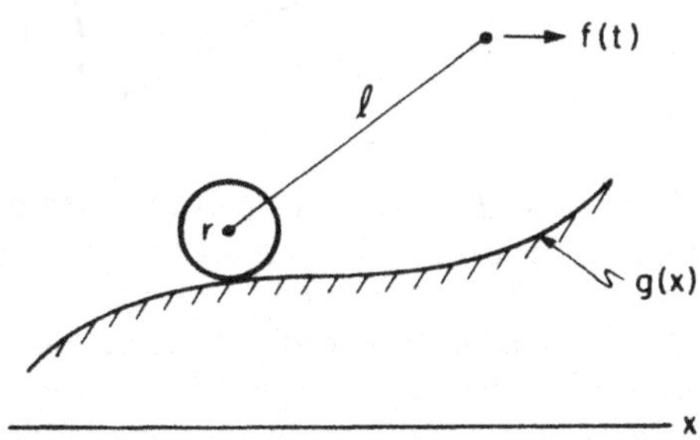

4.8. A particle moving in the xy plane is connected by an inextensible string of length l to a point P on the rim of a fixed disk of radius r, as shown. The line PO makes the angle θ with the x axis. What are the constraints on the finite and infinitesimal displacements of the point at the free end of the string having the position (x, y)?

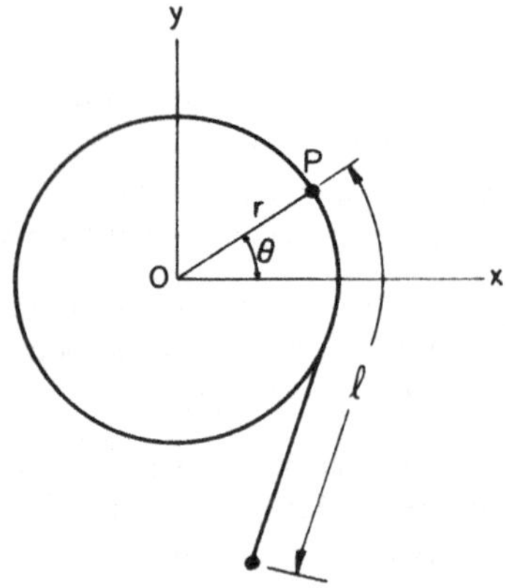

4.9. A circular shaft of variable radius $r(x)$ rotates with angular velocity $\omega(t)$ about its centerline, as shown. The shaft is translated along its centerline in a prescribed fashion $f(t)$. Two disks of radii r_1 and r_2, respectively, roll without slipping on the shaft. A mechanism permits the disks to rise and fall in such a way that the disk rims never lose contact with the shaft. Show that the relation, free of ω, between the angular displacements φ_1 and φ_2 of the disks is in general nonholonomic. State the general condition that must be satisfied in the exceptional case that the constraint is holonomic and give an example.

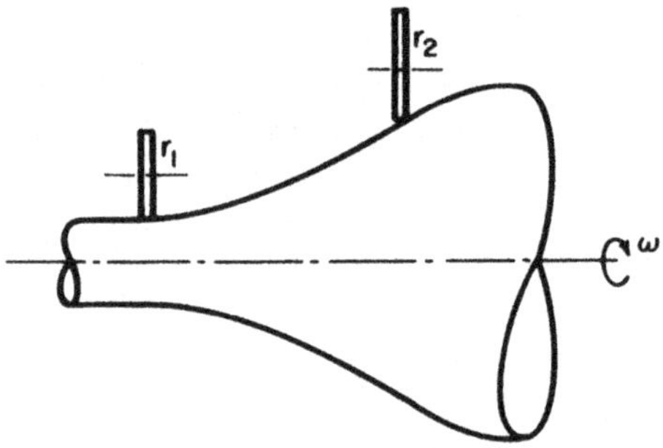

4.10. The position of a point P moving in a plane is given by the polar coordinates r and φ, where r is measured with respect to some fixed point O, and φ with respect to some fixed line L passing through O. We chose to express the position of P by means of the coordinates r, φ, and A, where A is double of the area swept out by the vector r, and $A = 0$ when $\varphi = 0$. Evidently, r, φ, and A must satisfy a constraint between them because two coordinates are sufficient to give the position of P. Find and classify this constraint.

4.11. A particle moving in the vertical plane is steered in such a way that the slope of its trajectory is proportional to its height. Formulate this constraint mathematically and classify it.

4.12. Write down and classify the equation of constraint of a particle moving in a plane if its slope is always proportional to the time.

4.13. A particle P moving in 3-space is steered in such a way that its velocity is directed for all time toward a point O which has a prescribed motion in space and time. Formulate and classify the equation(s) of constraint of the particle motion under the assumption that the positions of P and O never coincide.

4.14. A particle P can move on the bottom of a massless two-dimensional cage of width $2a$ and height b as shown. Let the position of the particle from the middle of the cage bottom be ξ. The cage is hinged at the middle of the top with a massless bar of length l, and the angle which the top of the cage makes with the normal to l is denoted by φ. Finally, the angle which the bar l makes with the vertical y axis is θ. The particle position may either be given

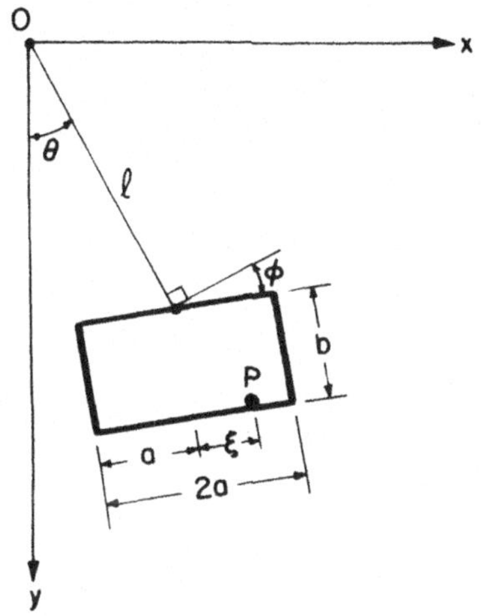

by its x and y components, or by the three coordinates θ, φ, and ξ. Establish the relation between θ, φ, and ξ on the one hand and x and y on the other. Since the position of the particle can be described by the two coordinates x and y, or by the three coordinates θ, φ, and ξ, it seems that the latter must satisfy a constraint relation between them. Can you find that relation? How many degrees of freedom does the particle have?

5

The Strictly Newtonian Mechanics Problem

5.1. General Remarks

In this section we define with some precision the problem of classical particle mechanics. As observed in Section 2.3, Newtonian systems have been divided into those called "strictly Newtonian" (SN), in which impulsive forces are excluded, and those called "Newtonian" (N), in which unbounded forces of bounded impulse are admitted. The strictly Newtonian systems are the more common; they are the ones on which a first course in dynamics is usually centered. Thus, we begin with formulation of (SN). Systems (N) which are not (SN) will be treated later in a chapter on impulsive motion (see Chapter 21).

5.2. The Given Quantities and Relations

We assume that a real number R is given which defines in $\mathscr{E}^3$ a sphere called the living space:

$$\mathscr{L}(R) = \left\{x \colon x \in \mathscr{E}^3;\ \sum_{i=1}^{3} x_i^2 \leq R^2\right\}. \tag{5.2.1}$$

At the time t the living space contains a nonzero particle population

$$\begin{aligned}\Pi_{\mathscr{L}}(t, R) = \{P_r[(x^r(t)), m_r] \colon x^r = (x_1^r, x_2^r, x_3^r) \in \mathscr{L}(R);\ m_r \in \mathscr{M};\\ t \in \mathscr{T};\ r = 1, 2, \ldots, n(t) \leq s\},\end{aligned} \tag{5.2.2}$$

where s and n are integers, and where the sets

$$\mathscr{M} = \{m_r: 0 < m_r < \infty\}, \qquad \mathscr{T} = \{t: -\infty < t < \infty\} \tag{5.2.3}$$

are defined on the line of reals. There are given, moreover, n *distinct* arbitrary vectors

$$x^r(0) = (x_1^r(0), x_2^r(0), x_3^r(0)) \qquad (r = 1, 2, \ldots, n), \tag{5.2.4}$$

called initial positions, and n vectors that may or may not be distinct:

$$\dot{x}^r(0) = (\dot{x}_1^r(0), \dot{x}_2^r(0), \dot{x}_3^r(0)) \qquad (r = 1, 2, \ldots, n), \tag{5.2.5}$$

called the initial velocities.

There are given $K \geq 1$ inequality relations

$$f^p(x_1^1, x_2^1, \ldots, x_2^n, x_3^n, t) \leq 0 \qquad (p = 0, 1, 2, \ldots, K-1) \tag{5.2.6}$$

with $f^0 \equiv 0$, and L equalities,

$$\sum_{j=1}^{3} \sum_{r=1}^{n} f_j^{qr}\, dx^j + f^q\, dt = 0 \qquad (q = 0, 1, 2, \ldots, L-1) \tag{5.2.7}$$

with $f_j^{0r} = f^0 \equiv 0$. The f_j^{qr} and f^q are everywhere continuous functions of the x_j^r, and they may depend on t as well.

The K relations (5.2.6) are nonholonomic configuration constraints. The L relations (5.2.7) are holonomic constraints if and only if the system (5.2.7) is integrable; otherwise they are nonholonomic equality constraints.

There are given, moreover, the n differential vector equations

$$m_r \ddot{x}^r = F^r(x_1^1, x_2^1, \ldots, x_3^n; \dot{x}_1^1, \ldots, \dot{x}_3^n; t) \quad (r = 1, 2, \ldots, n), \tag{5.2.8}$$

where

$$x^r = (x_1^r, x_2^r, x_3^r),$$
$$\dot{x}_i^r = dx_i^r/dt,$$
$$F^r = (F_1^r, F_2^r, F_3^r).$$

The F^r are bounded, piecewise continuous, and Lipschitzian functions in the x_i^r and $\dot{x}_i^r$, and they may depend on t as well.†

† Let $\| F \| = \Sigma_{r=1}^N | F^r |$ and $\| u \| = \Sigma_{r=1}^N | u_r |$. Moreover, let $u^1 = (u_1^1, u_2^1, \ldots, u_N^1)$ and $u^2 = (u_1^2, u_2^2, \ldots, u_N^2)$ be any two distinct points in a domain $U \in \mathscr{E}^N$, and let every

5.3. The First Problem

Given every $x^r(t)$ $(r = 1, 2, \ldots, n)$ having initial values (5.2.4) and initial time derivatives (5.2.5), satisfying all relations (5.2.6) and (5.2.7) as well as

$$x^p(t_i) \neq x^q(t_i) \qquad (p, q = 1, 2, \ldots, n;\ p \neq q) \tag{5.3.1}$$

for every $t_i \in \mathscr{T}$: *Find every* F^r.

The most common example of the first problem is the statics problem in which $x^r(t) \equiv x^r(0)$ $(r = 1, 2, \ldots, n)$. In other words, in the statics problem the prescribed motion is rest; it is desired to find forces such that rest persists for all time. One sees immediately that the statics problem must necessarily be subject to constraints, for an unconstrained particle will always accelerate with respect to an inertial frame when it is acted on by nonvanishing forces.

There are many meaningful problems of the first type which do involve motion rather than rest. An illustration of the first problem is furnished by:

Example 5.3.1. The ram of a metal shaper is to have a quick-return mechanism. Let its motion during the work-stroke and during the return-stroke be sinusoidal, but the return-stroke is twice as fast as the work-stroke. What force on the ram will achieve this if the load is constant during the work-stroke and zero during the return stroke, and if Coulomb friction acts?

In Fig. 5.3.1, whe show the time history of the motion of the ram with the midway position as the reference point. Then, the motion is given by

$$\begin{aligned} x &= -\,l \cos \frac{\pi}{T}\, t = x_1 \qquad &&\text{for } \dot{x} \geq 0, \\ x &= l \cos \frac{2\pi}{T}\, t = x_2 \qquad &&\text{for } \dot{x} \leq 0. \end{aligned} \tag{a}$$

This may be combined into a single equation by means of the sgn function. The function sgn u is $+1$ or -1 according as u is positive or negative. Its graph is shown in Fig. 5.3.2. With the aid of this function we may write (a) as

$$x = \tfrac{1}{2}(x_1 - x_2) \operatorname{sgn} \dot{x} + \tfrac{1}{2}(x_1 + x_2)$$

F^r be of the form $F^r(u_1, u_2, \ldots, u_N, t) = F^r(u, t)$. Then, F is said to be Lipschitzian on U in the variable u if

$$\| F(u^2, t) - F(u^1, t) \| \leq K \| u^2 - u^1 \|,$$

where K is the "Lipschitz constant."

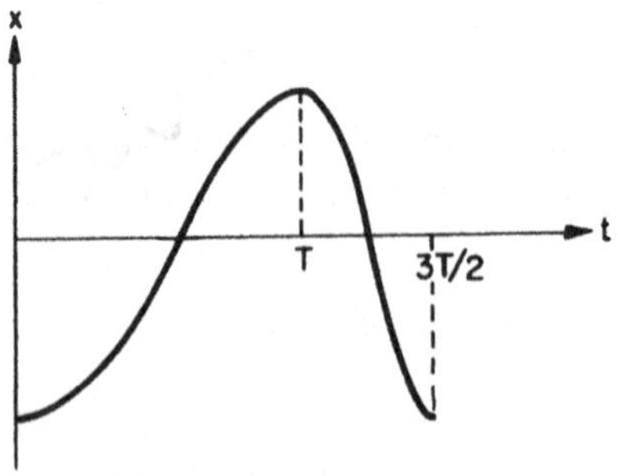

Fig. 5.3.1. Time history of the motion of Example 5.3.1.

or, after substitution and some simplification,

$$x = -\frac{l}{2}(1 + \operatorname{sgn}\dot{x})\cos\frac{\pi}{T}t + \frac{l}{2}(1 - \operatorname{sgn}\dot{x})\cos\frac{2\pi}{T}t. \tag{b}$$

From (a), the acceleration is

$$\begin{aligned} \ddot{x} &= l\left(\frac{\pi}{T}\right)^2 \cos\frac{\pi}{T}t = -\left(\frac{\pi}{T}\right)^2 x_1 = \ddot{x}_1 \qquad \text{for } \dot{x} \geq 0, \\ \ddot{x} &= -l\left(\frac{2\pi}{T}\right)^2 \cos\frac{2\pi}{T}t = -\left(\frac{2\pi}{T}\right)^2 x_2 = \ddot{x}_2 \qquad \text{for } \dot{x} \leq 0, \end{aligned} \tag{c}$$

or, using the sgn function,

$$\ddot{x} = \frac{l}{2}\left(\frac{\pi}{T}\right)^2(1 + \operatorname{sgn}\dot{x})\cos\frac{\pi}{T}t - \frac{l}{2}\left(\frac{2\pi}{T}\right)^2(1 - \operatorname{sgn}\dot{x})\cos\frac{2\pi}{T}t. \tag{d}$$

This may also be written, from (d), as

$$\ddot{x} = \frac{1}{2}\left(\frac{\pi}{T}\right)^2(3\operatorname{sgn}\dot{x} - 5)x. \tag{e}$$

The work load is

$$\begin{aligned} W &= C \neq 0 \qquad \text{for } \dot{x} \geq 0, \\ W &= 0 \qquad \text{for } \dot{x} \leq 0, \end{aligned}$$

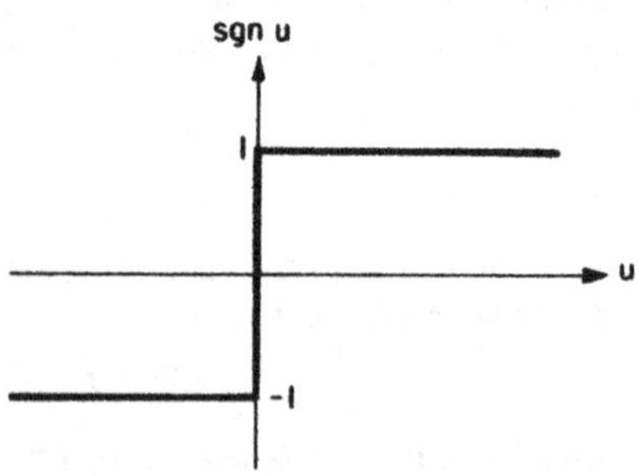

Fig. 5.3.2. The signum function used in Example 5.3.1.

or, using the sgn function,

$$W = \tfrac{1}{2}C(1 + \operatorname{sgn} \dot{x}). \tag{f}$$

The friction force always opposes the velocity, or

$$f = -c \operatorname{sgn} \dot{x}. \tag{g}$$

If P is the desired force,

$$m\ddot{x} = P + f - W$$

or, after some algebraic manipulations,

$$P = \frac{1}{2} m \left(\frac{\pi}{T}\right)^2 (3 \operatorname{sgn} \dot{x} - 5)x + c \operatorname{sgn} \dot{x} + \frac{1}{2} C(1 + \operatorname{sgn} \dot{x}). \tag{h}$$

This is the required answer.

A more involved example is that of finding the forces which must act on an airplane such that it moves on a given path and with prescribed time history of the motion from the take-off position in San Francisco to the terminal at Kennedy Airport in New York. One sees that problems of the first type may be very complex.

5.4. The Second Problem

Given every F^r satisfying the above properties (5.2.8), find every vector function $x^r(t)$ satisfying the initial conditions (5.2.4) and (5.2.5), the constraints (5.2.6) and (5.2.7), the impenetrability conditions (5.3.1), and which satisfy the set of differential equations (5.2.8) for all $t \in \mathscr{E}$.

The second problem is by far the more difficult and the more interesting of the two. It is essentially one of prediction, e.g., one wishes to find the particle motion for all future times when the initial state of each particle is given.

5.5. Other Problems

There exist many fascinating problems of dynamics not included among the above two. One of these is the so-called "identification problem." In it, all external forces acting on an unknown or partially known system are given. Certain features of the motion of the system are observed. Identify the entire system.

Another problem often encountered deals with systems "with memory." Here, the behavior of the system depends not only on its state at the time t but also on past states. In these systems, the forces are of the form $F^r(x_1{}^1, \ldots, x_3{}^n; \dot{x}_1{}^1, \ldots, \dot{x}_3{}^n; t; \tau)$, where $0 \leq \tau < t$.

Still another, and a very timely problem of dynamics, related remotely to the first problem, is the so-called "optimization problem." In it one wishes to determine forces such that the system will behave in an optimal manner. A typical example is this: The initial and terminal states of a particle are prescribed. The force acting on the particle is arbitrary except that its magnitude may not exceed some given amount. Find the force magnitude and direction for every value of t such that the system will attain the prescribed terminal state in the shortest possible time if it starts from the prescribed initial state.

5.6. Concluding Remarks

It is clear that the formulation of the first two problems is not exhaustive, nor does the list become complete when the other problems are added to it. In this book, we center our attention on the second problem.

6

Some Rigid Body Kinematics

6.1. The Rigid Body

Rigid bodies are, by definition, systems of particles in which the distance between any two particles remains constant for all time and for all configurations. In other words, the particles of a rigid body do not move with respect to a coordinate system fixed in it. For this reason it is sometimes convenient to fix a coordinate system in a moving body, and to describe the motion in terms of the components along the axes of this moving coordinate system. As the coordinate system may then not be a Galilean frame, the resulting expressions for the acceleration components along the axes of the moving frame, but relative to an inertial frame, may become quite complicated.

The system is called *rigid* if, and only if, the distance between any two particles remains constant for all time when nonvanishing forces act on some or all particles of a system. One can easily verify that there exist precisely $n(n-1)/2$ distinct distances between n particles. Therefore, a rigid system of n particles satisfies $n(n-1)/2$ holonomic constraints.

In the absence of these constraints, $3n$ coordinates are required to fix the configuration of the n particles. Nevertheless, the rigid system cannot have $3n - n(n-1)/2$ degrees of freedom (see Section 4.2) because that number is positive only for $n < 7$. One concludes that the $n(n-1)/2$ constraints are not all independent. In Fig. 6.1.1 a *two-dimensional* rigid system of four particles is shown. While there exist six distinct rigid connections between them, it is easily seen that any one of the six may be removed without destroying the rigidity, but, if more than one is removed, the system is no longer rigid.

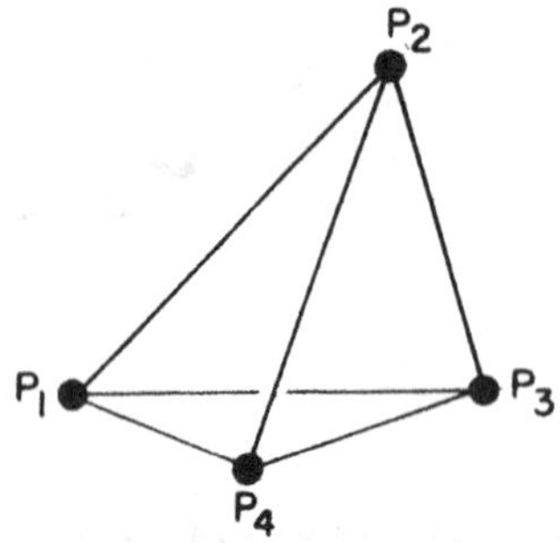

Fig. 6.1.1. Two-dimensional rigid system of four particles.

If the number of rigid connections exceeds that necessary for the system to be rigid, we call the system *over-rigid*. If the number of rigid connections is such that the removal of one of them destroys the rigidity of the system, we call the system *just-rigid*.

It is easy to see that a just-rigid system of two particles has one rigid connection, and a just-rigid system of three noncollinear particles has three rigid connections. A just-rigid system of n particles, with $n > 3$, requires $3n - 6$ rigid connections. This is readily demonstrated. Let m be the number of particles in excess of 3, or $n = 3 + m$. Each particle which is added to a rigid system requires, in general, three additional connections to be just-rigidly attached. Therefore, the m particles in excess of three require $3m$ rigid connections, while the original three particles have three connections. Thus, the least number of rigid connections between $n = 3 + m$ particles of a rigid system is $3 + 3m$. Since $m = n - 3$, the least number of connections is $3 + 3(n - 3) = 3n - 6$, as claimed.

If the particles are not on a straight line, the number of degrees of freedom of a just-rigid system of any number of particles is 6, i.e., it is $3n - (3n - 6)$. This result is unchanged for over-rigid systems because redundant rigid connections do not decrease the degrees of freedom of the system. Now, nine coordinates are necessary and sufficient to fix the configuration of three points, and if their distance is fixed, three constraints exist between them. Thus, the six degrees of freedom may be derived from the coordinates of three noncollinear points in a rigid body.

If all particles of a rigid system lie on a straight line, the configuration of the system is given by the position of two of its points; this requires six numbers. But there exists one constraint between these two points. Hence, a rigid system of particles on a straight line has only five degrees of freedom.

We have seen that a rigid system of particles has in general six degrees of freedom, regardless of the number of particles. Thus, the constraints have been utilized to reduce the dimensions of the configuration space from $3n$ to six. We know from an elementary study of mechanics that there exists a wide choice as to which six quantities may be used to describe the configuration. For instance, we may use the position of three points, as suggested above, but the most common method of defining the motion of a rigid body is based on Chasle's theorem:

The general motion of a rigid body may be described by a translation along a line and a rotation about that line.

With that description, the motion of a rigid body is given by three translational and three rotational components of the motion. The question of the best choice of coordinates to describe the motion of rigid bodies will be examined in more detail later on (see Chapter 11).

6.2. Finite Rotation

We wish to examine the motion of a rigid system of particles, called hereafter a *rigid body*, relative to an inertial reference frame. Let this inertial frame be given by the Cartesian triad of unit vectors $\hat{I}$, $\hat{J}$, $\hat{K}$ with origin at O. A convenient way to study this motion is to fix in the rigid body a Cartesian triad of unit vectors $\hat{i}$, $\hat{j}$, $\hat{k}$ with origin at O'. Then, the motion of the rigid body is that of the body-fixed Cartesian frame, and it in turn may be described by the translation of O' and by the rotation about a line through O. One sees, then, that if O' is fixed relative to the inertial frame, the only remaining motion is the rotation of the body-fixed triad about O'. For such a motion, one may fix O' at O without loss of generality.

Now, the concept of rotation is always studied in a first course in mechanics, but that study is usually carried out on an elementary level. In the balance of this chapter we study rotation in more general terms and by more advanced concepts and methods than commonly used in a first course in mechanics.

We first demonstrate the well-known fact that:

Finite rotation is not a vector quantity.

The three properties of vector quantities are:

(i) A vector has magnitude;
(ii) A vector has direction;
(iii) A vector may be added to another vector by the parallelogram law.

Finite rotation fails with respect to the third property, as we shall now show.

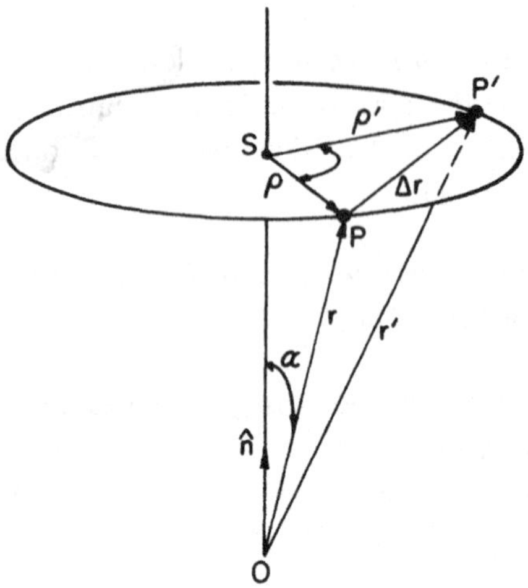

Fig. 6.2.1. Translation of P due to rotation about $\hat{n}$.

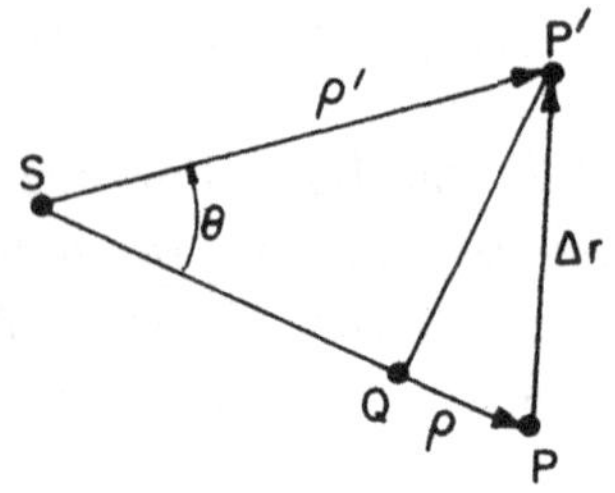

Fig. 6.2.2. The calculation of Δr.

Consider Fig. 6.2.1. Let a rigid body with point O fixed rotate about a line having the direction of the unit vector $\hat{n}$. Consider a point P on the body whose position vector with respect to O is r before rotation. Due to a rotation in the amount θ, the point is translated to P' with position vector r', and the change in the position of P is Δr. We denote the component of r normal to $\hat{n}$ by ϱ, and that of r' by ϱ', and the angle between $\hat{n}$ and r we call α. The vector Δr is decomposed into a component along ϱ, and one normal to it (see Fig. 6.2.2). The component normal to ϱ is $\overline{QP'}$ and is normal to the plane OSP defined by the vectors $\hat{n}$ and r; hence, it has the direction of $\hat{n}\times r$. An easy calculation shows this component to be equal to $(\hat{n}\times r)\sin\theta$. The component in the direction of ϱ is $\overline{PQ}$ and is normal to $\hat{n}\times r$, and it is also normal to $\hat{n}$ because it lies in the plane PSP'. Therefore, it has the direction of $\hat{n}\times(\hat{n}\times r)$. It is readily computed to be

$$\hat{n} \times (\hat{n} \times r)(1 - \cos\theta) = 2\hat{n} \times (\hat{n} \times r)\sin^2\frac{\theta}{2}.$$

It follows that

$$\Delta r = \sin\theta(\hat{n} \times r) + 2\sin^2\frac{\theta}{2}[\hat{n} \times (\hat{n} \times r)]. \tag{6.2.1}$$

Expanding (6.2.1) up to and including second-order terms in θ, one finds

$$\Delta r = \theta\hat{n} \times r + \frac{\theta^2}{2}[\hat{n} \times (\hat{n} \times r)] + \cdots . \tag{6.2.2}$$

Let us now consider two consecutive rotations $\theta_1\hat{n}_1$ and $\theta_2\hat{n}_2$, one in the amount θ_1 about a line of direction $\hat{n}_1$ and a second in the amount θ_2

about a line of direction $\hat{n}_2$, where $\hat{n}_1 \neq \hat{n}_2$, and neither $\hat{n}_1$ nor $\hat{n}_2$ are parallel to r, i.e., $\hat{n}_1 \times r \neq 0$, and $\hat{n}_2 \times r \neq 0$. Then the position of P after the first rotation is

$$r_1 = r + \theta_1 \hat{n}_1 \times r + \frac{\theta_1{}^2}{2} [\hat{n}_1 \times (\hat{n}_1 \times r)] + \cdots \tag{6.2.3}$$

and its position after the second rotation is

$$r_2 = r_1 + \theta_2 \hat{n}_2 \times r_1 + \frac{\theta_2{}^2}{2} [\hat{n}_2 \times (\hat{n}_2 \times r_1)] + \cdots . \tag{6.2.4}$$

Substituting (6.2.3) in (6.2.4) one finds up to and including second-order terms

$$\begin{aligned}(r_2 - r)_{1,2} = {} & \theta_1 \hat{n}_1 \times r + \theta_2 \hat{n}_2 \times r + \frac{\theta_1{}^2}{2} [\hat{n}_1 \times (\hat{n}_1 \times r)] \\ & + \frac{\theta_2{}^2}{2} [\hat{n}_2 \times (\hat{n}_2 \times r)] + \theta_1 \theta_2 [\hat{n}_1 \times (\hat{n}_1 \times r)] + \cdots .\end{aligned} \tag{6.2.5}$$

The subscript 1, 2 on the left-hand side of (6.2.5) indicates that $\theta_1 \hat{n}_1$ was performed first, and $\theta_2 \hat{n}_2$ subsequently. If this order of the rotations were reversed one would have (simply by exchanging subscripts 1 and 2)

$$\begin{aligned}(r_2 - r)_{2,1} = {} & \theta_2 \hat{n}_2 \times r + \theta_1 \hat{n}_1 \times r + \frac{\theta_2{}^2}{2} [\hat{n}_2 \times (\hat{n}_2 \times r)] \\ & + \frac{\theta_1{}^2}{2} [\hat{n}_1 \times (\hat{n}_1 \times r)] + \theta_2 \theta_1 [\hat{n}_1 \times (\hat{n}_2 \times r)] + \cdots .\end{aligned} \tag{6.2.6}$$

For rotations to add like vectors, it is necessary that $(r_2 - r_1)_{1,2} = (r_2 - r_1)_{2,1}$. Comparing (6.2.5) and (6.2.6), this requires that

$$\hat{n}_1 \times (\hat{n}_2 \times r) = \hat{n}_2 \times (\hat{n}_1 \times r). \tag{6.2.7}$$

But, by supposition $\hat{n}_1 \times r \neq 0$, $\hat{n}_2 \times r \neq 0$, and $\hat{n}_1 \neq \hat{n}_2$. Therefore, (6.2.7) is false, and the final position of P after two rotations does depend on the sequence in which they are made. A simple and well-known example where a change is the sequence of two 90° rotations leads to different end configurations is shown in Figs. 6.2.3 and 6.2.4.

While finite rotation is not a vector quantity, it turns out that

Infinitesimal rotation is a vector quantity

because when the second-order terms are deleted in (6.2.5) and (6.2.6),

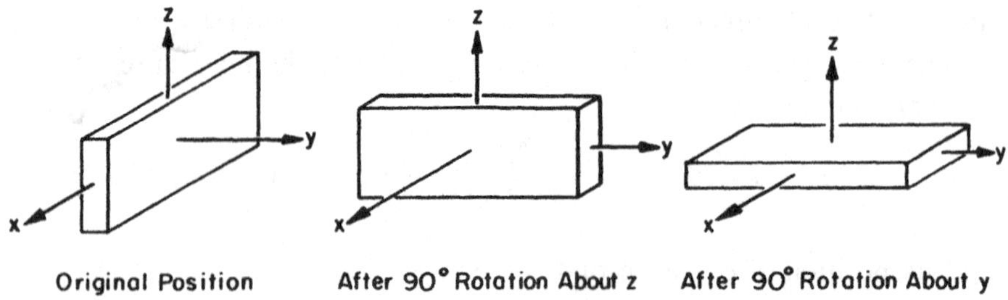

Fig. 6.2.3. Terminal configuration after two 90° rotations about two normal axes.

the equations become equal. In consequence, we also have:

Angular velocity is a vector quantity

because, if $d\theta$ is a vector, then $d\theta/dt$ is a vector.

6.3. The Direction Cosines

Let $\hat{\imath}$, $\hat{\jmath}$, $\hat{k}$ be a Cartesian reference triad with origin O, and let $\hat{\imath}'$, $\hat{\jmath}'$, $\hat{k}'$ be another Cartesian reference triad whose origin coincides with O. This is shown in Fig. 6.3.1.

Let α_1 be the angle between x' and x, α_2 that between x' and y, and α_3 that between x' and z. Then, the quantities

$$\begin{aligned} l_1 &= \cos\alpha_1 = \cos(x', x),\\ l_2 &= \cos\alpha_2 = \cos(x', y), \\ l_3 &= \cos\alpha_3 = \cos(x', z), \end{aligned} \tag{6.3.1}$$

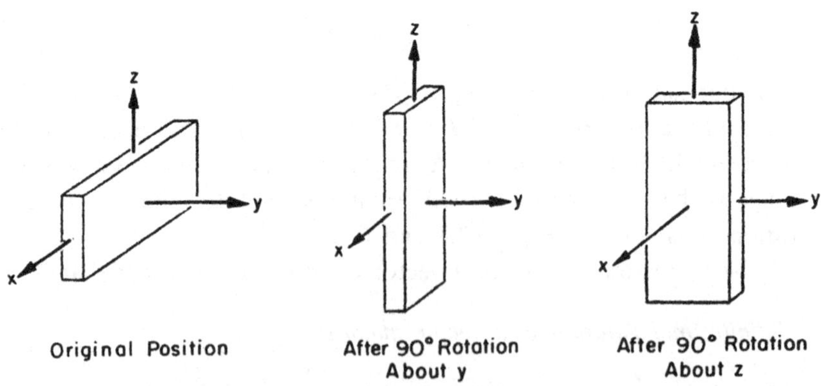

Fig. 6.2.4. Terminal configuration when the order of the rotations is reversed.

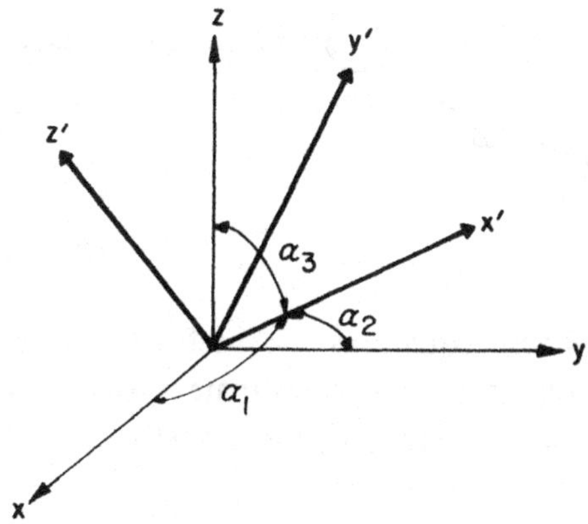

Fig. 6.3.1. Illustrations of the direction cosines.

are called the direction cosines of x' with respect to x, y, z. One has

$$\begin{aligned} l_1 &= \hat{\imath}' \cdot \hat{\imath}, \\ l_2 &= \hat{\imath}' \cdot \hat{\jmath}, \\ l_3 &= \hat{\imath}' \cdot \hat{k}. \end{aligned} \tag{6.3.2}$$

In a similar way we denote the direction cosines of y' with respect to x, y, z by m_1, m_2, m_3 and those of z' by n_1, n_2, n_3. Then

$$\begin{aligned} m_1 &= \hat{\jmath}' \cdot \hat{\imath}, \\ m_2 &= \hat{\jmath}' \cdot \hat{\jmath}, \\ m_3 &= \hat{\jmath}' \cdot k, \end{aligned} \tag{6.3.3}$$

and

$$\begin{aligned} n_1 &= \hat{k}' \cdot \hat{\imath}, \\ n_2 &= \hat{k}' \cdot \hat{\jmath}, \\ n_3 &= \hat{k}' \cdot \hat{k}. \end{aligned} \tag{6.3.4}$$

It follows that the primed vectors may be written in the unprimed system as

$$\begin{aligned} \hat{\imath}' &= l_1\hat{\imath} + l_2\hat{\jmath} + l_3\hat{k}, \\ \hat{\jmath}' &= m_1\hat{\imath} + m_2\hat{\jmath} + m_3\hat{k}, \\ \hat{k}' &= n_1\hat{\imath} + n_2\hat{\jmath} + n_3\hat{k}. \end{aligned} \tag{6.3.5}$$

Now, the cosine is an even function of its argument, or $\cos(x', x) = \cos(x, x')$. This is also clear because $\hat{\imath}' \cdot \hat{\imath} = \hat{\imath} \cdot \hat{\imath}'$. Thus, if we wish to express the unprimed unit vectors in the primed system, we also have

$$\begin{aligned} \hat{\imath} &= l_1\hat{\imath}' + m_1\hat{\jmath}' + n_1\hat{k}', \\ \hat{\jmath} &= l_2\hat{\imath}' + m_2\hat{\jmath}' + n_2\hat{k}', \\ \hat{k} &= l_3\hat{\imath}' + m_3\hat{\jmath}' + n_3\hat{k}'. \end{aligned} \tag{6.3.6}$$

These equations are useful to determine the vector components of any vector in one coordinate system when its components in the other are known. Thus, if, for a vector r,

$$\begin{aligned} r &= x\hat{\imath} + y\hat{\jmath} + z\hat{k} \\ &= x'\hat{\imath}' + y'\hat{\jmath}' + z'\hat{k}', \end{aligned} \tag{6.3.7}$$

one has

$$\begin{aligned} x' &= r \cdot \hat{\imath}' = l_1x + l_2y + l_3z, \\ y' &= r \cdot \hat{\jmath}' = m_1x + m_2y + m_3z, \\ z' &= r \cdot \hat{k}' = n_1x + n_2y + n_3z. \end{aligned} \tag{6.3.8}$$

If we consider $\hat{\imath}', \hat{\jmath}', \hat{k}'$ to be fixed in the rigid body, one of its points with position (x', y', z') has the position (x, y, z) relative to the $\hat{\imath}, \hat{\jmath}, \hat{k}$ system, and the nine quantities l_i, m_i, n_i $(i = 1, 2, 3)$ fix that position. But, clearly, these nine quantities cannot be independent because a body with one point fixed has only three degrees of freedom; therefore, three quantities must be sufficient to specify the position of one of its points. Indeed, the nine quantities given in (6.3.2) to (6.3.4) satisfy six relations between them. Three of these arise because, in each coordinate system, the unit vectors are orthogonal, and three more emerge from the fact that the vectors $\hat{\imath}, \hat{\jmath}, \hat{k}$ (or $\hat{\imath}', \hat{\jmath}', \hat{k}'$) have unit magnitude. These relations are

$$\begin{aligned} \hat{\imath} \cdot \hat{\jmath} = \hat{\jmath} \cdot \hat{k} = \hat{k} \cdot \hat{\imath} = 0, \\ \hat{\imath} \cdot \hat{\imath} = \hat{\jmath} \cdot \hat{\jmath} = \hat{k} \cdot \hat{k} = 1, \end{aligned} \tag{6.3.9}$$

and similar relations hold for the $\hat{\imath}', \hat{\jmath}', \hat{k}'$ system. If we form all possible dot products (6.3.9) for the $\hat{\imath}', \hat{\jmath}', \hat{k}'$, substitute (6.3.6), and make use in that result of the relations (6.3.9), we find

$$l_il_j + m_im_j + n_in_j = \delta_{ij} \qquad (i, j = 1, 2, 3), \tag{6.3.10}$$

where Kronecker's delta $\delta_{ij} = 1$ for $i = j$, and $\delta_{ij} = 0$ for $i \neq j$. These are the six relations satisfied by the nine direction cosines.

6.4. Orthogonal Transformations

The equations (6.3.8), which relate the quantities x, y, z to the quantities x', y', z', constitute a group of transformation equations, called a *linear vector transformation*:

$$x_i' = \sum_{j=1}^{3} a_{ij}x_j \qquad (i = 1, 2, 3) \tag{6.4.1}$$

with $x' = x_1'$, $y' = x_2'$, $z' = x_3'$ and similar relations for the unprimed symbols.

Now, the length of the vector r, given in (6.3.7), is independent of the coordinate system in which it is considered, or

$$\sum_{i=1}^{3} x_i'^2 = \sum_{i=1}^{3} x_i^2. \tag{6.4.2}$$

Substituting (6.4.1) in the left-hand side of (6.4.2),

$$\begin{aligned}\sum_{i=1}^{3} x_i'^2 = \sum_{i=1}^{3}\left[\sum_{j=1}^{3} a_{ij}x_j \sum_{k=1}^{3} a_{ik}x_k\right] &= \sum_{i=1}^{3}\sum_{j=1}^{3}\sum_{k=1}^{3} a_{ij}a_{ik}x_jx_k \\ &= \sum_{j=1}^{3}\sum_{k=1}^{3}\left(\sum_{i=1}^{3} a_{ij}a_{ik}\right)x_jx_k.\end{aligned} \tag{6.4.3}$$

If we substitute this quantity back into (6.4.2) we find

$$\sum_{j=1}^{3}\sum_{k=1}^{3}\left(\sum_{i=1}^{3} a_{ij}a_{ik}\right)x_jx_k = \sum_{i=1}^{3} x_i^2. \tag{6.4.4}$$

If (6.4.4) is to be an identity, one must have

$$\sum_{i=1}^{3} a_{ij}a_{ik} = \delta_{jk} \qquad (j, k = 1, 2, 3). \tag{6.4.5}$$

If we make the change of notation

$$\begin{aligned} l_1 &= a_{11}, & l_2 &= a_{12}, & l_3 &= a_{13}, \\ m_1 &= a_{21}, & m_2 &= a_{22}, & m_3 &= a_{23}, \\ n_1 &= a_{31}, & n_2 &= a_{32}, & n_3 &= a_{33}, \end{aligned}$$

Eq. (6.4.5) is identical with (6.3.10).

We shall write, for short,

$$\mathbf{A} = \begin{bmatrix} a_{11} & a_{12} & a_{13} \\ a_{21} & a_{22} & a_{23} \\ a_{31} & a_{32} & a_{33} \end{bmatrix} \tag{6.4.6}$$

and call $\mathbf{A}$ the *matrix of the linear vector transformation.* The a_{ij} are called the matrix elements.

Example 6.4.1. Consider a vector r in the plane. We represent it in the x_1, x_2 coordinate system and in the x_1', x_2' system, which is rotated by an angle θ with respect to x_1, x_2. This is shown in Fig. 6.4.1. We deduce from this diagram that

$$x_1' = x_1 \cos\theta + x_2 \sin\theta,$$
$$x_2' = -x_1 \sin\theta + x_2 \cos\theta.$$

This is a linear vector transformation with transformation matrix

$$\mathbf{A} = \begin{bmatrix} a_{11} & a_{12} \\ a_{21} & a_{22} \end{bmatrix},$$

where

$$a_{11} = \cos\theta, \qquad a_{12} = \sin\theta,$$
$$a_{21} = -\sin\theta, \qquad a_{22} = \cos\theta.$$

Equations (6.4.5) for $i, j, k = 1, 2$ become

$$a_{11}a_{11} + a_{21}a_{21} = 1,$$
$$a_{12}a_{12} + a_{22}a_{22} = 1,$$
$$a_{11}a_{12} + a_{21}a_{22} = 0,$$

and, substituting the values of the a_{ij} into these relations, one has

$$\cos^2\theta + \sin^2\theta = 1,$$
$$\sin^2\theta + \cos^2\theta = 1,$$
$$\cos\theta \sin\theta - \sin\theta \cos\theta = 0.$$

These are obviously true relations.

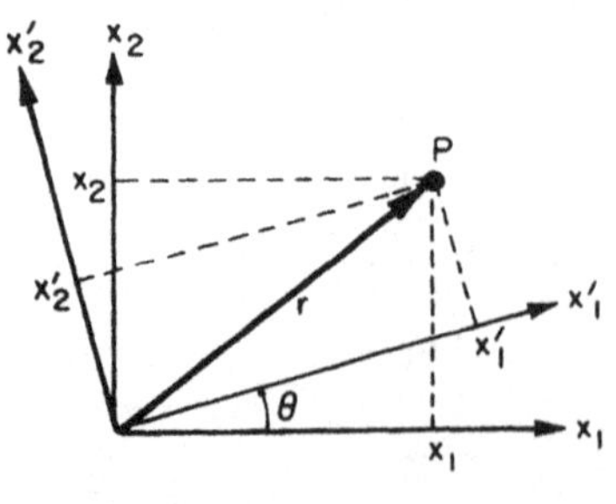

Fig. 6.4.1. Axis rotation of Example 6.4.1.

6.5. The Matrix Notation

Consider again the linear vector transformation (6.4.1), but written in expanded form

$$\begin{aligned} x_1{}' &= a_{11}x_1 + a_{22}x_2 + a_{13}x_3 , \\ x_2{}' &= a_{22}x_1 + a_{22}x_2 + a_{23}x_3 , \\ x_3{}' &= a_{31}x_1 + a_{32}x_2 + a_{33}x_3 . \end{aligned} \tag{6.5.1}$$

This set of equations is frequently written as the single matrix equation

$$\begin{bmatrix} x_1{}' \\ x_2{}' \\ x_3{}' \end{bmatrix} = \begin{bmatrix} a_{11} & a_{12} & a_{13} \\ a_{21} & a_{22} & a_{23} \\ a_{31} & a_{32} & a_{33} \end{bmatrix} \begin{bmatrix} x_1 \\ x_2 \\ x_3 \end{bmatrix} . \tag{6.5.2}$$

A matrix which contains only a single row, or only a single column is called a vector and, indeed, the elements of the first column in (6.5.2) are the components of the vector

$$x' = (x_1{}', x_2{}', x_3{}'); \tag{6.5.3}$$

those in the last column are the components of the vector

$$x = (x_1 , x_2 , x_3). \tag{6.5.4}$$

Then, in view of (6.4.6) we may write (6.5.2) in the abbreviated form

$$x' = \mathbf{A}x. \tag{6.5.5}$$

Since the left-hand side of (6.5.5) is a vector, so is the right-hand side. Moreover, two vectors are equal if and only if their components are equal. These components are the relations (6.5.1). Therefore, one has

$$\begin{bmatrix} x_1{}' \\ x_2{}' \\ x_3{}' \end{bmatrix} = \begin{bmatrix} a_{11}x_1 + a_{12}x_2 + a_{13}x_3 \\ a_{21}x_1 + a_{22}x_2 + a_{23}x_3 \\ a_{31}x_1 + a_{32}x_2 + a_{33}x_3 \end{bmatrix} . \tag{6.5.6}$$

A comparison between (6.5.2) and (6.5.6) shows by what rules the matrix product $\mathbf{A}x$ is computed.

A vector may either be written as a row or as a column matrix. However, when products are to be computed we always write x as a row vector if we want to calculate $x\mathbf{A}$, and we always write it as a column vector when calculating $\mathbf{A}x$. The first is called pre-multiplication, and the second postmultiplication by the vector. Matrix multiplication does not commute; in fact, we would have expected that $x\mathbf{A} \neq \mathbf{A}x$ because we know that vector multiplication does not commute, i.e., for two vectors r, R, we have, $r \times R \neq R \times r$.

We define the *unit matrix*

$$\mathbf{I} = \begin{bmatrix} 1 & 0 & 0 \\ 0 & 1 & 0 \\ 0 & 0 & 1 \end{bmatrix}, \tag{6.5.7}$$

and the rules of matrix multiplication show that a square matrix $\mathbf{A}$ and a vector x are unchanged when multiplied by the unit matrix. Thus,

$$\begin{aligned} \mathbf{AI} &= \mathbf{IA} = \mathbf{A}, \\ x\mathbf{I} &= \mathbf{I}x = x. \end{aligned} \tag{6.5.8}$$

We now ask whether a square matrix $\mathbf{B}$ exists such that $\mathbf{BA} = \mathbf{AB} = \mathbf{I}$. Such a matrix, if it exists, is called the *inverse* of $\mathbf{A}$ and it is written as $\mathbf{A}^{-1}$. Thus,

$$\mathbf{AA}^{-1} = \mathbf{A}^{-1}\mathbf{A} = \mathbf{I}. \tag{6.5.9}$$

(It is in general very laborious to compute the inverse of a matrix when that inverse exists.)

Next, we define the matrix which is obtained when the rows and columns are exchanged in $\mathbf{A}$. It is called the *transpose* of $\mathbf{A}$ and is denoted by $\tilde{\mathbf{A}}$. Thus, if $\mathbf{A}$ is as defined as in (6.4.6), the transpose is

$$\tilde{\mathbf{A}} = \begin{bmatrix} a_{11} & a_{21} & a_{31} \\ a_{12} & a_{22} & a_{32} \\ a_{13} & a_{23} & a_{33} \end{bmatrix}. \tag{6.5.10}$$

The matrix $\mathbf{A}$ is called *orthogonal*, if the transpose of $\mathbf{A}$ is equal to the inverse of $\mathbf{A}$. Therefore, if $\mathbf{A}$ is an orthogonal matrix,

$$\mathbf{A}^{-1} = \tilde{\mathbf{A}}. \tag{6.5.11}$$

We notice that, while it may be very laborious to calculate the inverse of a matrix $\mathbf{A}$ in general, it is certainly easy in the case of orthogonal matrices; all one needs to do is to exchange rows and columns in $\mathbf{A}$.

Combining (6.5.9) and (6.5.11), one sees that, if $\mathbf{A}$ is an orthogonal matrix,

$$\mathbf{A}\tilde{\mathbf{A}} = \tilde{\mathbf{A}}\mathbf{A} = \mathbf{I}. \tag{6.5.12}$$

The elements of matrices need not be real; they may be complex. Then, the conjugate complex of a matrix is found by changing each element to its conjugate complex. The conjugate complex of a matrix is denoted by an asterisk. Thus, if

$$\mathbf{A} = \begin{bmatrix} a_{11} + ib_{11} & a_{12} + ib_{12} & a_{13} + ib_{13} \\ a_{21} + ib_{21} & a_{22} + ib_{22} & a_{23} + ib_{23} \\ a_{31} + ib_{31} & a_{32} + ib_{32} & a_{33} + ib_{33} \end{bmatrix}, \tag{6.5.13}$$

then one has

$$\mathbf{A}^* = \begin{bmatrix} a_{11} - ib_{11} & a_{12} - ib_{12} & a_{13} - ib_{13} \\ a_{21} - ib_{21} & a_{22} - ib_{22} & a_{23} - ib_{23} \\ a_{31} - ib_{31} & a_{32} - ib_{32} & a_{33} - ib_{33} \end{bmatrix}. \tag{6.5.14}$$

It is clear that if $\mathbf{A}$ is real, $\mathbf{A} = \mathbf{A}^*$ (because all $b_{ij} = 0$).

Finally, when a matrix $\mathbf{A}$ has complex elements, the conjugate complex of its transpose is called the *adjoint* matrix, denoted by $\mathbf{A}^\dagger$, or

$$\mathbf{A}^\dagger = \tilde{\mathbf{A}}^*. \tag{6.5.15}$$

When

$$\mathbf{A}^\dagger = \mathbf{A} \tag{6.5.16}$$

the matrix is called *self-adjoint*. When

$$\mathbf{A}^\dagger \mathbf{A} = \mathbf{I}$$

which is the same as writing

$$\mathbf{A}^\dagger = \mathbf{A}^{-1}, \tag{6.5.17}$$

$\mathbf{A}$ is said to be a *unitary* matrix.

6.6. Properties of the Rotation Matrix

We show here that the transformation matrix $\mathbf{A}$ of (6.5.5) defined in (6.4.6) is orthogonal. Thus, we must show in accordance with (6.5.11) that the inverse of that matrix is equal to its transpose.

Let the elements of the inverse $\mathbf{A}^{-1}$ of $\mathbf{A}$ be denoted by α_{ij}, or

$$\mathbf{A}^{-1} = \begin{bmatrix} \alpha_{11} & \alpha_{12} & \alpha_{13} \\ \alpha_{21} & \alpha_{22} & \alpha_{23} \\ \alpha_{31} & \alpha_{32} & \alpha_{33} \end{bmatrix}. \tag{6.6.1}$$

Now pre-multiply (6.5.5) by $\mathbf{A}^{-1}$; then

$$x = \mathbf{A}^{-1}x', \tag{6.6.2}$$

and the elements of x are

$$x_i = \sum_{j=1}^{3} \alpha_{ij}x_j' \qquad (i = 1, 2, 3), \tag{6.6.3}$$

while the elements of x' are, from (6.5.6),

$$x_k' = \sum_{i=1}^{3} a_{ki}x_i \qquad (i = 1, 2, 3). \tag{6.6.4}$$

Therefore (6.6.4) must be the explicit solution of (6.6.3) for the x_k'. If we substitute (6.6.3) in (6.6.4) we may regard this as having transformed x' first into x, and then transformed x back into x'; thus we must get an identity. The substitution gives

$$\begin{aligned} x_k' &= \sum_{i=1}^{3} a_{ki} \sum_{j=1}^{3} \alpha_{ij}x_j' \\ &= \sum_{j=1}^{3}\left(\sum_{i=1}^{3} a_{ki}\alpha_{ij}\right)x_j'. \end{aligned} \tag{6.6.5}$$

Since this must be an identity, the coefficient of x_j' must be zero for $j \neq k$, and unity for $j = k$; in equation form,

$$\sum_{i=1}^{3} a_{ki}\alpha_{ij} = \delta_{kj}. \tag{6.6.6}$$

But, this equation is nothing more than the definition of the inverse in terms of the identity; written in matrix form, it is simply

$$\mathbf{A}\mathbf{A}^{-1} = \mathbf{I} \tag{6.6.7}$$

where $\mathbf{I}$ is the unit matrix defined in (6.5.7).

Consider now the double sum

$$\sum_{k=1}^{3}\sum_{i=1}^{3} a_{kl}a_{ki}\alpha_{ij};$$

we may write it in the two forms:

$$\begin{aligned} \sum_{k=1}^{3}\sum_{i=1}^{3} a_{kl}a_{ki}\alpha_{ij} &= \sum_{i=1}^{3}\left(\sum_{k=1}^{3} a_{kl}a_{ki}\right)\alpha_{ij} \\ &= \sum_{k=1}^{3}\left(\sum_{i=1}^{3} a_{ki}\alpha_{ij}\right)a_{kl}. \end{aligned} \tag{6.6.8}$$

Substituting the relation (6.4.5) in the first form, we find

$$\sum_{i=1}^{3}\left(\sum_{k=1}^{3} a_{kl}a_{ki}\right)\alpha_{ij} = \sum_{i=1}^{3} \delta_{il}\alpha_{ij} = \alpha_{lj}, \tag{6.6.9}$$

whereas, substituting (6.6.6) into the second form, we have

$$\sum_{k=1}^{3}\left(\sum_{i=1}^{3} a_{ki}\alpha_{ij}\right)a_{kl} = \sum_{k=1}^{3} \delta_{kj}a_{kl} = a_{jl}. \tag{6.6.10}$$

Since (6.6.9) and (6.6.10) are merely two different forms of the same quantity, we have

$$\alpha_{lj} = a_{jl} \qquad (j, l = 1, 2, 3). \tag{6.6.11}$$

In words: Consider the element a_{lj} of the matrix $\mathbf{A}$. The corresponding element of its inverse is by definition α_{lj}; the above relation (6.6.11) states that element is the same as that of $\mathbf{A}$ with the position of the indices reversed, or it is the corresponding element of the transpose of $\mathbf{A}$. Therefore (6.6.11) leads to the matrix equation

$$\mathbf{A}^{-1} = \tilde{\mathbf{A}}, \tag{6.6.12}$$

which was to be shown.

This result has an important consequence. It implies that, if a linear orthogonal vector transformation $x' \rightarrow x$ is given by

$$x' = \mathbf{A}x, \tag{6.6.13}$$

we may solve that system for x by writing

$$x = \tilde{\mathbf{A}}x'. \tag{6.6.14}$$

Not every square matrix $\mathbf{A}$ possesses an inverse, but every such matrix possesses a transpose. Thus, our result shows that the transformation (6.6.13) is one-to-one. Moreover, we have now a very simple method for solving a set of equations like (6.6.3) for the x_j'.

6.7. The Composition of Rotations

Let us suppose that a rigid body is subjected to two successive rotations. Let the first be that from x to x', so that

$$x_k' = \sum_{j=1}^{3} b_{kj}x_j \qquad (k = 1, 2, 3) \tag{6.7.1}$$

or, in matrix notation,

$$x' = \mathbf{B}x. \tag{6.7.2}$$

The second rotation is from x' to x'', or

$$x_i'' = \sum_{k=1}^{3} a_{ik} x_k' \qquad (i = 1, 2, 3) \tag{6.7.3}$$

or, in matrix notation,

$$x'' = \mathbf{A}x'. \tag{6.7.4}$$

Substituting (6.7.1) in (6.7.3) gives

$$x_i'' = \sum_{k=1}^{3} a_{ik} \sum_{j=1}^{3} b_{kj} x_j = \sum_{j=1}^{3} \left(\sum_{k=1}^{3} a_{ik} b_{kj} \right) x_j \qquad (i = 1, 2, 3). \tag{6.7.5}$$

In matrix form, the substitution of (6.7.2) in (6.7.4) gives

$$x'' = \mathbf{AB}x. \tag{6.7.6}$$

If we denote the product of the transformation matrices by $\mathbf{C}$, or

$$\mathbf{AB} = \mathbf{C}, \tag{6.7.7}$$

we find that its elements are

$$c_{ij} = \sum_{k=1}^{3} a_{ik} b_{kj} \qquad (i, j = 1, 2, 3), \tag{6.7.8}$$

in agreement with (6.7.5). Therefore:

> *The composition of two successive linear vector transformations with transformation matrices* $\mathbf{A}$ *and* $\mathbf{B}$*, respectively, is equivalent to a simple linear vector transformation with transformation matrix* $\mathbf{C} = \mathbf{AB}$.

It is evident that this result is easily extended to the composition of more than two rotations.

6.8. Applications

Consider a rigid body, one of whose points is fixed at the origin of some x, y, z coordinate system; that coordinate system is *not* fixed in the body. Then a sequence of rotations of the body produces a sequence of displacements of every one of its points (except those lying on the axes of rotation). Thus, after the rotations are completed, every general point is in a new position. The position change has components x, y, z in the coordinate system which is not fixed in the body. We now fix an x', y', z' coordinate system in the body, and we ask: What are the components of the position change relative to some fixed frame along the body-fixed axes? The preceding theory has yielded that answer in terms of a transformation matrix whose elements are the direction cosines of the body-fixed axes with respect to the axis-system which is not fixed in the body. We have found two results:

(i) The transformation matrix is orthogonal;

(ii) the transformation matrix of a sequence of several rotations is equal to the product of the individual transformation matrices, multiplied together in the inverse sequence of that in which the rotations were made. Since rotations do not add like vectors, a different sequence results in general in a different transformation matrix.

We shall now apply this theory to two well-known descriptions of a rotation which is the composition of a sequence of three rotations. It is clear that a general displacement of a point on a rigid body with one point fixed is describable by three independent rotations since a rigid body with one point fixed has three degrees of freedom.

These applications differ only in the manner in which the rotation is decomposed into three components, and the quantities to be determined are the direction cosines of the body-fixed axes. The best-known description of a rotation is that in terms of the Euler angles.

(a) The Euler Angles

The reader who has met the Euler angles in his or her first course in mechanics may have been frustrated by the fact that the final formulas are not the same in all books on mechanics. The Euler angles themselves are the same in all of them, and the symbols used to denote them are usually the same as well. But, the sequence of carrying out the Euler rotation is not always the same and, to date, no standard sequence has been agreed upon. We shall follow the most widely adopted sequence, which is that given in Goldstein (pp. 107-109). At first, consider an x_1, x_2, x_3 coordinate system not fixed in the body, and an x_1', x_2', x_3' system fixed in the body, as shown in Fig. 6.8.1. Before rotation, these systems coincide. Now, let

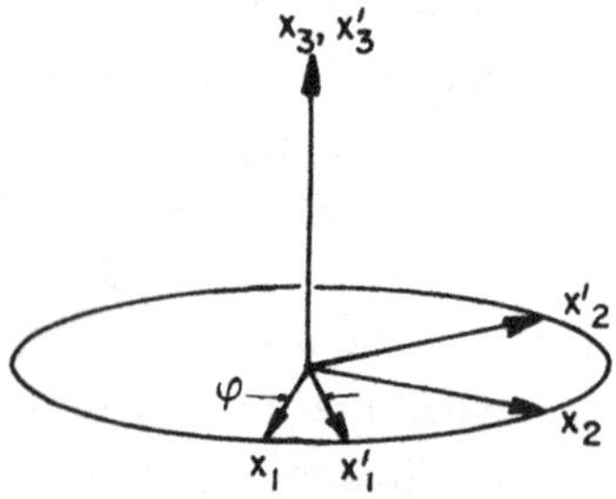

Fig. 6.8.1. First Euler angle, φ.

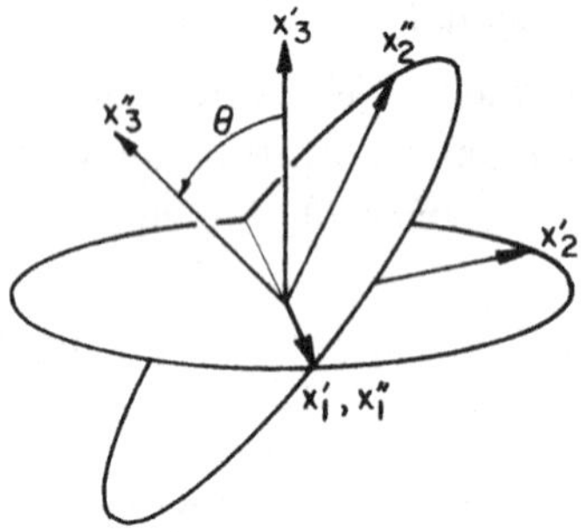

Fig. 6.8.2. Second Euler angle, θ.

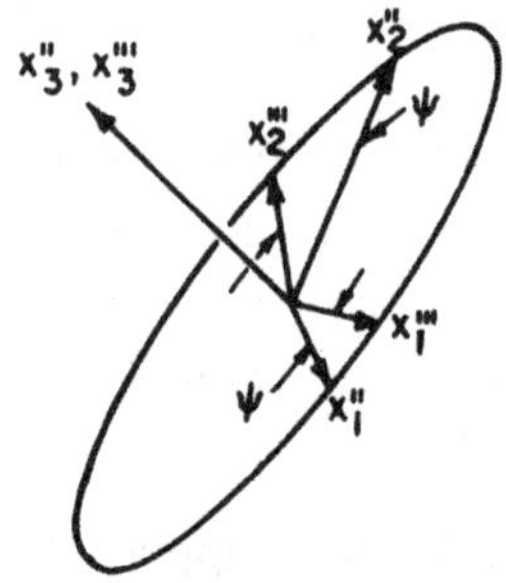

Fig. 6.8.3. Third Euler angle, ψ.

there be a right-handed rotation in the amount φ about the x_3 axis. By our theory, we have

$$x' = \mathbf{D}x \tag{6.8.1}$$

and $\mathbf{D}$ is the matrix of the direction cosines

$$\begin{aligned}
d_{11} &= \cos(x_1', x_1) = \cos\varphi, \\
d_{12} &= \cos(x_1', x_2) = \sin\varphi, \\
d_{13} &= \cos(x_1', x_3) = 0, \\
d_{21} &= \cos(x_2', x_1) = -\sin\varphi, \\
d_{22} &= \cos(x_2', x_2) = \cos\varphi, \\
d_{23} &= \cos(x_2', x_3) = 0, \\
d_{31} &= \cos(x_3', x_1) = 0, \\
d_{32} &= \cos(x_3', x_2) = 0, \\
d_{33} &= \cos(x_3', x_3) = 1.
\end{aligned}$$

Therefore,

$$\mathbf{D} = \begin{bmatrix} \cos\varphi & \sin\varphi & 0 \\ -\sin\varphi & \cos\varphi & 0 \\ 0 & 0 & 1 \end{bmatrix}. \tag{6.8.2}$$

Next we consider the x_1', x_2', x_3' system as not fixed in the body, and we call the body-fixed system the x_1'', x_2'', x_3'' system, as shown in Fig. 6.8.2. Before rotation, the two systems coincide. Then we execute a right-handed rotation θ about the x_1' axis as shown; that axis lies in the "line of nodes." The transformation is

$$x'' = \mathbf{C}x'. \tag{6.8.3}$$

By the method used above, we find

$$\mathbf{C} = \begin{bmatrix} 1 & 0 & 0 \\ 0 & \cos\theta & \sin\theta \\ 0 & -\sin\theta & \cos\theta \end{bmatrix}. \tag{6.8.4}$$

Finally, we consider the x_1'', x_2'', x_3'' system not fixed in the body while the x_1''', x_2''', x_3''' system is fixed in the body. Before rotation, the two systems coincide. We now produce a right-handed rotation ψ about the x_3'' axis as shown in Fig. 6.8.3. The transformation is

$$x''' = \mathbf{B}x'', \tag{6.8.5}$$

and the transformation matrix is

$$\mathbf{B} = \begin{bmatrix} \cos\psi & \sin\psi & 0 \\ -\sin\psi & \cos\psi & 0 \\ 0 & 0 & 1 \end{bmatrix}. \tag{6.8.6}$$

By the rule of the composition of rotations, the transformation from x to x''' is

$$x''' = \mathbf{A}x, \tag{6.8.7}$$

where

$$\mathbf{A} = \mathbf{BCD}, \tag{6.8.8}$$

and the matrix multiplication results in

$$\mathbf{A} = \begin{bmatrix} \cos\psi\cos\varphi - \cos\theta\sin\varphi\sin\psi & \cos\psi\sin\varphi + \cos\theta\cos\varphi\sin\psi & \sin\psi\sin\theta \\ -\sin\psi\cos\varphi - \cos\theta\sin\varphi\cos\psi & -\sin\psi\sin\varphi + \cos\theta\cos\varphi\cos\psi & \cos\psi\sin\theta \\ \sin\theta\sin\varphi & -\sin\theta\cos\varphi & \cos\varphi \end{bmatrix} \tag{6.8.9}$$

If we write

$$x_1''' = x', \qquad x_2''' = y', \qquad x_3''' = z',$$
$$x_1 = x, \qquad x_2 = y, \qquad x_3 = z,$$

the elements of the vector x''' are

$$x' = (\cos\psi\cos\varphi - \cos\theta\sin\varphi\sin\psi)x + (\cos\psi\sin\varphi + \cos\theta\cos\varphi\sin\psi)y + (\sin\psi\sin\theta)z, \tag{6.8.10}$$

$$y' = (-\sin\psi\cos\varphi - \cos\theta\sin\varphi\cos\psi)x + (-\sin\psi\sin\varphi + \cos\theta\cos\varphi\cos\psi)y + (\cos\psi\sin\theta)z, \tag{6.8.11}$$

$$z' = (\sin\theta\sin\varphi)x - (\sin\theta\cos\varphi)y + (\cos\theta)z. \tag{6.8.12}$$

Infinitesimal rotations cannot be described by three independent Euler angle rotations. This is seen by expanding (6.8.10) to (6.8.12) to first-order terms in infinitesimal quantities. They reduce to

$$x' = x + (\varphi + \psi)y,$$
$$y' = -(\varphi + \psi)x + y + \theta z,$$
$$z' = -\theta y + z.$$

These transformations involve only the two angles θ and $\varphi + \psi$, rather than three angles. A geometrical interpretation of this result evolves from Figs. 6.8.1 to 6.8.3. When the angle θ in Fig. 6.8.2 is infinitesimal, a small rotation φ (Fig. 6.8.1) and a subsequent small rotation ψ (Fig. 6.8.3) have the same effect as a single rotation $\varphi + \psi$.

The interpretation of (6.8.10) to (6.8.12) is this: Consider a rigid body one of whose points O is fixed in some reference frame. Let (x, y, z) be a triad† with origin at O; this is also fixed in the reference frame. Finally, let x', y', z' be a body-fixed triad which initially coincides with x, y, z. Now, let the body execute a rotation dn about a line of direction $\hat{n}$ which passes through O. Then a general point P of the body has coordinates (x, y, z) in the fixed triad, and (x', y', z') in the body-fixed triad. If the rotation about $\hat{n}$ is decomposed into the three Euler rotations φ, θ, and ψ, the equations (6.8.10) to (6.8.12) give the transformation of coordinates from x, y, z to x', y', z'.

† In 6.3, we used the term triad to denote the set of orthogonal unit vectors $\hat{i}, \hat{j}, \hat{k}$. We now extend the use of this term to the orthogonal x, y, z system of axes, as is usually done.

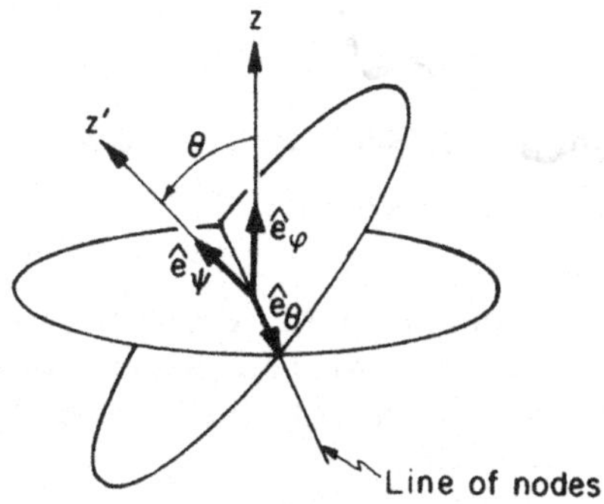

Fig. 6.8.4. Angular velocity unit vectors of the Euler angles.

Suppose the body-fixed coordinate system has an orientation given by the Euler angles φ, θ, and ψ. If the body is now given an infinitesimal rotation dn, what will be the changes $d\varphi$, $d\theta$, $d\psi$ in Euler angles?

In Fig. 6.8.4 we show the body-fixed x', y', z' axes and the unit vectors of the Euler rotations.

The vector $\hat{e}_\varphi$ lies in the z axis. Therefore its components on the body-fixed coordinates are found from the transformation **A** of (6.8.8), e.g., from the last column of (6.8.9). If we write

$$d\varphi\hat{k} = d\varphi_{x'}\hat{i}' + d\varphi_{y'}\hat{j}' + d\varphi_{z'}\hat{k}'$$

we have, therefore,

$$\begin{aligned} d\varphi_{x'} &= \sin\theta \sin\psi \, d\varphi, \\ d\varphi_{y'} &= \sin\theta \cos\psi \, d\varphi, \\ d\varphi_{z'} &= \cos\theta \, d\varphi. \end{aligned}$$

The vector $\hat{e}_\theta$ lies in the line of nodes. Consequently, its components on the body-fixed axes are those of x_1'', formed from the transformation **B**, defined in (6.8.5). They are [see the first column of (6.8.6)]

$$\begin{aligned} d\theta_{x'} &= \cos\psi \, d\theta, \\ d\theta_{y'} &= -\sin\psi \, d\theta, \\ d\theta_{z'} &= 0. \end{aligned}$$

The unit vector $\hat{e}_\psi$ lies in the body-fixed z' axis so that

$$\begin{aligned} d\psi_{x'} &= d\psi_{y'} = 0, \\ d\psi_{z'} &= d\psi. \end{aligned}$$

Combining these,

$$dn = (\sin\theta \sin\psi \, d\varphi + \cos\psi \, d\theta)\hat{i}' + (\sin\theta \cos\psi \, d\varphi - \sin\psi \, d\theta)\hat{j}' + (\cos\theta \, d\varphi + d\psi)\hat{k}'. \tag{6.8.13}$$

Example 6.8.1. A rigid body rotates about a point with angular velocity $\omega = \hat{n}(d\theta/dt)$. What are the components of ω on a body-fixed x', y', z' coordinate system in terms of the Euler angles if the origin of the body-fixed triad lies at the fixed point?

From (6.8.13) we have

$$\begin{aligned} \omega_{x'} &= \dot\varphi \sin\theta \sin\psi + \dot\theta \cos\psi, \\ \omega_{y'} &= \dot\varphi \sin\theta \cos\psi - \dot\theta \sin\psi, \\ \omega_{z'} &= \dot\varphi \cos\theta + \dot\psi. \end{aligned} \tag{6.8.14}$$

If the x, y, z system is that from which the Euler angle displacements were executed, one can show in a similar way that the angular velocity components of ω along these axes are

$$\begin{aligned} \omega_x &= \dot\psi \sin\theta \sin\varphi + \dot\theta \cos\varphi, \\ \omega_y &= -\dot\psi \sin\theta \cos\varphi + \dot\theta \sin\varphi, \\ \omega_z &= \dot\psi \cos\theta + \dot\varphi. \end{aligned}$$

The third of these may be deduced by inspection from Fig. 6.8.4. The demonstration of these formulas is left as an exercise.

A certain lack of symmetry is evident in these formulas. For instance, in examining the matrix **A** in (6.8.9), it is evident that the diagonal elements a_{ii} are very unlike each other. While the elements a_{ij}, for $i, j = 1, 2$, are similar in structure, those of the last row and column are quite different. This lack of symmetry is due to the particular choice of component rotations φ, θ, and ψ. These rotations were chosen by Euler in order to discuss the gyroscope, where φ is called the precession, θ the nutation, and ψ the spin. There exist other descriptions of rotation which do not have this lack of symmetry. One of these is called the Rodrigues formulas. We shall discuss them next.

(b) The Rodrigues Formulas

The derivation of the Rodrigues formulas begins with the (exact) formula (6.2.1):

$$\Delta r = \sin\theta(\hat{n} \times r) + 2\sin^2\frac{\theta}{2}\,[\hat{n} \times (\hat{n} \times r)]. \tag{6.8.15}$$

We now introduce the parameter (vector)

$$\lambda = \tan\frac{\theta}{2}\,\hat{n}. \tag{6.8.16}$$

Then

$$\sin\theta = 2\sin\frac{\theta}{2}\cos\frac{\theta}{2} = \frac{2\tan(\theta/2)}{1+\tan^2(\theta/2)} = \frac{2\,|\,\lambda\,|}{1+\lambda^2},$$

and similarly

$$\sin^2\frac{\theta}{2} = \frac{\tan^2(\theta/2)}{1+\tan^2(\theta/2)} = \frac{\lambda^2}{1+\lambda^2}.$$

In this way, (6.8.15) becomes

$$\Delta r = r' - r = \frac{2}{1+\lambda^2}\,[\lambda \times r + \lambda \times (\lambda \times r)]. \tag{6.8.17}$$

But, we also may write the triple vector product in (6.8.17) as

$$\lambda \times (\lambda \times r) = \lambda(\lambda \cdot r) - \lambda^2 r. \tag{6.8.18}$$

Therefore, the relation between r' and r is

$$r' = \frac{2}{1+\lambda^2}\,[\lambda \times r + \lambda(\lambda \cdot r)] + \frac{1-\lambda^2}{1+\lambda^2}\,r. \tag{6.8.19}$$

This formula gives the position of P after rotation, i.e., r' in terms of the position r before rotation, and of the parameter λ.

Let us now write

$$\begin{aligned} r &= x = x_1\hat{i} + x_2\hat{j} + x_3\hat{k},\\ r' &= x' = x_1'\hat{i} + x_2'\hat{j} + x_3'\hat{k},\\ \lambda &= \lambda_1\hat{i} + \lambda_2\hat{j} + \lambda_3\hat{k}. \end{aligned} \tag{6.8.20}$$

The substitution of these formulas in (6.8.19) gives the transformation equations, whose matrix form is

$$x' = \frac{1}{1+\lambda^2}\,\mathbf{A}x \tag{6.8.21}$$

with

$$\mathbf{A} = \begin{bmatrix} 1+\lambda_1^2-\lambda_2^2-\lambda_3^2 & 2(\lambda_1\lambda_2-\lambda_3) & 2(\lambda_1\lambda_3+\lambda_2) \\ 2(\lambda_1\lambda_2+\lambda_3) & 1-\lambda_1^2+\lambda_2^2-\lambda_3^2 & 2(\lambda_2\lambda_3-\lambda_1) \\ 2(\lambda_1\lambda_3-\lambda_2) & 2(\lambda_2\lambda_3+\lambda_1) & 1-\lambda_1^2-\lambda_2^2+\lambda_3^2 \end{bmatrix} \tag{6.8.22}$$

This transformation matrix has the symmetries which are lacking in that of the Euler angles.

Still another representation, of particular interest in quantum mechanics, is that of the so-called Cayley–Klein parameters. It shows the same dissymmetries as the Euler angle transformation, and it is a complex representation utilizing the Neumann sphere. The main difference between the Cayley–Klein parametric representation and the two given here is that the former uses four parameters, i.e., one more than the number of degrees of freedom. Hence, the Cayley–Klein parameters must satisfy a holonomic equation of constraint. The reader interested in the Cayley–Klein parameters is referred to Goldstein (pp. 109-118).

6.9. Problems

6.1. Let

$$\mathbf{A} = \begin{bmatrix} 0 & 1 \\ 1 & 0 \end{bmatrix}, \quad \mathbf{B} = \begin{bmatrix} -1 & 0 \\ 0 & 1 \end{bmatrix}.$$

Show that

$$\mathbf{AB} \neq \mathbf{BA}.$$

6.2. Let

$$\mathbf{A} = [a_{ij}] = \begin{bmatrix} 0 & 1 \\ -1 & 0 \end{bmatrix}.$$

Determine by computing $\mathbf{A}^{-1}$ whether $\mathbf{A}$ is orthogonal. (Note: The inverse is $\mathbf{A}^{-1} = \mathbf{C} = [c_{ji}]$, where $c_{ji} = \alpha_{ji}/a$, where α_{ji} is the cofactor of a_{ij}, and where $a = | a_{ij} |$ is the determinant of $\mathbf{A}$.)

6.3. Let $\mathbf{A} = [a_{ij}]$, $i, j = 1, 2, 3$, be a rotation matrix. Prove that $\mathbf{B} = \mathbf{A} - \tilde{\mathbf{A}}$ is skew-symmetric (i.e., $b_{ii} = 0$, $b_{ij} = -b_{ji}$). Is $\mathbf{B}$ a rotation matrix?

6.4. Prove that, if $\mathbf{C} = \mathbf{AB}$, then $\tilde{\mathbf{C}} = \tilde{\mathbf{B}}\tilde{\mathbf{A}}$.

6.5. Show that the sum of two orthogonal matrices is not, in general, an orthogonal matrix but their product is.

6.6. Consider rotation matrices $\mathbf{C}$ and $\mathbf{B}$ given in (6.8.4) and (6.8.6), respectively, with $| \theta |$ and $| \psi |$ so small that terms of second order in small quantities can be ignored. Show that the composite rotation $\mathbf{BC}$ is a skew matrix.

6.7. Show that the rotations in Exercise 6.6 may be regarded as vectors, i.e., show that $\mathbf{BC} = \mathbf{CB}$ in Exercise 6.6.

6.8. Show that under the assumption of Exercise 6.6, the elements off the diagonal of $\mathbf{BC}$ are the same as those of $\mathbf{B} + \mathbf{C}$, and the elements on the diagonal of $\mathbf{BC}$ and $\mathbf{B} + \mathbf{C}$ differ by a factor of 2.

6.9. Starting with the left-handed coordinate system shown, consider the following sequence of rotations: (i) a rotation θ about the y axis into the x', y', z' system; (ii) a rotation φ about the x' axis into the x'', y'', z'' system; (iii) a

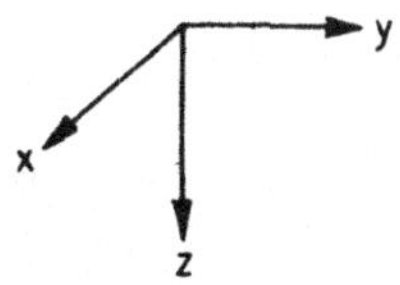

rotation ψ about the z'' axis into the x''', y''', z''' system. Construct the rotation matrix of the composite rotation analogous to (6.8.9).

6.10. Calculate the vector dn for Problem 6.9 above analogous to (6.8.13).

7

Some Rigid Body Kinetics

7.1. Introductory Remarks

The subject of rotation of rigid bodies and of mass moments and products of inertia is usually treated in a first course of mechanics. The inertial parameters arise in the following way:

Let a rigid body be composed of n particles m_i $(i = 1, 2, \ldots, n)$, with position vectors

$$r_i = (x_i, y_i, z_i) \tag{7.1.1}$$

where x, y, z is a triad *fixed in the rigid body*. If the angular velocity vector of the body is

$$\omega = (\omega_x, \omega_y, \omega_z), \tag{7.1.2}$$

i.e., the body rotates about a line with angular velocity ω, then the angular momentum of the ith particle of mass m_i and position r_i, denoted by the vector

$$L_i = (L_{x_i}, L_{y_i}, L_{z_i}) \tag{7.1.3}$$

is, by definition,

$$\begin{aligned} L_i &= r_i \times m_i \dot{r}_i \\ &= m_i[r_i \times (\omega \times r_i)] \end{aligned}$$

or, using the expansion formula of the triple vector product,

$$L_i = m_i[(r_i \cdot r_i)\omega - (r_i \cdot \omega)r_i]. \tag{7.1.4}$$

The angular momentum L of the entire body is then simply the sum

$$L = \omega \sum_{i=1}^{n} m_i r_i^2 - \sum_{i=1}^{n} m_i (r_i \cdot \omega) r_i, \tag{7.1.5}$$

and the Cartesian components of the momentum in the direction of the body-fixed axes are found by substituting (7.1.1) and (7.1.3) in (7.1.5); they are

$$\begin{aligned} L_x &= \omega_x I_{xx} + \omega_y I_{xy} + \omega_z I_{xz}, \\ L_y &= \omega_x I_{yx} + \omega_y I_{yy} + \omega_z I_{yz}, \\ L_z &= \omega_x I_{zx} + \omega_y I_{zy} + \omega_z I_{zz}, \end{aligned} \tag{7.1.6}$$

where

$$I_{xx} = \sum_{i=1}^{n} m_i (y_i^2 + z_i^2), \quad I_{yy} = \sum_{i=1}^{n} m_i (x_i^2 + z_i^2), \quad I_{zz} = \sum_{i=1}^{n} m_i (y_i^2 + x_i^2) \tag{7.1.7}$$

are called the moments of inertia of the body relative to the x, y, z system of coordinates, and

$$I_{xy} = -\sum_{i=1}^{n} m_i x_i y_i, \quad I_{xz} = -\sum_{i=1}^{n} m_i x_i z_i, \quad I_{yz} = -\sum_{i=1}^{n} m_i y_i z_i \tag{7.1.8}$$

are called the products of inertia.[†] As is evident from (7.1.8),

$$I_{xy} = I_{yx}, \quad I_{xz} = I_{zx}, \quad I_{yz} = I_{zy}. \tag{7.1.9}$$

The array

$$\begin{bmatrix} I_{xx} & I_{xy} & I_{xz} \\ I_{yx} & I_{yy} & I_{yz} \\ I_{zx} & I_{zy} & I_{zz} \end{bmatrix}$$

is called the inertia tensor of the rigid system of particles.

If the rigid body is a continuous solid, the summations are replaced by integrations over the volume and, provided $\varrho = dm/dv$ exists, the particle masses m_i are replaced by

$$dm = \varrho \, dv, \tag{7.1.10}$$

where $\varrho(x, y, z)$ is the local density, and dv is the volume of the mass element

[†] Our departure from the usual notation is to call I_{xy} what is normally denoted by $-I_{xy}$; this has simplifying effects in later developments.

dm having the position vector (x, y, z). Thus, for instance, the first of equations (7.1.7) becomes the Riemann integral

$$I_{xx} = \int_V (y^2 + z^2)\varrho \, dv \tag{7.1.11}$$

and the first of equations (7.1.9) becomes

$$I_{xy} = -\int_V xy\varrho \, dv. \tag{7.1.12}$$

To establish the validity of the procedures used in replacing sums like those in (7.1.7) and (7.1.8) by integrals like those in (7.1.11) and (7.1.12) we shall now show that both the limit of the finite sum and the integral which replaces it are equal to (so-called) Stieltjes integrals of the form

$$I = \int_V f(x, y, z) \, dm(x, y, z). \tag{7.1.13}$$

Before proceeding with this demonstration, a few words regarding Stieltjes integrals may be in order. Consider a real-valued, bounded, nondecreasing function $F(t)$ defined on the closed interval

$$[a, b] = a \leq t \leq b.$$

We call P a partition of $[a, b]$ if P is a finite collection of nonoverlapping subintervals whose union is $[a, b]$. Denoting a partition by the partition points we write

$$P = (t_1, t_2, \ldots, t\ ), \qquad t_k < t_{k+1}. \tag{7.1.14}$$

Now, let ξ_k be intermediate points, or

$$t_k \leq \xi_k \leq t_{k+1}. \tag{7.1.15}$$

Then, Q is also a partition of $[a, b]$ where

$$Q = (t_1, \xi_1, t_2, \xi_2, \ldots, \xi_{n-1}, t_n). \tag{7.1.16}$$

We say that Q is a refinement of P, or Q is finer than P, because every subinterval of Q is contained in some subinterval of P; this implies that every partition point of P is also a partition point of Q.

Let $\varphi(t)$ be a continuous, real-valued, bounded function defined on

$[a, b]$. Then it can be shown† that the Stieltjes sum

$$S(P; \varphi, F) = \sum_{k=1}^{n} \varphi(\xi_k)[F(t_{k+1}) - F(t_k)] \tag{7.1.17}$$

converges to the Stieltjes integral

$$I = \int_a^b \varphi(t)\, dF(t) \tag{7.1.18}$$

as the partition is made finer and finer in the sense that

$$\max |\, t_{k+1} - t_k \,| \to 0. \tag{7.1.19}$$

Usually, $\varphi(t)$ is called the integrand, and $F(t)$ the integrator.

Consider now the real-valued, bounded function $m(x) > 0$, defined only at a finite number of distinct points x_K ($K = 1, 2, \ldots, N$) on $[a, b]$. Then, we may also regard $m(x)$ as a nonnegative function defined everywhere on $[a, b]$ whose value is zero for every $x \neq x_K$, and equal to $m(x_K) > 0$ for every $x = x_K$. We denote the sum of the $m(x_K)$ by

$$M = \sum_{K=1}^{N} m(x_K). \tag{7.1.20}$$

Our problem is resolved if we can show that M is also given by a Stieltjes integral

$$M = \int_{x=a}^{x=b} dF(x). \tag{7.1.21}$$

To do this, consider the Stieltjes sum

$$S(P; 1, F) = \sum_{k=1}^{n} [F(t_{k+1}) - F(t_k)] \tag{7.1.22}$$

with the partition

$$P = (t_1, t_2, \ldots, t_n)$$

such that $t_k = x_k$ for all k and K, and at most one x_K can be interior to any subinterval $[t_k, t_{k+1}]$. Moreover, let $F(t)$ be given by

$$F(t) = \sum_{x_K \leq t} m(x_K). \tag{7.1.23}$$

† For a proof of this statement refer to any text on real analysis. See, for instance, Bartle, R. G., *The Elements of Real Analysis*, John Wiley and Sons, Inc., New York (1964), pp. 275–277.

This function is called a "discrete distribution corresponding to $m(x)$ and x_K." Evidently, $F(t)$ is defined everywhere on $[a, b]$; it is bounded and nondecreasing; it has a jump of value $m(x_K)$ at each x_K and it is constant between these values.

The differences are given by

$$\begin{aligned} F(t_{k+1}) - F(t_k) &= m(x_K), \\ &= 0, \end{aligned} \tag{7.1.24}$$

according as the interval $[t_k, t_{k+1}]$ does or does not include a point x_K. It follows that the Stieltjes sum

$$\sum_{k=1}^{n} [F(t_{k+1}) - F(t_k)] = \sum_{K=1}^{N} m(x_K). \tag{7.1.25}$$

In view of the properties of $F(t)$ [and of $\varphi(t) \equiv 1$], the Stieltjes sum converges to the Stieltjes integral, or

$$\sum_{K=1}^{N} m(x_K) = \int_{x=a}^{x=b} dF(x), \tag{7.1.26}$$

which was to be shown.

Now, if we wish to consider a continuous function $m(x)$ defined on $[a, b]$, we have

$$M = \int_{x=a}^{x=b} dm(x), \tag{7.1.27}$$

which is a Stieltjes integral. If $\varrho = dm/dx$ exists on $[a, b]$, (7.1.27) may be replaced by the Riemann integral

$$M = \int_a^b \varrho(x)\, dx. \tag{7.1.28}$$

Example 7.1.1. A homogeneous solid cube of edge length a and density ϱ is constrained so that one vertex is fixed. Calculate the angular momentum components under an angular velocity $\omega = (1, 2, 3)$.

The cube and the coordinate system are shown in Fig. 7.1.1. By direct integration we obtain

$$\begin{aligned} I_{xx} = \quad I_{yy} = \quad I_{zz} &= \tfrac{2}{3}\varrho a^5 = \tfrac{2}{3}Ma^2, \\ -I_{xy} = -I_{xz} = -I_{yz} &= \tfrac{1}{4}\varrho a^5 = \tfrac{1}{4}Ma^2, \end{aligned}$$

where $M = \varrho a^3$ is the cube mass.

Then,

$$L_x = -\tfrac{7}{12}Ma^2, \qquad L_y = \tfrac{1}{3}Ma^2, \qquad L_z = \tfrac{5}{4}Ma^2.$$

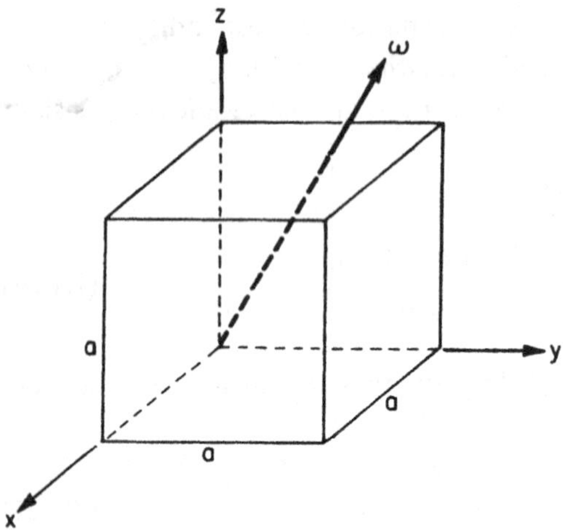

Fig. 7.1.1. Cube of Example 7.1.1.

7.2. The Inertial Parameters in Rotated Axes

From the previous section it is clear that the magnitudes of the inertial parameters depend on the orientation of the body-fixed axes and will change when this orientation changes. As is well known, for every choice of origin of body-fixed axes, there exists one preferred orientation for which the products of inertia of the body vanish, and the moments of inertia are stationary with respect to neighboring orientations. These preferred directions are called the "principal axes"; they are preferred because the vanishing of the products of inertia greatly simplifies the equations of rotational motion.

In order to find the principal axes, one must study the effect of axis rotation on the inertial parameters. This study is brought into conformity with Sections 6.4–6.6 when x, y, and z are replaced by x_1, x_2, and x_3, respectively. Also, we use the notation

$$I_{x_ix_j} = I_{ij}$$

because it does not give rise to confusion.

Now, a moment of inertia such as I_{xx} in (7.1.11) may be written as

$$I_{xx} = \int_V (r^2 - x^2)\, dm$$

because of the Pythagorean theorem. Hence, in general, the moment of inertia is

$$I_{ii} = \int_V (r^2 - x_ix_i)\, dm, \tag{7.2.1}$$

and the product of inertia is

$$I_{ij} = -\int_V x_i x_j \, dm \qquad (i \neq j).$$

Therefore, we may write for all $i, j = 1, 2, 3$,

$$I_{ij} = \int_V (r^2 \, \delta_{ij} - x_i x_j) \, dm,$$

where δ_{ij} is Kronecker's delta, defined in connection with (6.3.10). Now, if the coordinates after rotation are x_i' ($i = 1, 2, 3$), then the inertial parameters with respect to these rotated coordinates are evidently, from (7.2.1),

$$I'_{ij} = \int_V (r^2 \, \delta_{ij} - x_i' x_j') \, dm. \tag{7.2.2}$$

But the new axes are related to the old ones by (6.6.4), which are (with a suitable change of index notation)

$$x_i' = \sum_{k=1}^{3} a_{ik} x_k , \qquad x_j' = \sum_{l=1}^{3} a_{jl} x_l . \tag{7.2.3}$$

The substitution of these in (7.2.2) gives

$$I'_{ij} = \delta_{ij} \int_V r^2 \, dm - \sum_{k=1}^{3} \sum_{l=1}^{3} a_{ik} a_{jl} \int_V x_k x_l \, dm. \tag{7.2.4}$$

The inertial parameter I'_{ij} after rotation is related to the one before axis-rotation in a simple way. From (7.2.1), the quantity

$$\int_V x_k x_l \, dm = \delta_{kl} \int_V r^2 \, dm - I_{kl} ,$$

and the substitution of this expression in (7.2.4) gives

$$I'_{ij} = \delta_{ij} \int_V r^2 \, dm + \sum_{k=1}^{3} \sum_{l=1}^{3} a_{ik} a_{jl} \left(I_{kl} - \delta_{kl} \int_V r^2 \, dm \right).$$

The last term contains the coefficient

$$\sum_{k=1}^{3} \sum_{l=1}^{3} a_{ik} a_{jl} \, \delta_{kl} = \sum_{k=1}^{3} a_{ik} a_{jk} = \delta_{ij} ,$$

where the last equality is the result of (6.6.6) and (6.6.11). Therefore, the inertial parameters with respect to rotated coordinate systems in terms of those before rotation are simply

$$I'_{ij} = \sum_{k=1}^{3} \sum_{l=1}^{3} a_{ik} a_{jl} I_{kl}. \tag{7.2.5}$$

Moreover, as the rotation matrix is orthogonal (i.e., its inverse is its transpose) we also find

$$I_{ij} = \sum_{k=1}^{3} \sum_{l=1}^{3} a_{ki} a_{lj} I'_{kl}. \tag{7.2.6}$$

7.3. Angular Momentum and Principal Axes

Let us write the angular momentum equations (7.1.6) in the matrix form

$$\begin{bmatrix} L_x \\ L_y \\ L_z \end{bmatrix} = \begin{bmatrix} I_{xx} & I_{xy} & I_{xz} \\ I_{yx} & I_{yy} & I_{yz} \\ I_{zx} & I_{zy} & I_{zz} \end{bmatrix} \begin{bmatrix} \omega_x \\ \omega_y \\ \omega_z \end{bmatrix}, \tag{7.3.1}$$

and let us suppose that the angular momentum vector L and the angular velocity vector ω have the same direction, or

$$L = I\omega, \tag{7.3.2}$$

where I is a *scalar* constant of proportionality.

When (7.3.2) is written in matrix form, one finds

$$\begin{bmatrix} L_x \\ L_y \\ L_z \end{bmatrix} = \begin{bmatrix} I & 0 & 0 \\ 0 & I & 0 \\ 0 & 0 & I \end{bmatrix} \begin{bmatrix} \omega_x \\ \omega_y \\ \omega_z \end{bmatrix}. \tag{7.3.3}$$

It follows that, when the directions of angular momentum and angular velocity coincide, (7.1.6) becomes

$$\begin{aligned} I_{xx}\omega_x + I_{xy}\omega_y + I_{xz}\omega_z &= I\omega_x, \\ I_{yx}\omega_x + I_{yy}\omega_y + I_{xz}\omega_z &= I\omega_y, \\ I_{zx}\omega_x + I_{zy}\omega_y + I_{zz}\omega_z &= I\omega_z. \end{aligned} \tag{7.3.4}$$

Nontrivial solutions of (7.3.4) require that the determinant of the coefficients of the $\omega_x, \omega_y, \omega_z$ vanish, or that

$$\begin{bmatrix} I_{xx} - I & I_{xy} & I_{xz} \\ I_{yx} & I_{yy} - I & I_{yz} \\ I_{zx} & I_{zy} & I_{zz} - I \end{bmatrix} = 0. \tag{7.3.5}$$

This is a cubic equation in I which can be shown to have always three real roots I_x, I_y, and I_z. [The proof of this statement is not given here. It is based on the fact that the matrix of the inertia tensor is self-adjoint, as defined in (6.5.16).]

By comparing (7.3.1) and (7.3.3) it is evident that the latter corresponds to the former when the products of inertia vanish, and this means that the three roots I_x, I_y, and I_z of (7.3.5) are the principal moments of inertia.

It is, of course, *not* true in general that, for any rotational motion, the angular velocity and momentum vectors coincide in direction, and our method for finding principal moments of inertia did not suppose that they always do. Rather, we made use of the invariance of a vector under coordinate rotation. Explained more fully, we utilized the fact that a given vector does not change in magnitude or direction when the coordinate system is rotated; all that changes is the magnitude and direction of the components. Therefore, if angular velocity and momentum vectors do coincide, that fact is not altered by a rotation of the coordinate system. This observation together with the one that the angular momentum vector is always given by (7.3.1) regardless of the direction of the coordinate system establishes our procedure.

If the x, y, z system is any coordinate system fixed in the body, its origin O fixed in some point in the body, then the principal moments of inertia are found as the roots I_x, I_y, and I_z of the determinantal equation (7.3.5).

The directions of the principal axes are found as follows:

If we substitute a root I_i ($i = x, y, z$) in (7.3.4) we obtain the three equations

$$\begin{aligned} (I_{xx} - I_i)\omega_x^{(i)} + I_{xy}\omega_y^{(i)} + I_{xz}\omega_z^{(i)} &= 0, \\ I_{yx}\omega_x^{(i)} + (I_{yy} - I_i)\omega_y^{(i)} + I_{yz}\omega_z^{(i)} &= 0, \\ I_{zx}\omega_x^{(i)} + I_{zy}\omega_y^{(i)} + (I_{zz} - I_i)\omega_z^{(i)} &= 0. \end{aligned} \tag{7.3.6}$$

For every i, there are three equations in the three unknowns $\omega_x^{(i)}, \omega_y^{(i)}, \omega_z^{(i)}$. Then, a point $P^{(i)}$ having coordinates $x_i = \omega_x^{(i)}$, $y_i = \omega_y^{(i)}$, $z_i = \omega_z^{(i)}$ lies on that principal axis passing through the origin about which the (principal) moment of inertia is I_i.

Example 7.3.1. Calculate the principal moments of inertia of the cube of Example 7.1.1 about a vertex.

Substitution in (7.3.5) gives

$$\begin{vmatrix} 8A - I & -3A & -3A \\ -3A & 8A - I & -3A \\ -3A & -3A & 8A - I \end{vmatrix} = 0$$

where $A = Ma^2/12$. Expansion gives

$$I^3 - 24AI^2 + 165A^2I - 242A^3 = 0,$$

which has the roots

$$I_1 = 2A, \qquad I_2 = I_3 = 11A.$$

Therefore, the principal moments of inertia are

$$I_x = \tfrac{1}{6}Ma^2, \qquad I_y = I_z = \tfrac{11}{12}Ma^2.$$

To find the direction of the principal axis about which the moment of inertia is I_x, substitute $I = I_x$ and the other inertial parameters in (7.3.6). One finds $y_1/x_1 = 1$ and $z_1/x_1 = 1$, or the x_1 axis is the diagonal passing through the fixed vertex. The other two principal axes are orthogonal to the x_1 axis and to each other. Their orientation in the plane normal to the x_1 axis is immaterial because $I_y = I_z$.

7.4. The Ellipsoids of Cauchy and Poinsot

A geometrical interpretation of principal axes, frequently discussed in a first course in mechanics, is provided by Cauchy's ellipsoid of inertia.

Suppose the inertial parameters of a rigid body with respect to an x, y, z triad are given; let them have the magnitudes $I_{xx}, I_{yy}, I_{zz}, I_{xy}, I_{xz}, I_{yz}$. Then, if an x' axis passing through the origin of the x, y, z triad has direction cosines a_{11}, a_{12}, a_{13}, the moment of inertia $I_{x'x'}$ about that axis is, from (7.2.5),

$$I_{x'x'} = \sum_{k=1}^{3} \sum_{l=1}^{3} a_{ik}a_{jl}I_{kl}$$

or, written out (with $j = 1, 2, 3$; $k, l = 1, 2, 3$, and/or $k, l = x, y, z$),

$$I_{x'x'} = I'_{xx} = a_{11}^2 I_{xx} + a_{12}^2 I_{yy} + a_{13}^2 I_{zz} + 2a_{11}a_{12}I_{xy} + 2a_{11}a_{13}I_{xz} + 2a_{12}a_{13}I_{yz}. \tag{7.4.1}$$

In Fig. 7.4.1(a) an x, y, z triad is shown as well as the line Ox'. If this line subtends the angles α, β, and γ with the x, y, and z axes, respectively, we have

$$\cos\alpha = a_{11}, \qquad \cos\beta = a_{12}, \qquad \cos\gamma = a_{13}.$$

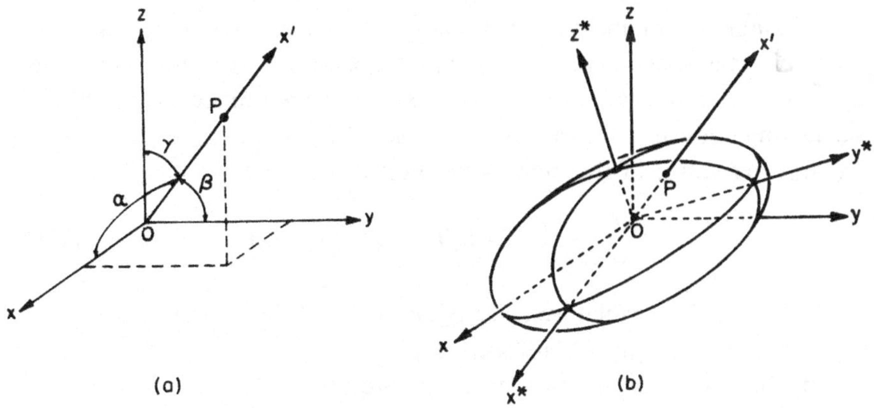

Fig. 7.4.1. Arbitrary triad and Cauchy's ellipsoid.

On the x' axis, we mark a point P whose distance $\overline{OP}$ from O is numerically equal to $1/(I'_{xx})^{1/2}$, as calculated from (7.4.1). Then, it is evident from that diagram that the coordinates of P are

$$\begin{aligned} x &= a_{11}\overline{OP} = a_{11}/(I'_{xx})^{1/2}, \\ y &= a_{12}/(I'_{xx})^{1/2}, \\ z &= a_{13}/(I'_{xx})^{1/2}. \end{aligned} \tag{7.4.2}$$

If we solve these equations for the a_{ij} and substitute them into (7.4.1), that equation becomes

$$I_{xx}x^2 + I_{yy}y^2 + I_{zz}z^2 + 2I_{xy}xy + 2I_{xz}xz + 2I_{yz}yz = 1. \tag{7.4.3}$$

This is the equation of an ellipsoid centered on the origin of the x, y, z triad, called Cauchy's ellipsoid of inertia. Therefore, for any rigid body:

> *Cauchy's ellipsoid defined by* (7.4.3) *is the locus of points P such that the square of the distance between O and P is inversely proportional to the mass moment of inertia about the line connecting O and P.*

This observation is illustrated in Fig. 7.4.1(b). In that figure, the x^*, y^*, and z^* axes are also shown. These are the principal axes of the ellipsoid as well as those of the rigid body. In this way, the principal axes of inertia have been interpreted in a simple geometrical way.

Related to Cauchy's ellipsoid is Poinsot's ellipsoid, which gives a geometrical interpretation of torque-free rotation of rigid bodies.

The angular momentum components of a rigid body with one point fixed in inertial space were calculated in (7.1.6). It is important to remember that these components are along the axes of a coordinate system fixed in the moving body and, hence, *not fixed in inertial space*, in general. Therefore, the time derivative of the momentum vector is

$$\frac{dL}{dt} = \dot{L}_x\hat{\imath} + \dot{L}_y\hat{\jmath} + \dot{L}_z\hat{k} + \omega \times L, \tag{7.4.4}$$

in which ω is the angular velocity of the rotating body and, hence, of the x, y, z axes relative to a Galilean frame.

If the rigid body is acted on by a torque

$$M = (M_x, M_y, M_z), \tag{7.4.5}$$

Newton's law for its rotation is

$$\frac{dL}{dt} = M_z \tag{7.4.6}$$

When (7.4.4) and (7.4.5) are substituted in (7.4.6), the components of this equation are found to be

$$\begin{aligned} \dot{L}_x + \omega_y L_z - \omega_z L_y &= M_x, \\ \dot{L}_y + \omega_z L_x - \omega_x L_z &= M_y, \\ \dot{L}_z + \omega_x L_y - \omega_y L_x &= M_z, \end{aligned} \tag{7.4.7}$$

where the momentum components are given in (7.1.6). Evidently, their substitution in (7.4.7) will result in lengthy equations. However, when the x, y, z axes are the principal axes (so that the products of inertia all vanish), there results

$$\begin{aligned} I_x\dot{\omega}_x - (I_y - I_z)\omega_y\omega_z &= M_x, \\ I_y\dot{\omega}_y - (I_z - I_x)\omega_z\omega_x &= M_y, \\ I_z\dot{\omega}_z - (I_x - I_y)\omega_x\omega_y &= M_z. \end{aligned} \tag{7.4.8}$$

These are the well-known Euler equations of rotation of a rigid body about a point. Their simplicity is due to the use of principal axes, and this is, in fact, the main advantage accruing from principal axes.

Under torque-free motion, the Euler equations are

$$\begin{aligned} I_x\dot{\omega}_x - (I_y - I_z)\omega_y\omega_z &= 0, \\ I_y\dot{\omega}_y - (I_z - I_x)\omega_z\omega_x &= 0, \\ I_z\dot{\omega}_z - (I_x - I_y)\omega_x\omega_y &= 0. \end{aligned} \tag{7.4.9}$$

But, when no torque acts, it follows from (7.4.6) that the momentum vector L is a constant, both in magnitude and direction; its direction, evidently fixed in inertial space, is the so-called "invariable line." The magnitude of the momentum vector is given by

$$L \cdot L = L^2 = I_x^2\omega_x^2 + I_y^2\omega_y^2 + I_z^2\omega_z^2. \tag{7.4.10}$$

Moreover, when no torque acts, it is simple to integrate Euler's equations of motion. Multiplying the first of (7.4.9) by ω_x, the second by ω_y, the third by ω_z, and adding them gives

$$I_x\dot{\omega}_x\omega_x + I_y\dot{\omega}_y\omega_y + I_z\dot{\omega}_z\omega_z = 0,$$

and this integrates to

$$I_x\omega_x^2 + I_y\omega_y^2 + I_z\omega_z^2 = 2T = \text{const}, \tag{7.4.11}$$

where T is the kinetic energy of rotation.

If we let the end point of the angular velocity vector have the position (x, y, z), (7.4.10) and (7.4.11) become

$$\begin{aligned} I_x^2x^2 + I_y^2y^2 + I_z^2z^2 &= L^2 = \text{const}, \\ I_xx^2 + I_yy^2 + I_zz^2 &= 2T = \text{const}. \end{aligned} \tag{7.4.12}$$

Both of these are seen to be the equations of ellipsoids referred to principal coordinates, and both ellipsoids are rigidly attached to the rotating body. Evidently, both equations must be satisfied by torque-free rotation, or the end point of the angular velocity vector must lie on the intersection of these two ellipsoids, i.e., a curve. If an observer were fixed in the body (and he could actually see the angular velocity vector) he would observe its end point moving along this curve; it is called the "polhode."

We will now show that the angular velocity vector ω, fixed at some point O on the line of action of the momentum vector L, terminates on a plane which is fixed in inertial space.

In Fig. 7.4.2 we show the constant angular momentum vector L and the angular velocity vector ω; the endpoint of ω is the point P. The line OQ is the invariable line. Now, let the normal from P on the invariable line intersect it at R, as shown. Then, OR is the component of ω along the invariable line, or

$$OR = \omega \cdot L/|L|, \tag{7.4.13}$$

where $L/|L|$ is the unit vector along L. But

$$\omega = \omega_x\hat{\imath} + \omega_y\hat{\jmath} + \omega_z\hat{k},$$

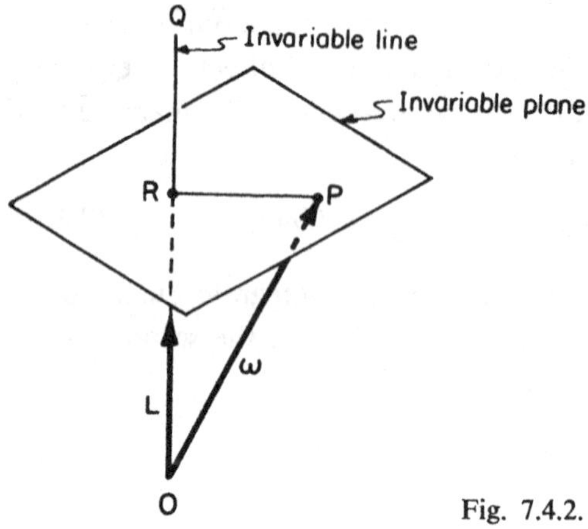

Fig. 7.4.2. Moment vector and invariable plane.

and

$$L = I_x\omega_x\hat{i} + I_y\omega_y\hat{j} + I_z\omega_z\hat{k}$$

when principal axes are used. Therefore,

$$\omega \cdot L = I_x\omega_x^2 + I_y\omega_y^2 + I_z\omega_z^2 = 2T = \text{const} \tag{7.4.14}$$

in virtue of the second equation of (7.4.12). The substitution of (7.4.14) into (7.4.13) gives

$$OR = 2T/|L| = \text{const.} \tag{7.4.15}$$

It follows that R is a fixed point in inertial space, i.e., it does not move up and down along the invariable line during the motion. But then, any perpendicular to OQ meeting the invariable line at R must lie in a plane which is also fixed in inertial space; this is what we wished to show. This plane is called the "invariable plane," and it is perpendicular to the invariable line.

The second equation of (7.4.12) is the equation of the so-called "Poinsot ellipsoid." As we saw, this ellipsoid is the locus of the endpoints of the angular velocity vector. But, if that vector terminates both on the surface of the Poinsot ellipsoid and on the invariable plane, then, during the motion, the Poinsot ellipsoid (fixed in the rotating body) must roll on the invariable plane, which is fixed in inertial space. Therefore, the polhode is the locus of points on the ellipsoid which have contact with the invariable plane during this rolling. The locus of contact points on the invariable plane is called the "herpolhode"; it is the locus of the endpoints of the angular velocity vector seen by an observer who is fixed in inertial space.

The interpretation of the rotation of rigid bodies in terms of the Poinsot ellipsoid rolling on the invariable plane is called Poinsot's representation.

Example 7.4.1.† Discuss the motion of a homogeneous circular disk supported at its center on a needle point when it is set spinning about a line of known orientation.

Let the radius of the disk be r, its mass M, locate an x, y, z triad in the disk so that the origin is at the disk center, and $\hat{k}$ is normal to the disk, let the initial angular velocity be ω_0, and let the vector ω_0 make an angle α with the normal to the disk. Let the angular velocity of the disk be ω (so that $\omega = \omega_0$ when $t = 0$). This is shown in Fig. 7.4.3.

Clearly, the $\hat{i}$, $\hat{j}$, $\hat{k}$ triad lies in the principal axes, and the principal moments of inertia are

$$I_x = I_y = \tfrac{1}{4}Mr^2, \qquad I_z = 2I_x = \tfrac{1}{2}Mr^2.$$

Let us now fix the $\hat{i}$, $\hat{j}$ system in the disk such that ω_0 lies in the y, z plane. Thus, if

$$\omega = \omega_x\hat{i} + \omega_z\hat{j} + \omega_z\hat{k},$$

we have at $t = 0$

$$\omega_x(0) = 0, \qquad \omega_y(0) = \omega_0 \sin\alpha, \qquad \omega_z(0) = \omega_0 \cos\alpha.$$

The motion is torque-free; hence, (7.4.9) are applicable, and the third of these is

$$I_z\dot{\omega}_z = 0$$

because $I_x = I_y$. This equation implies

$$\omega_z = \text{const} = \omega_0 \cos\alpha.$$

The remaining two equations are

$$\dot{\omega}_x + \omega_0 \cos\alpha\,\omega_y = 0,$$
$$\dot{\omega}_y - \omega_0 \cos\alpha\,\omega_x = 0,$$

because $I_z = 2I_x$.

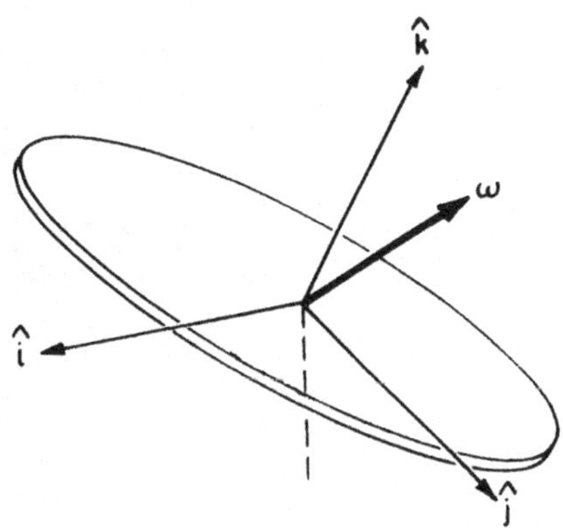

Fig. 7.4.3. Disk of Example 7.4.1.

† Synge, J. L., and Griffith, B. A., *Principles of Mechanics*, 3rd ed., McGraw-Hill Book Company, Inc., New York (1959), pp. 316–317.

These equations are easily uncoupled and integrate to

$$\omega_x = -\omega_0 \sin\alpha \sin[(\omega_0 \cos\alpha)t],$$
$$\omega_y = \omega_0 \sin\alpha \cos[(\omega_0 \cos\alpha)t].$$

Example 7.4.2. (Lainé, E., p. 47). A homogeneous lamina under no forces has the form of an isosceles triangle OAB, as shown in Fig. 7.4.4. The line OH, which bisects the angle at O, has the length h, and the base AB has the length $6^{1/2}h$. The lamina can turn freely about the fixed point O. Initially, it has the angular velocity 2ω about a line whose projection on the OAB plane falls on OH, and which makes an angle of 30° with OH. Discuss the motion and determine the polhode and herpolhode.

Let a right-handed x, y, z triad be fixed in the lamina at O. The x axis is parallel to AB and positive in the direction of B, the y axis is along OH positive in the direction of H, and the z axis is normal to the x and y axes. It follows from symmetry that the y and z axes are principal axes and, hence, the same is true for the x axis. The principal mass moments of inertia are easily found to be

$$I_x = \tfrac{1}{2}Mh^2, \qquad I_y = \tfrac{1}{4}Mh^2, \qquad I_z = \tfrac{3}{4}Mh^2, \tag{a}$$

where M is the lamina mass.

As the motion is force-free, the integrals (7.4.10) and (7.4.11) apply. We write them as

$$\begin{aligned} I_x^2\omega_x^2 + I_y^2\omega_y^2 + I_z^2\omega_z^2 &= D^2k^2, \\ I_x\omega_x^2 + I_y\omega_y^2 + I_z\omega_z^2 &= Dk^2, \end{aligned} \tag{b}$$

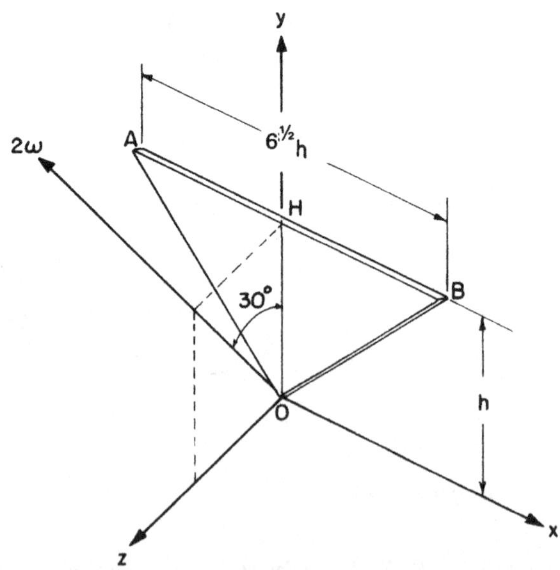

Fig. 7.4.4. Triangular lamina of Example 7.4.2.

where D and k are constants that are determined from the initial conditions

$$\omega_x(0) = 0, \qquad \omega_y(0) = 3^{1/2}\omega, \qquad \omega_z(0) = \omega \tag{c}$$

with $\omega > 0$.

Substituting (a) and (c) in (b), one finds

$$\tfrac{3}{2}Mh^2\omega^2 = Dk^2, \qquad \tfrac{3}{4}M^2h^4\omega^2 = D^2k^2,$$

from which the constants are found to be

$$D = \tfrac{1}{2}Mh^2, \qquad k^2 = 3\omega^2. \tag{d}$$

Then, the first of the Euler equations (7.4.9) and the two integrals (b) constitute the system of equations

$$\begin{aligned} \dot{\omega}_x + \omega_y\omega_z &= 0, \\ 2\omega_x^2 + \omega_y^2 + 3\omega_y^2 &= 6\omega^2, \\ 4\omega_x^2 + \omega_y^2 + 9\omega_z^2 &= 12\omega^2, \end{aligned} \tag{e}$$

where use was made of (a) and (d). These equations are sufficient to determine the three components of the angular velocity.

From the last two equations of (e) one finds

$$\omega_y^2 = 3\omega^2 - \omega_x^2, \qquad \omega_z^2 = \tfrac{1}{3}(3\omega^2 - \omega_x^2),$$

which shows that these velocity components both become zero when $\omega_x^2 = 3\omega^2$. Hence, the angular velocity can never exceed $3^{1/2}\omega$ in magnitude.

It is evident from Fig. 7.4.4 and from (c) that, initially, ω_y and ω_z are positive; thus, they remain positive during some time interval and are given by

$$\omega_y = (3\omega^2 - \omega_x^2)^{1/2}, \qquad \omega_z[(3\omega^2 - \omega_x^2)/3]^{1/2}. \tag{f}$$

The substitution of (f) in the first equation of (e) shows that, during this time interval, ω_x satisfies

$$3^{1/2}\dot{\omega}_x + 3\omega^2 - \omega_x^2 = 0,$$

which integrates under $\omega_x(0) = 0$ to

$$\frac{3^{1/2}\omega - \omega_x}{3^{1/2}\omega + \omega_x} = e^{2\omega t}.$$

Solving this equation for ω_x and substituting it in (f), one finds

$$\begin{aligned} \omega_x &= -\,3^{1/2}\omega \tanh \omega t, \\ \omega_y &= 3^{1/2}\omega/\cosh \omega t, \\ \omega_z &= \omega/\cosh \omega t. \end{aligned} \tag{g}$$

These equations give the angular velocity components along the moving axes. They show that, as $t \to \infty$, $|\omega_x| \to 3^{1/2}\omega$ and ω_y, $\omega_z \to 0$.

To determine the velocity components relative to an inertial X, Y, Z system with origin at O, let us choose the Z axis to lie in the direction of the angular momentum vector L, i.e., in the direction of the invariable line. The orientation of the X and Y axes will be left open for the present.

We denote by θ the Euler angle between the positive z and Z axis, and by ψ the angle which the projection of the Z axis on the xy plane makes with the positive y axis. Then, the angular momentum vector has the length Dk, and its components $I_x\omega_x$, $I_y\omega_y$, and $I_z\omega_z$ along the moving axis are, respectively,

$$\begin{aligned} I_x\omega_x &= Dk \sin\theta \sin\psi, \\ I_y\omega_y &= Dk \sin\theta \cos\psi, \\ I_z\omega_z &= Dk \cos\theta. \end{aligned}$$

Combining this with (d) and (g), one has

$$\begin{aligned} \sin\theta \sin\psi &= -\tanh\omega t, \\ \sin\theta\cos\psi &= 1/(2\cosh\omega t), \\ \cos\theta &= 3^{1/2}/(2\cosh\omega t). \end{aligned} \tag{h}$$

From the last of (h) one sees that $\cos\theta(0) = 3^{1/2}/2$, so that we may choose $\theta(0) = \pi/6$. Hence, $\theta(t)$ increases from its initial value $\pi/6$ to $\pi/2$ as $t \to \infty$.

As one may rewrite (h) in the form

$$\begin{aligned} \cos\theta &= 3^{1/2}/(2\cosh\omega t), \qquad & \sin\theta &= H^{1/2}/(2\cosh\omega t), \\ \sin\psi &= -2\sinh\omega t/H^{1/2}, \qquad & \cos\psi &= 1/H^{1/2}, \end{aligned} \tag{i}$$

where

$$H(t) = 1 + 4\sinh^2\omega t,$$

it is evident that the Euler angles θ and ψ are now fully determined as functions of time.

To find the third Euler angle we utilize the first two equations of (6.8.14):

$$\begin{aligned} \omega_x &= \dot\varphi \sin\theta\sin\psi + \dot\theta\cos\psi, \\ \omega_y &= \dot\varphi\sin\theta\cos\psi - \dot\theta\sin\psi. \end{aligned}$$

Eliminating $\dot\theta$ between them, one has

$$\dot\varphi\sin\theta = \omega_x\sin\psi + \omega_y\cos\psi$$

or, substituting (g) and (i) into this relation, one finds the differential equation for φ as

$$\dot\varphi = \frac{2(3)^{1/2}\omega(2\sinh^2\omega t + 1)}{H} = 3^{1/2}\omega + \frac{3^{1/2}\omega}{H}.$$

It follows that

$$\begin{aligned} \varphi &= 3^{1/2}\omega t + \int \frac{3^{1/2}\omega\, dt}{1 + 4\sinh^2\omega t} \\ &= 3^{1/2}\omega t + \int \frac{3^{1/2}\omega\, dt}{\cosh^2\omega t + 3\sinh^2\omega t} \\ &= 3^{1/2}\omega t + \int \frac{3^{1/2}\omega\, dt}{\cosh^2\omega t(1 + 3\tanh^2\omega t)} \end{aligned}$$

and this last form is readily integrated by the substitution $\tanh^2 \omega t = u$. There results

$$\varphi = 3^{1/2}\omega t + \tan^{-1}(3^{1/2} \tanh \omega t) + \varphi(0), \tag{j}$$

where the value of $\varphi(0)$ depends on the choice of orientation of the X axis. Since $\tanh \omega t$ increases from 0 to 1 as t goes from 0 to $+\infty$, the arctangent increases from 0 to $\pi/3$ as $t \to \infty$. Equations (i) and (j) determine the time history of the three Euler angles.

Since $\theta \to \pi/2$ with $t \to \infty$, the z axis tends to the X, Y plane, or the lamina tends to a plane containing the Z axis; at the same time, the lamina tends to a uniform angular velocity of magnitude $3^{1/2}\omega$.

To find the polhode, we observe that the endpoint $P = P(x, y, z)$ of the angular velocity vector relative to the moving triad is, because of (g),

$$\begin{aligned} x &= -\, 3^{1/2}\omega \tanh \omega t, \\ y &= 3^{1/2}\omega/\cosh \omega t, \\ z &= \omega/\cosh \omega t. \end{aligned} \tag{k}$$

The last two of (k) may be combined to give

$$y = 3^{1/2} z$$

for all t. Thus, the polhode must lie in the plane defined by this equation. As it must also lie on the Poinsot ellipsoid, it is, in fact, an arc of the ellipse which is formed by the intersection of the Poinsot ellipsoid with this plane.

To discuss the herpolhode, we note that this curve lies in the invariable plane and is the locus of the endpoint of the angular velocity vector. Thus, the herpolhode is the translation of the locus of the point (X, Y) from the plane $Z = 0$ into the invariable plane.

Now, the locus of the endpoint P of the angular velocity vector relative to the inertial axes is defined by (see Example 6.8.1)

$$\begin{aligned} X &= \dot\psi \sin\theta \sin\varphi + \dot\theta \cos\varphi, \\ Y &= -\, \dot\psi \sin\theta \cos\varphi + \dot\theta \sin\varphi, \\ Z &= \dot\psi \cos\theta + \dot\varphi. \end{aligned}$$

Let us rewrite the first two as

$$\begin{aligned} X &= \dot\theta\left(\cos\varphi + \frac{\dot\psi \sin\theta}{\dot\theta} \sin\varphi\right), \\ Y &= \dot\theta\left(\sin\varphi - \frac{\dot\psi \sin\theta}{\dot\theta} \cos\varphi\right). \end{aligned} \tag{l}$$

Now, from (i) we have by direct differentiation

$$\begin{aligned} \dot\psi &= -\frac{2\omega \cosh \omega t}{H}, \\ \dot\theta &= \frac{3^{1/2}\omega \tanh \omega t}{H^{1/2}}. \end{aligned} \tag{m}$$

Then, if we define an angle α by

$$\tan \alpha = -\frac{\dot{\psi} \sin \theta}{\dot{\theta}},$$

we find with the aid of (i) and (m)

$$\tan \alpha = 1/(3^{1/2} \tanh \omega t). \tag{n}$$

Thus, as t grows from zero to ∞, $\tan \alpha$ decreases from $+\infty$ to $3^{-1/2}$, or α may be regarded as decreasing from $\pi/2$ to $\pi/6$.

Consistent with (n), we may write

$$\sin \alpha = \cosh \omega t / H^{1/2},$$
$$\cos \alpha = 3^{1/2} \sinh \omega t / H^{1/2},$$

which permits us to rewrite (l) as

$$\begin{aligned} X &= \frac{\dot{\theta} \cos(\varphi + \alpha)}{\cos \alpha}, \\ Y &= \frac{\dot{\theta} \sin(\varphi + \alpha)}{\cos \alpha}, \end{aligned} \tag{o}$$

where θ, φ, and α are all functions of time.

To find X and Y explicitly as functions of time consider (j) with the initial value $\varphi(0) = -\pi/2$. Then,

$$\varphi = 3^{1/2}\omega t - [\pi/2 - \tan^{-1}(3^{1/2} \tanh \omega t)],$$

or because of (n),

$$\varphi = 3^{1/2}\omega t - \alpha,$$

so that the argument of the trigonometric functions in (o) is

$$\varphi + \alpha = 3^{1/2}\omega t. \tag{p}$$

Therefore, the substitution of (m) and (p) in (o) gives the desired result:

$$\begin{aligned} X &= \frac{\omega \cos(3^{1/2}\omega t)}{\cosh \omega t}, \\ Y &= \frac{\omega \sin(3^{1/2}\omega t)}{\cosh \omega t}. \end{aligned} \tag{q}$$

The herpolhode is found by translating the curve defined by (q) from the plane $Z = 0$ into the invariable plane $Z = \text{const}$.

If we write

$$\varrho = \omega / \cosh \omega t, \qquad \beta = 3^{1/2}\omega t$$

we find

$$\varrho = \omega / \cosh(\beta / 3^{1/2}).$$

Thus we find for the herpolhode a sort of spiral tending to zero as $t \to \infty$.

7.5. The General Motion of Rigid Bodies

It was shown in Section 6.1 that an unconstrained rigid body has six degrees of freedom, i.e., it requires six numbers to specify its configuration. One of the more common ways to define the sequence of its configurations is to give the motion of the mass center, and the motion about the mass center, all relative to a Galilean frame. If F is the resultant of the external forces acting on the body, and M is the moment of these forces about the mass center, the motion of the strictly Newtonian problem is governed by

$$\begin{aligned} m\frac{dv}{dt} &= F, \\ \frac{dL}{dt} &= M, \end{aligned} \tag{7.5.1}$$

where v is the velocity of the mass center relative to the Galilean frame, L is the angular momentum relative to the same frame, and m is the total mass of the rigid body; it is necessarily constant.

If we fix in the rigid body a triad of principal axes with origin at the mass center we may write the vector equations of motion (7.5.1) in component form along the body-fixed, and hence *not* Galilean, reference frame. If the velocity of the mass center is

$$v = v_x\hat{\imath} + v_y\hat{\jmath} + v_z\hat{k},$$

the angular velocity of the body is

$$\omega = \omega_x\hat{\imath} + \omega_y\hat{\jmath} + \omega_z\hat{k},$$

the angular momentum is

$$L = L_x\hat{\imath} + L_y\hat{\jmath} + L_z\hat{k},$$

the force is

$$F = F_x\hat{\imath} + F_y\hat{\jmath} + F_z\hat{k},$$

and the moment is

$$M = M_x\hat{\imath} + M_y\hat{\jmath} + M_z\hat{k},$$

we find for the derivatives

$$\begin{aligned} \frac{dv}{dt} &= \dot{v}_x\hat{\imath} + \dot{v}_y\hat{\jmath} + \dot{v}_z\hat{k} + \omega \times v, \\ \frac{dL}{dt} &= \dot{L}_x\hat{\imath} + \dot{L}_y\hat{\jmath} + \dot{L}_z\hat{k} + \omega \times L, \end{aligned}$$

where the cross products arise because of the motion of the unit vectors $\hat{i}$, $\hat{j}$, and $\hat{k}$. The substitution of all these in (7.5.1) gives

$$\begin{aligned} m[\dot{v}_x - (v_y\omega_z - v_z\omega_y)] &= F_x, \\ m[\dot{v}_y - (v_z\omega_x - v_x\omega_z)] &= F_y, \\ m[\dot{v}_z - (v_x\omega_y - v_y\omega_x)] &= F_z, \\ I_x\dot{\omega}_x - (I_y - I_z)\omega_y\omega_z &= M_x, \\ I_y\dot{\omega}_y - (I_z - I_x)\omega_z\omega_x &= M_y, \\ I_z\dot{\omega}_z - (I_x - I_y)\omega_x\omega_y &= M_z. \end{aligned} \tag{7.5.2}$$

These are the six component equations of motion of a rigid body parallel to body-fixed principal coordinates with origin at the mass center.

Example 7.5.1. Formulate the dynamical problem of a heavy, homogeneous disk which rolls without slipping on an inclined plane in such a way that the plane of the disk remains always perpendicular to the inclined plane. This is the condition of a wheel on a carriage with two or more wheels.

Let us chose coordinate systems as shown in Fig. 7.5.1. The X, Y, Z system is inertial. The X axis is inclined by α to the horizontal and points down the plane, while the Y axis is horizontal; hence, the X, Y plane is the inclined plane. The x, y, z coordinates are not fixed in intertial space. The origin of the x, y, z sys-

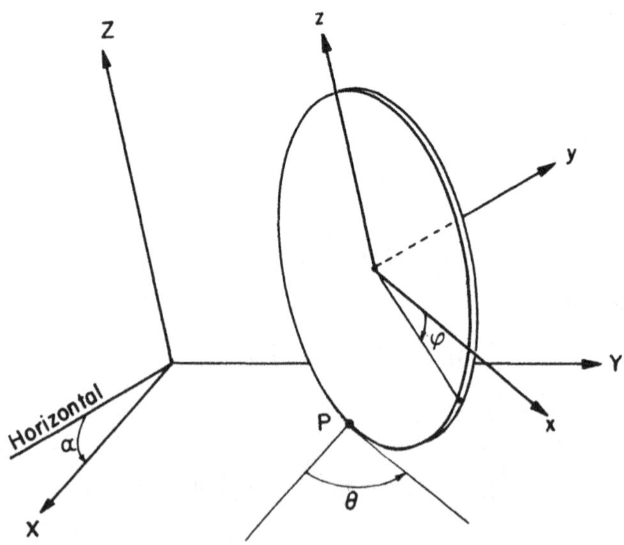

Fig. 7.5.1. Rolling disk on inclined plane of Example 7.5.1.

tem is at the center of the disk. The disk is always in the x, z plane, and the x, y plane is always parallel to the X, Y plane. It is evident that x, y, z are principal axes. Let φ be the angle through which the disk has rotated about the y axis, and let θ be the angle from the positive X axis to the tangent to the trajectory of the disk contact point P. Let the disk have mass m and radius r.

The external forces acting on the disk are the reaction at P, the gravitational force, and the forces which keep the disk plane normal to the inclined X, Y plane. We write for the reaction at P

$$F = F_x\hat{\imath} + F_y\hat{\jmath} + F_z\hat{k}, \tag{a}$$

where the $\hat{\imath}$, $\hat{\jmath}$, $\hat{k}$ are the unit vectors along the x, y, z axes, and the gravitational force at the disk center is

$$W = mga \sin\alpha \sin\theta\hat{\imath} - mga \sin\alpha \cos\theta\hat{\jmath} - mga \cos\alpha\hat{k}. \tag{b}$$

Let the forces which keep the disk plane normal to the inclined plane be forces F_1 and $-F_1$, both in the y, z plane parallel to the y axis, and acting a unit distance from the disk center. Thus, their sum is zero, and their moment about the disk center is

$$M_1 = \hat{k}\times F_1\hat{\jmath} + (-\hat{k})\times(-F_1)\hat{\jmath} = -2F_1\hat{\imath}. \tag{c}$$

If the position vector of the disk center is given by

$$R = X\hat{I} + Y\hat{J} + r\hat{K},$$

where $\hat{I}$, $\hat{J}$, $\hat{K}$ are unit vectors along the X, Y, and Z axes respectively, the vector equation of translation of the mass center is

$$\begin{aligned} m(\ddot{X}\hat{I} + \ddot{Y}\hat{J}) = {} & (F_x + mga \sin\alpha \sin\theta)\hat{\imath} \\ & + (F_y - mga \sin\alpha \cos\theta)\hat{\jmath} \\ & + (F_z - mga \cos\alpha)\hat{k}. \end{aligned}$$

But

$$\begin{aligned} \hat{\imath} &= \cos\theta\hat{I} + \sin\theta\hat{J}, \\ \hat{\jmath} &= -\sin\theta\hat{I} + \cos\theta\hat{J}, \\ \hat{k} &= \hat{K}. \end{aligned}$$

Therefore, the equations of mass center translation in component form are

$$\begin{aligned} m\ddot{X} &= F_x \cos\theta - F_y \sin\theta + 2\, mga \sin\alpha \sin\theta \cos\theta, \\ m\ddot{Y} &= F_x \sin\theta + F_y \cos\theta + mga \sin\alpha(\sin^2\theta - \cos^2\theta), \\ 0 &= F_z - mga \cos\alpha \end{aligned}$$

or, utilizing well-known trigonometric identities,

$$\begin{aligned} m\ddot{X} &= F_x \cos\theta - F_y \sin\theta + mga \sin\alpha \sin 2\theta, \\ m\ddot{Y} &= F_x \sin\theta + F_y \cos\theta - mga \sin\alpha \cos 2\theta, \\ 0 &= F_z - mga \cos\alpha. \end{aligned} \tag{d}$$

These are the first three equations of (7.5.2) applied to the present problem.

To obtain the other three we note that

$$\begin{aligned} \omega &= \omega_x \hat{\imath} + \omega_y \hat{\jmath} + \omega_z \hat{k} \\ &= \dot{\varphi} \hat{\jmath} + \dot{\theta} \hat{k}, \end{aligned} \tag{e}$$

and

$$\dot{\hat{\jmath}} = -\dot{\theta}\hat{\imath}, \qquad \dot{\hat{k}} = 0. \tag{f}$$

The principal moments of inertia of the disk are

$$I_x = I_z = \tfrac{1}{4}mr^2, \qquad I_y = \tfrac{1}{2}mr^2. \tag{g}$$

The angular momentum is, in view of (e),

$$L = I_y \dot{\varphi} \hat{\jmath} + I_z \dot{\theta} \hat{k},$$

and the time rate of change of the momentum is, because of (f),

$$\dot{L} = -I_y \dot{\theta} \dot{\varphi} \hat{\imath} + I_y \ddot{\varphi} \hat{\jmath} + I_z \ddot{\theta} \hat{k}. \tag{h}$$

The moment of the applied forces about the disk center is, in view of (c),

$$\begin{aligned} M &= -2F_1 \hat{\imath} + (-r)\hat{k} \times (F_x \hat{\imath} + F_y \hat{\jmath} + F_z \hat{k}) \\ &= (-2F_1 + rF_y)\hat{\imath} - rF_x \hat{\jmath}. \end{aligned} \tag{i}$$

Equating (h) and (i) and utilizing (g), the equations of rotation about the disk center are

$$\begin{aligned} -\tfrac{1}{2}mr^2 \dot{\varphi} \dot{\theta} &= -2F_1 + rF_y, \\ \tfrac{1}{2}mr^2 \ddot{\varphi} &= -rF_x, \\ \tfrac{1}{4}mr^2 \ddot{\theta} &= 0. \end{aligned} \tag{j}$$

These are the second three equations of (7.5.2) for this particular problem.

The condition of pure rolling is that the contact point of the disk be at rest, or $\dot{R} = \omega \times r\hat{k}$. Thus

$$\begin{aligned} \dot{X}\hat{I} + \dot{Y}\hat{J} &= (\dot{\varphi}\hat{\jmath} + \dot{\theta}\hat{k}) \times r\hat{k} \\ &= r\dot{\varphi}\hat{\imath} = r\dot{\varphi}(\cos\theta \hat{I} + \sin\theta \hat{J}), \end{aligned}$$

or in component form

$$\dot{X} = r\dot{\varphi}\cos\theta, \qquad \dot{Y} = r\dot{\varphi}\sin\theta. \tag{k}$$

Equations (d), (j), and (k) formulate the problem and constitute the required answer. Equations (k) imply

$$\sin\theta \, dX = \cos\theta \, dY,$$

which is of the form (4.6.1) and, hence, nonholonomic. Nevertheless, the problem is readily soluble. From the last of (j) we obtain

$$\dot{\theta} = A = \text{const}, \qquad \theta = At + B, \tag{l}$$

where B is also a constant. Next, we eliminate F_y between the first two equations of (d). This gives

$$m(\ddot{X}\cos\theta + \ddot{Y}\sin\theta) = F_x + mga\sin\alpha\sin\theta. \tag{m}$$

We can now eliminate F_x by means of the second equation of (j), $\ddot{X}$ and $\ddot{Y}$ by means of (k), and θ by the second equation of (l). This results in

$$\ddot{\varphi} = C \sin(At + B), \tag{n}$$

where

$$C = (2g \sin \alpha)/3r.$$

Equation (n) integrates to

$$\begin{aligned}\dot{\varphi} &= -\frac{C}{A}\cos(At + B) + \dot{\varphi}_0,\\ \varphi &= -\frac{C}{A^2}\sin(At + B) + \dot{\varphi}_0 t + \varphi_0,\end{aligned} \tag{o}$$

where $\dot{\varphi}_0$ and φ_0 are constants. The substitution of the second equation of (l) and the first equation of (o) into (k) permits the computation of X and Y as functions of t, (m) is then used to compute F_x, and the first or second equation of (d) to compute F_y, both as functions of t, F_z is known from the third relation of (d) and is constant, and the first equation of (j) is used to find F_1.

7.6. Problems

7.1. Show that the kinetic energy of a uniform, straight rigid rod of mass m is given by

$$T = \tfrac{1}{6}m(u \cdot u + u \cdot v + v \cdot v),$$

where u is the velocity of one of its extremities and v is that of the other extremity.

7.2. Five particles of mass m, $2m$, $3m$, $4m$, and $5m$ respectively are interconnected by eight rigid, massless bars of equal length a, as shown, so as to form a pyramid. A right-handed Cartesian coordinate system is fixed in the rigid

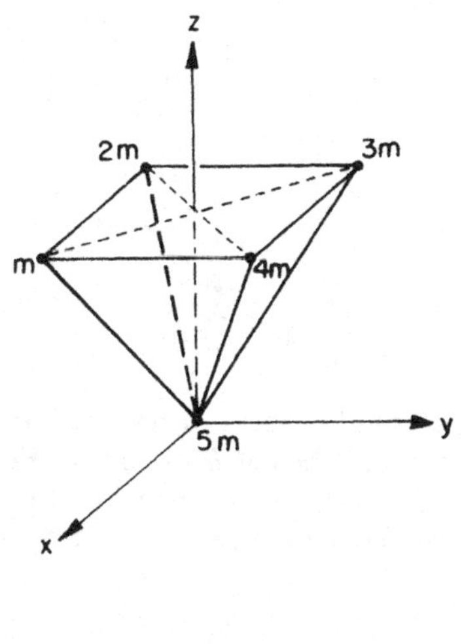

body with origin at the vertex of the pyramid; the x axis is parallel to the line connecting m and $2m$, and the y axis is parallel to the line connecting m and $4m$. The pyramid rotates with angular velocity ω about a line connecting $5m$ and $4m$, as shown. Calculate the x, y, and z components of the angular momentum of the pyramid.

7.3. A homogeneous rectangular lamina of mass per unit area m has sides a and $2a$. If the origin of the x, y system of Cartesian coordinates is at one corner, and the axes are parallel to the sides of the rectangle, as shown, calculate

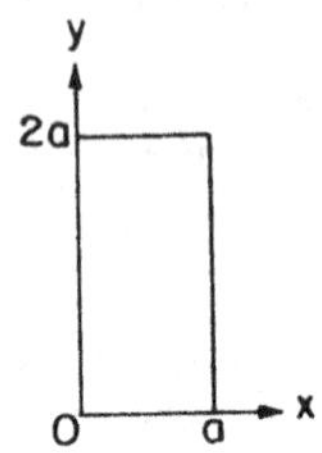

the inertial parameters relative to the x and y axes and the principal moments of inertia about O. Give the directions of the principal axes relative to the positive x axis.

7.4. Show that if a point O lies in a plane of mass symmetry of a rigid body, then one of the principal axes through O must lie in that plane of mass symmetry. In consequence, the principal moments of inertia are found by solving a quadratic equation.

7.5. A homogeneous solid has the form of a tetrahedron as shown. The angles AOB, AOC, and BOC are all right angles; the distances of A, B, and C from O are, respectively, $OA = a$, $OB = b$, $OC = c$. If the total mass of the solid is M, find the ellipsoid of inertia with respect to O and calculate the principal moments of inertia with respect to O when $a = b = c$.

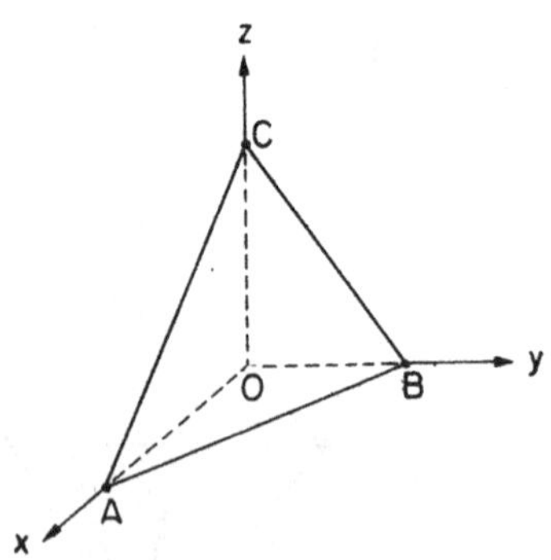

7.6. A homogeneous solid of mass M has the form of a right cone of revolution of base radius R and height h as shown. Find (a) the principal moments of inertia relative to O, (b) the moment of inertia relative to a side, and (c) the moment of inertia relative to a base diameter.

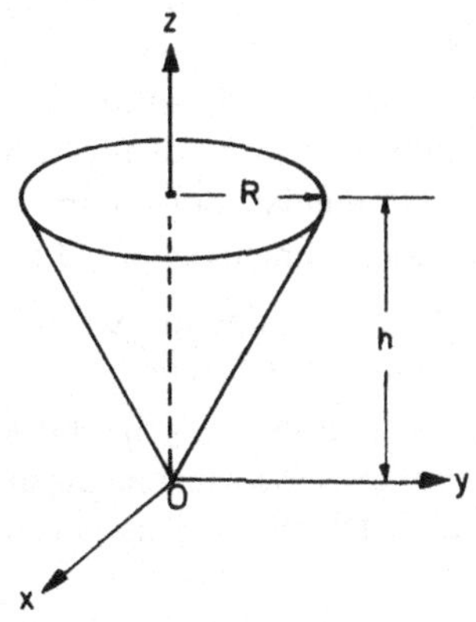

7.7. A homogeneous tetrahedron of mass density ϱ is bounded by the faces ABC, ACD, BCD, and ABD, as shown. The coordinates of A are $(-a/2, 0, 0)$, those of B are $(a/2, 0, 0)$, those of C are $(0, 0, 3^{1/2}/2)$, and those of D are $(0, 3^{1/2}a/2, 0)$. Find the principal axes and moments of inertia about O.

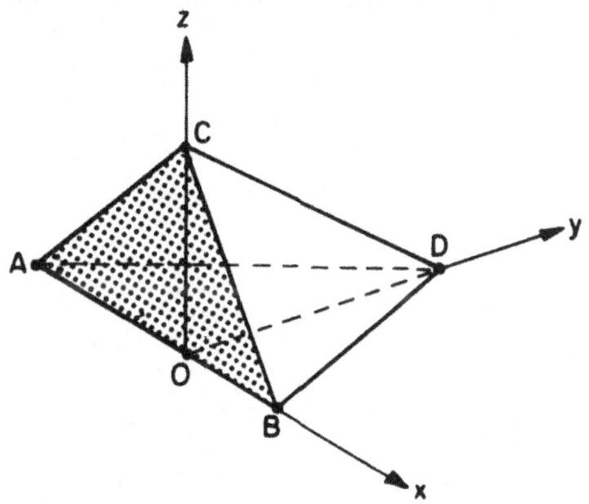

7.8. A homogeneous, right circular cylinder of radius R and height R rests with a flat side on a plane. Find the principal axes and moments of inertia of the cylinder with respect to a point on the circle of intersection between the cylinder surface and the plane.

7.9. A homogeneous, right circular cone with vertex half-angle α rolls without slipping on a horizontal plane, as shown. The angular speed of the cone about

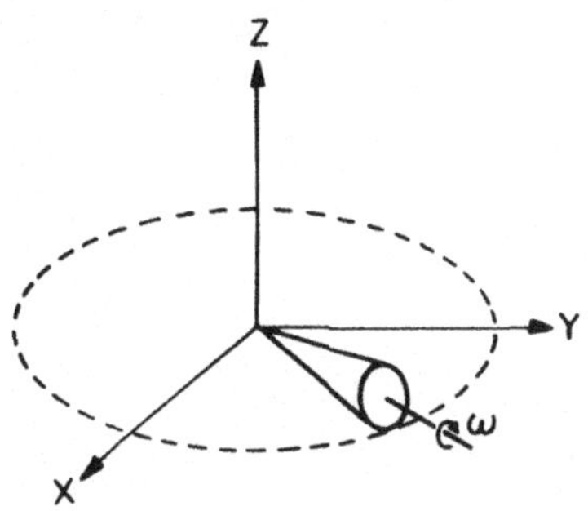

an axis through the apex and normal to the base is $\omega = \text{const.}$

(a) Write down the equations of motion.

(b) Compute the angular velocity of the cone about the Z axis.

(c) Compute the total angular momentum of the cone.

(d) Find the forces and moments required to maintain the motion.

7.10. A homogeneous rod has its midpoint fixed. Find the force-free motion of the rod under arbitrary initial conditions.

7.11. A heavy rod is so oriented that at some instant t_0 its mass center is higher than one of its extremities. Show that one can always move the lower extremity in a horizontal plane such that the mass center of the rod remains above the lower extremity.

7.12. A heavy, homogeneous sphere of mass m and radius r rolls on a perfectly rough, horizontal turntable that rotates with constant angular velocity ω about a fixed, vertical line. Find the equations of motion of the center of the sphere in an inertial coordinate system and show that every trajectory of the center of the sphere is a circular arc with center and radius dependent on initial conditions.

8

The Nature of Lagrangean Mechanics

8.1. General Remarks

In the preceding chapters, the basic concepts underlying Newtonian mechanics have been discussed. Thus, these chapters constitute essentially a review of the material treated in a first course of mechanics. We now sketch in broad outline the essential generalizations made by Lagrange.

8.2. The Generalizations by Lagrange

Lagrangean mechanics extends and generalizes Newtonian mechanics in two basic ways:

(i) Lagrangean mechanics is a general dynamical theory of discrete, constrained systems, provided only that the constraints belong to the class defined in (4.3.4) and (4.3.6). In particular, one need not know the precise form of the constraints before formulating the problem completely.

(ii) The introduction of "generalized coordinates" (also often called Lagrange coordinates) reduces both the number of equations of motion and the number of equations of constraint to the smallest possible number. Yet, one need not know any particular set of generalized coordinates in order to formulate the problem completely.

Lagrange himself wrote†:

> I have set out to reduce the theory of that science [mechanics], and the art of solving problems in that domain, to general formulas whose simple development yields all equations necessary for the solution of every problem... Those who love analysis will note with pleasure that mechanics has now become one of its branches and will be grateful to me for having thus extended its domain.

Historically, Lagrangean mechanics evolved in a natural way as a consequence of extending the notions of "virtual displacement" and "virtual work" from the domain of statics to that of dynamics. These concepts are not essential to the derivation of the Lagrangean equations of motion for holonomic systems; however, the Lagrangean equations of nonholonomic systems cannot be derived without introducing either the notion of virtual work or some equivalent concept.

The introduction of the notion of virtual work into Lagrangean mechanics arises from the manner in which forces are classified in that theory. In Lagrangean mechanics every bounded force is either a "constraint force" or it is not (in the latter case it is called a "given force,"), and constraint forces are systems of forces which do no work in a virtual displacement or, simply, which do no virtual work. In Newtonian mechanics, on the other hand, forces are either internal or external to the system of particles, and this distinction is intrinsic to Newtonian mechanics because the internal forces satisfy the third law while the external forces do not. Either of them may include both constraint forces and given forces, so that the classifications of forces in the Newtonian and Lagrangean mechanics are not equivalent.

An entirely separate, but equally important feature of Lagrangean mechanics is the notion of generalized coordinates and generalized forces. Every set of coordinates just large enough to describe the configuration of a dynamical system is called a set of generalized coordinates.‡ These have the property of satisfying all holonomic constraints. Thus, only nonholonomic constraints remain in the problem formulation when generalized coordinates are used. The Lagrangean formulation of the dynamics problem is made in terms of an undefined set of generalized coordinates; no prior selection of a specific set is necessary to formulate the problem. It is this feature which has given rise to the name "generalized" coordinates.

† Author's translation of excerpt from Lagrange, J. L., *Mécanique analytique*, new edition, Mme. Ve. Courcier, Paris, 1811.

‡ This definition will be made precise later on.

Similarly, the given forces are expressed in components along the generalized coordinates and are called "generalized" forces. These are generalized in quite another sense as well: For instance, a generalized coordinate may be an angle; in that case, the corresponding generalized force component is, in physical terms, a moment, not a force.

Finally, in Lagrangean mechanics, the so-called inertia terms, those which arise from the acceleration of massive elements, are derived entirely from the kinetic energy. This obliterates the essential distinction between rigid systems of particles on the one hand and rigid, continuous bodies on the other (see Section 7.1). Thus, if one knows how to compute the kinetic energy of rigid, continuous bodies in terms of any coordinate system describing its configuration, one can readily incorporate it in the framework of the Lagrangean equations of motion.

It should not be concluded from the observations relative to generalized coordinates that Newtonian mechanics is restricted to the use of Cartesian coordinates. In fact, we showed in Section 2.3 that any set of coordinates defined by a Galilean frame $g = (\hat{e}_1, \hat{e}_2, \hat{e}_3)$ may be utilized, where the $\hat{e}_i$ are unit vectors spanning Euclidean 3-space. However, when the equations of motion are written down from Newton's laws, it is necessary to specify at the outset which coordinates are to be used in the problem formulation.

The introduction of generalized coordinates leads directly to the so-called Lagrangean equations of motion. Some scientists feel, therefore, that Lagrangean mechanics consists of no more than the derivation of these equations. It is our view that such an interpretation does injustice to Lagrange's monumental contribution. While Lagrange was not the discoverer of the concept of virtual displacements, we owe to him the bold step of lifting it from the domain of statics and introducing it into dynamics by utilizing d'Alembert's principle. Once this was accomplished, the general theory of constraints and that of generalized coordinates followed in a natural way.

9

Virtual Displacement and Virtual Work

9.1. General Observations

The central concepts permitting the extension of mechanics from the Newtonian point of view to the Lagrangean are the notions of *virtual displacement* and of *virtual work*. These concepts were formulated early in the development of mechanics insofar as their application to *statics* is concerned. Johann Bernoulli (1667-1748) discovered them (1717) for the case of holonomic configuration constraints, and Gabriel Fourier extended them in 1798 to inequality constraints.

In the problems of statics, the configuration does not change with time; hence, virtual displacements do not involve time. But, dynamics is the science of motion, i.e., of configuration changes *in* time. Nevertheless, a virtual displacement does not involve time in *dynamics* either. This has led to the unfortunate phrase that virtual displacements "take place in zero time." However, even if one abandons that phrase, it must be explained how one can obtain information useful in the science of dynamics from the concept of a displacement not involving time. The rationale for this procedure will be found in d'Alembert's principle.

9.2. Classification of Displacements

The strictly Newtonian (second) problem of classical particle mechanics, subject to equality constraints only, is one of finding continuous scalar functions $u_s(t)$ $(s = 1, 2, \ldots, N)$ which have prescribed initial values $u_s(0)$, $\dot{u}_s(0)$, and which satisfy the set of differential equations

$$m_s\ddot{u}_s = F_s(u_1, u_2, \ldots, u_N; \dot{u}_1, \dot{u}_2, \ldots, \dot{u}_N; t) \qquad (s = 1, 2, \ldots, N) \tag{9.2.1}$$

as well as the constraint equations

$$\sum_{s=1}^{N} A_{rs}\, du_s + A_r\, dt = 0 \qquad (r = 1, 2, \ldots, L < N). \tag{9.2.2}$$

We now define three different classes of displacements: *actual*, *possible*, and *virtual* displacements.

Actual displacements, as the name implies, give the actual motion. Thus:

> *The class of functions $u_s(t)$, piecewise of class C^2, which satisfy the equations of motion and the equations of constraint is called the class of actual displacements. The vector $u(t) = (u_1(t), u_2(t), \ldots, u_N(t))$ is called an* actual displacement vector.

If displacements satisfy all equations of constraint but not necessarily the equations of motion they are called "possible"; in other words:

> *The set of infinitesimal quantities du_s $(s = 1, 2, \ldots, N)$ which satisfy the system of equations*
>
> $$\sum_{s=1}^{N} A_{rs}\, du_s + A_r\, dt = 0 \qquad (r = 1, 2, \ldots, L < N) \tag{9.2.3}$$
>
> *are called the class of* possible displacements. *The vector*
>
> $$du = (du_1, du_2, \ldots, du_N)$$
>
> *is called a* possible displacement vector.

It is clear that the infinitesimals of the actual displacements are members of the class of possible displacements, but the converse is not necessarily true. The above definition states that every vector of infinitesimal length which lies in the tangent plane defined by the constraint equations is a possible displacement vector.

Finally, we define "virtual" displacements which play such a central role in Lagrangean mechanics:

The set of infinitesimal quantities δu_s $(s = 1, 2, \ldots, N)$ *which satisfy the system of equations*

$$\sum_{s=1}^{N} A_{rs}\, \delta u_s = 0 \qquad (r = 1, 2, \ldots, L < N) \tag{9.2.4}$$

are called the class of virtual displacements. *The vector*

$$\delta u = (\delta u_1, \delta u_2, \ldots, \delta u_N)$$

is called a virtual displacement vector.

In other terms, every vector of infinitesimal length which lies in the tangent plane defined by (9.2.4) is a virtual displacement vector. Since the tangent plane (9.2.3) is different from (9.2.4), the class of possible displacements and the class of virtual displacements have, in general, no members in common. However, when all constraints are catastatic, these two classes are identical.

9.3. D'Alembert's Principle

D'Alembert's principle is often said to reduce the problem of dynamics to one of statics; in a certain sense, about to be explained, this statement is true.

In (2.5.1), the forces acting on the particle P_r were written as

$$F^r = \sum_\alpha \Theta_\alpha{}^r + \sum_\beta \Phi_\beta{}^r + \sum_j \Psi_j{}^r, \tag{9.3.1}$$

where

$$F_i{}^r = \sum_\alpha \Theta_\alpha{}^r$$

is the resultant of the internal forces, and

$$F_e{}^r = \sum_\beta \Phi_\beta{}^r + \sum_j \Psi_j{}^r$$

is the resultant of the external forces. While the division of forces into internal and external ones is intrinsic to Newtonian mechanics because of the third law of motion [see (2.5.4)], it is nevertheless known that among both the internal and the external forces, there are some which arise from the constraints, and others which do not.

It is clear that constraints give rise to forces. We prove this assertion as follows: The presence of constraints either changes the motion, or it does not. If it does not, the constraints may be altogether ignored in the problem formulation. If it does, the constraints must give rise to forces because, by Newton's second law, forces are the only agents which can change the motion of particles. Thus, we either have no constraints, or we have constraints and forces due to them. We call the latter *constraint forces*, and we denote the resultant of the constraint forces acting on the rth particle by

$$F^{r\prime} = \sum_{\gamma=1}^{L} C_{\gamma}{}^{L}, \tag{9.3.2}$$

where $C_\gamma{}^r$ is the force from the γth constraint acting on the rth particle.

The above definition does not tell us how to recognize constraint forces. The property which distinguishes them from other forces will be stated a little further on, and that property will then serve as a precise, formal definition of the notion of "constraint force."

All forces which are not constraint forces are called *given forces*, and the resultant of the given forces acting on the rth particle is denoted by

$$F^r = \sum_{\delta=1}^{D} \Delta_{\delta}{}^{r}. \tag{9.3.3}$$

With this notation, Newton's second law for the motion of the n particles may be written as

$$m_r \ddot{x}^r - F^r = F^{r\prime} \qquad (r = 1, 2, \ldots, n), \tag{9.3.4}$$

or, in component form,

$$m_s \ddot{u}_s - F_s = F_s{}' \qquad (s = 1, 2, \ldots, N = 3n). \tag{9.3.5}$$

When these equations are summed, one has in vector form

$$\sum_{r=1}^{n} (m_r \ddot{x}^r - F^r) = \sum_{r=1}^{n} F^{r\prime}, \tag{9.3.6}$$

or, in component form,

$$\sum_{s=1}^{N} (m_s \ddot{u} - F_s) = \sum_{s=1}^{N} F_s{}'. \tag{9.3.7}$$

D'Alembert's principle states:

The totality of the constraint forces may be disregarded in the dynamics problem of systems of particles.

This principle is of far-reaching importance in Lagrangean mechanics. It is, in fact, the counterpart to the third law in the Newtonian mechanics of systems of particles. There, the separation of forces into external and internal ones gave rise to a special axiom with respect to the internal forces. Similarly, in the Lagrangean mechanics of constrained systems, the separation of forces into given and constraint forces requires a special axiom with respect to the latter. That axiom is d'Alembert's principle.

This principle will be interpreted here in terms of the geometrical significance of constraints (only equality constraints will be considered for the present). These define either a surface or a tangent plane in the configuration space. D'Alembert's principle states that the totality of the constraint forces does not contribute to the acceleration of the particles of the system. Because these forces are lost to this effort, they are sometimes referred to as "lost forces."

D'Alembert's principle can be given an interesting dynamical interpretation. Let us write (9.3.6) in components *in* and *normal to* the tangent plane or surface defined by the constraints, i.e.,

$$\sum_{r=1}^{n}(m_r\ddot{x}^r - F^r)_{\text{norm}} + \sum_{r=1}^{n}(m_r\ddot{x}^r - F^r)_{\text{tan}} = \sum_{r=1}^{n}(F^{r'})_{\text{norm}} + \sum_{r=1}^{n}(F^{r'})_{\text{tan}}.$$

Then, because the motion is constrained to the tangent plane, the normal component of $\ddot{x}^r$ relative to the surface vanishes, and we find the two equations

$$\sum_{r=1}^{n}(m_r\ddot{x}^r - F^r)_{\text{tan}} = \sum_{r=1}^{n}(F^{r'})_{\text{tan}}, \tag{a}$$

$$\sum_{r=1}^{n}(F^r)_{\text{norm}} + \sum_{r=1}^{n}(F^{r'})_{\text{norm}} = 0. \tag{b}$$

But, by d'Alembert's principle, the right-hand side of (a) is set equal to zero. Hence, d'Alembert's principle may be stated in two parts:

(a) The force components of the given forces tangent to the constraint surface (or in the tangent plane) are the only ones which contribute to the particle acceleration, or

$$\sum_{r=1}^{n}[m_r\ddot{x}^r - (F^r)_{\text{tan}}] = 0 \tag{9.3.8a}$$

and

(b) The normal components of the given forces are in equilibrium with the constraint forces, or

$$\sum_{r=1}^{n}[(F^r)_{\text{norm}} + F^{r'}] = 0. \tag{9.3.8b}$$

Comparing (a) and (9.3.8a) we find

$$\sum_{r=1}^{n} (F^{r'})_{\text{tan}} = 0, \tag{c}$$

and that result could also have been deduced from (9.3.8b). We see that d'Alembert's principle may be expressed in terms of (c), from which (9.3.8a) and (9.3.8b) emerge as consequences. But (9.3.8b) and (c) are equations of forces in static equilibrium. In our view, this is the rationale for the statement that d'Alembert's principle reduces a problem in dynamics to one in statics.

If we had carried out the above considerations on (9.3.7) we would have found instead of (9.3.8a) and (9.3.8b) the equations

$$\sum_{s=1}^{N} [m_s \ddot{u}_s - (F_s)_{\text{tan}}] = 0, \tag{9.3.8c}$$

and

$$\sum_{s=1}^{N} [(F_s)_{\text{norm}} + F_s{'}] = 0. \tag{9.3.8d}$$

[Because of a widespread misunderstanding of d'Alembert's principle, it is necessary to add some critical remarks.

One frequently finds d'Alembert's principle "applied" to the unconstrained motion of a single particle. Authors who do this proceed as follows: They rewrite Newton's second law for a single particle

$$m\ddot{r} = F \tag{a}$$

in the form

$$F - m\ddot{r} = 0, \tag{b}$$

where F is the resultant of all forces acting on the particle, and they call (a) Newton's principle, and (b) d'Alembert's principle. They argue that, if F in (b) is a force and it may be added to $-m\ddot{r}$, then it follows from homogeneity requirements that $-m\ddot{r}$ is also a force (usually called the "reversed effective force" while $m\ddot{r}$ is called the "inertia force"). Thus, (b) states that the sum of two forces vanishes. This is the statement of a statics problem; hence, the dynamics problem (a) has been reduced to the statics problem (b).

Now, it is evident that (a) and (b) are the same equations, their only difference being that in (b) all nonzero terms have been transferred to the same side of the equal sign. Certainly, (b) does not involve any new "principle" not contained in (a). Hamel (p. 220) called this interpretation of d'Alembert's principle an insult to d'Alembert.]

Bernoulli is the originator of the principle of virtual work in *statics*. Let us consider (with Bernoulli) a problem of statics in which a concurrent force system is in equilibrium, or $\sum_{r=1}^{n} F^r = 0$. Then, Bernoulli introduced "imagined displacements not violating the constraints." He called them "virtual displacements," and we shall denote them here by δx^r. Bernoulli called the inner product $F^r \cdot \delta x^r$ the work done by the force F^r in a virtual displacement δx^r, or briefly, the virtual work done by F^r. His principle of virtual work states that

$$\sum_{r=1}^{n} F^r \cdot \delta x^r = 0$$

or a force system in static equilibrium is one which does no work in a virtual displacement.

[Bernoulli's vocabulary is unfortunate because no actual displacements occur, and no real work is being done. We retain it here out of respect for tradition and for the masters of the past. However, Crandall, Karnopp, Kurtz and Pridmore-Brown in their excellent book[†] have departed from this tradition. They call δx^r an "admissible variation," and $F^r \cdot \delta x^r$ the "variational indicator."]

It was Lagrange, following the ideas of Bernoulli, who utilized the fact that the totality of the constraint forces does no virtual work. Multiplying (9.3.4) by δx^r, or (9.3.5) by δu_s, and summing over the index, one obtains

$$\sum_{r=1}^{n} (m_r \ddot{x}^r - F^r) \cdot \delta x^r = \sum_{r=1}^{n} F^{r'} \cdot \delta x^r, \tag{9.3.9}$$

or

$$\sum_{s=1}^{N} (m_s \ddot{u}_s - F_s)\, \delta u_s = \sum_{s=1}^{N} F_s'\, \delta u_s. \tag{9.3.10}$$

Then, by the principles of d'Alembert and Bernoulli, the constraint forces do no virtual work, or

$$\sum_{r=1}^{n} (m_r \ddot{x}^r - F^r) \cdot \delta x^r = 0, \tag{9.3.11}$$

or

$$\sum_{s=1}^{N} (m_s \ddot{u}_s - F_s)\, \delta u_s = 0. \tag{9.3.12}$$

[†] S. H. Crandall, D. C. Karnopp, E. F. Kurtz, Jr., and D. C. Pridmore-Brown, *Dynamics of Mechanical and Electromechanical Systems*, McGraw-Hill Book Co., New York (1968).

These equations are usually referred to in the literature as "Lagrange's form of d'Alembert's principle." They are so fundamental to the development of Lagrangean mechanics that Pars refers to either as *the fundamental equation.*

A word may be in order concerning the difference between the forces occurring in (9.3.9) and (9.3.10), respectively. They are of the forms

$$F^r = F^r(x^1, x^2, \ldots, x^{3n}; \dot{x}^1, \dot{x}^2, \ldots, \dot{x}^{3n}; t)$$

and

$$F_s = F_s(u_1, u_2, \ldots, u_N; \dot{u}_1, \dot{u}_2, \ldots, \dot{u}_N; t).$$

There is a fundamental difference between these; F^r is a vector in 3-space while F_s is the component, in the direction of the unit vector $\hat{e}_s$, of a vector in N-space. Expressed differently, F^r is a force acting on a particle of mass m_r having position vector x^r; both the force and position vectors have, in general, three Cartesian components in physical space. Therefore, we shall call F^r a "physical force." F_s, on the other hand, is defined by means of the vector sum

$$F = (F_1, F_2, \ldots, F_N) = \sum_{s=1}^{N} F_s\hat{e}_s,$$

where the N-space has three times as many coordinates (or base vectors) as the system has particles. Hence, when there is more than one particle in the system, the N-dimensional configuration space has no counterpart in the world of observables.

We use the definition:

The virtual work δW *done by a force*

$$F = (F_1, F_2, \ldots, F_N)$$

in a virtual displacement

$$\delta u = (\delta u_1, \delta u_2, \ldots, \delta u_N)$$

is the inner product

$$\delta W = F \cdot \delta u = \sum_{s=1}^{N} F_s \, \delta u_s.$$

As stated heretofore, Lagrangean mechanics utilizes the classification of forces into those doing, and not doing, virtual work. We shall say:

If one can find forces acting on the $n = N/3$ *particles of the system which have the resultant*

$$F' = \sum_{s=1}^{N} F_s' \hat{e}_s$$

and which satisfy the relation

$$F' \cdot \delta u = 0$$

for all u_s, $\dot{u}_s$, and t, *where* δu *is an arbitrary virtual displacement, then the component forces of* F' *are* constraint forces. *All forces which are not constraint forces are called* given forces.

One immediate consequence of this definition is:

If a physical force $F^{r'}$ *does no virtual work it is a constraint force.*

It is interesting to note that the Lagrangean separation of forces into given and constraint forces does not obliterate the Newtonian separation into internal and external ones. To see this, consider a constraint force which is an internal force in the Newtonian sense. For each such force there exists an equal and opposite force because, by Newton's third law, internal forces vanish in pairs. Hence, each such pair is a force system in equilibrium and, by Bernoulli's principle, that force system can do no virtual work. However, each individual force of such a pair may do virtual work, as seen from

Example 9.3.1.† Consider two particles m_1 and m_2, connected by a rigid massless bar of length l. They are constrained to move in the x axis. Let the position of m_1 be x^1, and that of m_2 be x^2. Let the force exerted by m_2 on m_1 be $F^{1'}$, and that exerted by m_1 on m_2 be $F^{2'}$. By Newton's third law we have

$$F^{1'} = - F^{2'}. \tag{a}$$

The constraint is

$$(x^1 - x^2)^2 = l^2$$

or

$$(x^1 - x^2)(\delta x^1 - \delta x^2) = 0. \tag{b}$$

Since $x^1 \neq x^2$, (b) implies

$$\delta x^1 = \delta x^2. \tag{c}$$

The virtual work of this pair of forces is

$$\delta W = F^{1'} \cdot \delta x^1 + F^{2'} \cdot \delta x^2. \tag{d}$$

† Leitmann, G., personal communication.

Substituting (a) into (d), one has

$$\delta W = F^{1'}(\delta x^{1'} - \delta x^{2}),$$

and this vanishes because of (c). However, $F^{1'} \cdot \delta x^1 \neq 0$ and $F^{2'} \cdot \delta x^2 \neq 0$ because the $F^{i'}$ are neither zero nor normal to the δx^i.

Next, consider a constraint force which is *external.* In Lagrangean mechanics the only forces contemplated are those acting *on* particles; hence, this force is exerted by a surface on the particle. By the principles of statics, the particle exerts an equal and opposite force on the surface, so that these two forces are also a force system in equilibrium and, hence, do no virtual work. But the force exerted on the surface is not a member of the set of forces exerted on the particles of the system; therefore, it does not contribute to the sum of forces which occur in the fundamental equation (9.3.9). It follows that no external constraint force can do virtual work.

9.4. The Nature of the Forces of Constraint

We have defined forces of constraint as forces which do no work in a virtual displacement, but we have not discussed their existence, nor have we explained the choice of the words "constraint force" used to name them.

An equality constraint has been interpreted as a surface in which the C, E, S, or T trajectory of every motion lies.

For simplicity, let us begin by considering a single particle moving in the smooth surface

$$f(u_1, u_2, u_3) = 0. \tag{9.4.1}$$

The word "smooth" implies that all first partial derivatives of f are defined everywhere.

Suppose the particle tends to move out of the surface. Then, a force intrinsic to the surface must exist which prevents this motion. This force which ensures that the constraint is satisfied must be completely determined by the geometry of the surface; in particular, the direction of this force should coincide with a preferred direction of the surface.

Surfaces may, in general, have more than one preferred direction, but spherical surfaces and planes have one and only one such direction at every point—the *normal.* Since the constraint force direction must be defined for any surface (including plane and spherical ones) the only reasonable assumption for the constraint force direction is that it coincide with the direction of the gradient of the surface, i.e. with

$$\operatorname{grad} f = \frac{\partial f}{\partial u_1}\hat{\imath} + \frac{\partial f}{\partial u_2}\hat{\jmath} + \frac{\partial f}{\partial u_3}\hat{k}, \tag{9.4.2}$$

where $\hat{\imath}$, $\hat{\jmath}$, and $\hat{k}$ are the unit vectors along the u_1, u_2, and u_3 axes, respectively. This assumption is, in fact, implicit in d'Alembert's principle, as is evident from (9.3.8b) and (9.3.8d).

Then, if we write the constraint force as

$$F' = F_1'\hat{\imath} + F_2'\hat{\jmath} + F_3'\hat{k},$$

we must have

$$F' = -\lambda \operatorname{grad} f, \tag{9.4.3}$$

where the parameter λ is an undetermined scalar quantity; it is called a *Lagrangean multiplier* (because this formulation of F' is due to Lagrange).

Now, every virtual displacement

$$\delta u = \delta u_1\hat{\imath} + \delta u_2\hat{\jmath} + \delta u_3\hat{k}$$

lies *in* a tangent plane of the surface $f = 0$, while $\operatorname{grad} f$ is normal to it. It follows that

$$\delta W = -\lambda \operatorname{grad} f \cdot \delta u = 0, \tag{9.4.4}$$

or the virtual work done by the constraint force in a virtual displacement vanishes.

Consider, next, a particle which is constrained to move along a smooth space curve. As before (see Section 5.2) we consider the curve as the intersection of two smooth surfaces:

$$f_1(u_1, u_2, u_3) = 0,$$
$$f_2(u_1, u_2, u_3) = 0,$$

such that $\operatorname{grad} f_1$ and $\operatorname{grad} f_2$ exist but are not collinear anywhere.

One would expect that the constraint force intrinsic to the curve is in a plane normal to that curve, but it is clear that that condition does not fix its direction; there are forces of infinitely many directions lying in this plane. This expectation is borne out by analysis.

In Fig. 9.4.1 we show two intersecting surfaces $f_1 = 0$ and $f_2 = 0$, and the curve C which is formed by their intersection. The two vectors $-\lambda_1 \operatorname{grad} f_1$ and $-\lambda_2 \operatorname{grad} f_2$ are, in general, oblique vectors, the first normal

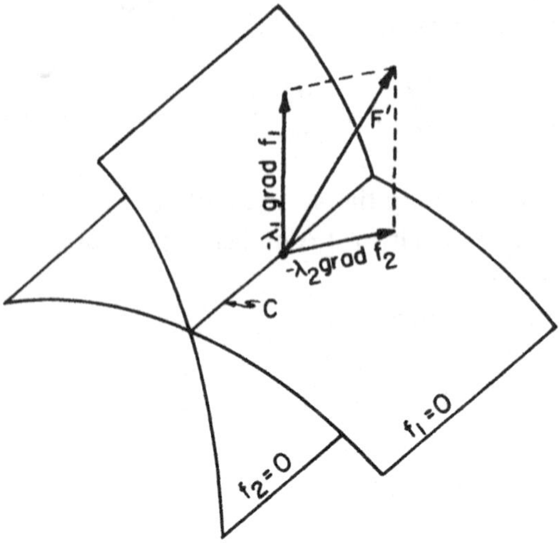

Fig. 9.4.1. Gradient vectors of intersecting surfaces.

to the surface $f_1 = 0$, and the second to $f_2 = 0$. The constraint force F' is the vector sum of these oblique vectors, or

$$F' = -\lambda_1 \operatorname{grad} f_1 - \lambda_2 \operatorname{grad} f_2.$$

This force lies in a plane defined by the two gradient vectors which is normal to the curve, but its direction is not fixed unless the ratio λ_1/λ_2 of the Lagrange multipliers λ_1 and λ_2 is known.

It will be readily seen how this concept of the constraint forces, produced by holonomic configuration constraints, is extended, both to $3n = N$ dimensions, and to $L < N$ equations of constraint. Moreover, the extension of these notions to differential equations of constraint (catastatic, or not) follows readily if one replaces "surface" by "element of a tangent plane to the surface."

A word may be in order regarding the treatment of forces that arise from the presence of equality constraints (i.e., surfaces or tangent surfaces), but that *do* virtual work. An example of such a force is Coulomb friction. This is a force which is proportional to the normal force to the surface, i.e., it is a function of a constraint force, but its line of action lies in the tangent plane to the surface; thus, it does virtual work.

When forces are present which are produced by surfaces but which do virtual work, they are usually due to properties of surfaces other than their geometries such as surface "lubrication" or "roughness," etc.

When the given forces include some which are independent of the presence of surfaces, and others which exist only because of the existence of these surfaces, it is also necessary to divide the constraint forces into two subclasses: Those that exist only because of the existence of the surfaces, and those which would continue to act even if the surfaces were made to disappear. This subject is discussed more fully at the end of this chapter (see Sections 9.9 and 9.10).

The foregoing discussion of constraint forces has left open the question whether there exist forces not due to surfaces which do no virtual work. Let us suppose that such forces exist. In that case, they are certainly normal to every virtual displacement. Thus, we may assume the existence of a surface normal to these forces in which the virtual displacements lie, and this supposition has no influence on the problem formulation. It is for this reason that *all* forces doing no virtual work are called constraint forces.

If there are L constraints, the virtual work done by all constraint forces is

$$\delta W = -\sum_{s=1}^{N}\sum_{r=1}^{L} \lambda_r A_{rs}\,\delta u_s \tag{9.4.5}$$

(where the

$$A_{rs} = \partial f_r/\partial u_s \tag{9.4.6}$$

if the rth constraint is holonomic, but not otherwise). Also, by definition, $\delta W = 0$. One sees from (9.4.5) that the total constraint force components in the u_s direction arising from all constraints is

$$F_s' = -\sum_{r=1}^{L} \lambda_r A_{rs}. \tag{9.4.7}$$

If we now rewrite (9.3.5) in accordance with (9.4.7) we have

$$m_s\ddot{u}_s - F_s = -\sum_{r=1}^{L} \lambda_r A_{rs} \qquad (s = 1, 2, \ldots, N). \tag{9.4.8}$$

Each of these equations contains, in general, $N + L$ unknowns: the u_s (N in number) and the λ_r (L in number). There are $N + L$ equations which these unknowns must satisfy; they are the N equations (9.4.8) and the L constraint equations

$$\sum_{s=1}^{N} A_{rs}\dot{u}_s + A_r = 0 \qquad (r = 1, 2, \ldots, L). \tag{9.4.9}$$

Let us now consider an *arbitrary* vector

$$\delta u = (\delta u_1, \delta u_2, \ldots, \delta u_N), \tag{9.4.10}$$

whose components need *not* satisfy the relations

$$\sum_{s=1}^{N} A_{rs}\,\delta u_s = 0 \qquad (r = 1, 2, \ldots, L). \tag{9.4.11}$$

Then, multiplying each of the equations (9.4.8) with its corresponding component of (9.4.10) and adding them, one finds

$$\sum_{s=1}^{N} \left(m_s \ddot{u}_s - F_s + \sum_{r=1}^{L} \lambda_r A_{rs} \right) \delta u_s = 0. \tag{9.4.12}$$

It is instructive to compare (9.4.12) with the fundamental equation (9.3.12). In (9.3.12) the term $\sum_{s=1}^{N} \lambda_r A_{rs}$ is absent, and the δu_s *must* be virtual displacements, e.g., they must satisfy (9.2.4), while in (9.4.12) the δu_s are completely arbitrary. This leads to the so-called:

Lagrange Multiplier Rule. *The equation*

$$\sum_{s=1}^{N} (m_s \ddot{u}_s - F_s)\,\delta u_s = 0,$$

in which the δu_s must satisfy

$$\sum_{s=1}^{N} A_{rs}\,\delta u_s = 0 \qquad (r = 1, 2, \ldots, L),$$

is completely equivalent to the equation

$$\sum_{s=1}^{N} \left(m_s \ddot{u}_s - F_s + \sum_{r=1}^{L} \lambda_r A_{rs} \right) \delta u_s = 0,$$

in which the δu_s are arbitrary.

The Lagrange multiplier rule will be discussed in greater detail in Section 13.2.

Because of the arbitrariness of the δu_s in (9.4.12), we have

$$m_s \ddot{u}_s - F_s + \sum_{r=1}^{L} \lambda_r A_{rs} = 0 \qquad (s = 1, 2, \ldots, N), \tag{9.4.13}$$

and this is, in fact, (9.4.8), from which (9.4.12) was obtained. Thus, it is not a new conclusion.

We shall now give some examples of the preceding theory. Some of these involve rigid bodies rather than particles. This will present no difficulty to the reader who has taken a first course in dynamics.

The first example is a problem of static equilibrium of a constrained system of two particles.

Example 9.4.1. One end of a massless rigid rod of length $2a$ carries a heavy mass m_0; a second heavy mass m_1 is fixed at the center of the rod. The mass m_0 is constrained to move on a smooth quarter-circle of radius r in the vertical plane, and the other end of the rod moves on a smooth horizontal line a distance h below the center of the circle and lying in the same vertical plane. Find the equilibrium positions or, stated differently, when is the motion that of rest?

Let x, y be the vertical plane with origin of the coordinate system at the center of the circle, as shown in Fig. 9.4.2(a). Let the coordinates of m_0 be x_0, y_0, and those of m_1 be x_1, y_1. The constraint which ensures that m_0 moves on the circle is

$$x_0^2 + y_0^2 - r^2 = 0. \tag{a}$$

The constraint which ensures that the distance between the two masses is a is

$$(x_1 - x_0)^2 + (y_1 - y_0)^2 - a^2 = 0. \tag{b}$$

The constraint that the lower end of the rod moves on the horizontal line a distance h below the y axis is

$$x_0 + 2(x_1 - x_0) - h = 0. \tag{c}$$

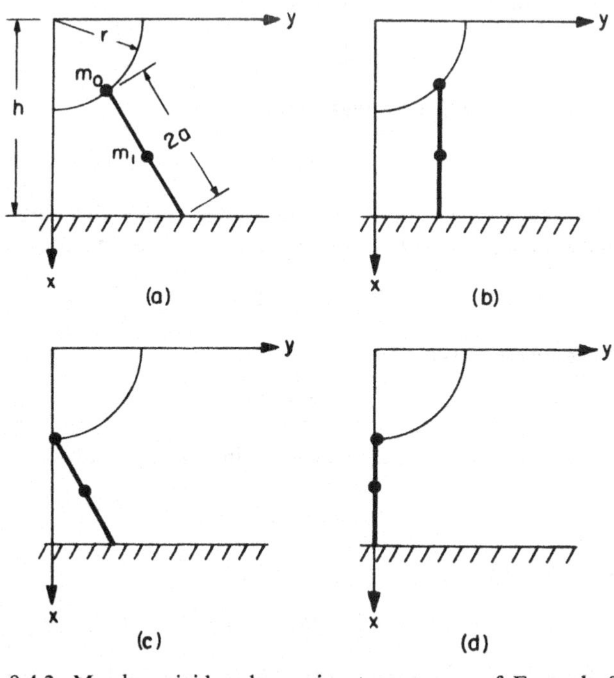

Fig. 9.4.2. Massless rigid rod carrying two masses of Example 9.4.1.

From these, the constraints on the virtual displacements are found as

$$\begin{gathered} x_0\,\delta x_0 + y_0\,\delta y_0 = 0, \\ (x_0 - x_1)\,\delta x_0 + (x_1 - x_0)\,\delta x_0 + (y_0 - y_1)\,\delta y_0 + (y_1 - y_0)\,\delta y_1 = 0, \\ -\,\delta x_0 + 2\,\delta x_1 = 0. \end{gathered} \tag{d}$$

The given forces on m_0 and m_1, respectively, are

$$\begin{aligned} X_0 &= m_0 g, \qquad & Y_0 &= 0, \\ X_1 &= m_1 g, \qquad & Y_1 &= 0. \end{aligned} \tag{e}$$

In writing down (9.4.12) for this example, all acceleration terms vanish because we seek the state of rest. Using Lagrange multipliers λ, μ, and ν, we find

$$\begin{aligned} &-m_0 g\,\delta x_0 - m_1 g\,\delta x_1 + \lambda(x_0\,\delta x_0 + y_0\,\delta y_0) + \mu[(x_0 - x_1)\,\delta x_0 + (x_1 - x_0)\,\delta x_1 \\ &\quad + (y_0 - y_1)\,\delta y_0 + (y_1 - y_0)\,\delta y_1] + \nu(-\,\delta x_0 + 2\,\delta x_1) = 0. \end{aligned} \tag{f}$$

Since the δx_0, δx_1, δy_0, δy_1 in (f) are completely arbitrary their coefficients must vanish, or

$$\begin{aligned} -m_0 g + \lambda x_0 + \mu(x_0 - x_1) - \nu &= 0, \\ \lambda y_0 + \mu(y_0 - y_1) &= 0, \\ -m_1 g + \mu(x_1 - x_0) + 2\nu &= 0, \\ \mu(y_1 - y_0) &= 0. \end{aligned} \tag{g}$$

In view of the last of these, the second may be rewritten as

$$\lambda y_0 = 0. \tag{h}$$

From the last equation of (g) we find, when $\mu \neq 0$,

$$y_1 - y_0 = 0. \tag{i}$$

This means that the massless rod is vertical, which is, therefore, one of the equilibrium configurations; it is shown in Fig. 9.4.2(b). We find from (b)

$$x_1 - x_0 = \pm\, a, \tag{j}$$

and substituting this into (c),

$$x_0 = h - 2a, \tag{k}$$

which shows that the plus sign must be chosen in (j). Then, substituting (j) in (k), we obtain

$$x_1 = h - a, \tag{l}$$

and, substituting (k) in (a),

$$y_0 = [r^2 - (h - 2a)^2]^{1/2}, \tag{m}$$

which is also the value of y_1 because of (i). Hence, all coordinates are determined.

The Lagrange multipliers are now also determined. Substituting the equilibrium coordinates in (g), we find

$$-m_0 g + \lambda(h - 2a) - \mu a - \nu = 0,$$
$$\lambda[r^2 - (h - 2a)^2]^{1/2} = 0,$$
$$-m_1 g + \mu a + 2\nu = 0.$$

Since $[r^2 - (h - 2a)^2]^{1/2} \neq 0$ in general, $\lambda = 0$, and the above equations become

$$-m_0 g - \mu a - \nu = 0,$$
$$-m_1 g + \mu a + 2\nu = 0.$$

When these are added, we find

$$\nu = (m_0 + m_1)g \tag{n}$$

and, subsequently,

$$\mu = -(2m_0 + m_1)(g/a). \tag{o}$$

Thus, the multipliers are also determined. The exceptional case when $r^2 - (h - 2a^2) = 0$ will be discussed later.

A second equilibrium configuration is found from (h). When $\lambda \neq 0$, we have

$$y_0 = 0. \tag{p}$$

This means that m_0 is at the lowest point on the circle, or from (a)

$$x_0 = \pm r, \tag{q}$$

and from (c)

$$x_1 = \tfrac{1}{2}(h + r); \tag{r}$$

this implies that the plus sign must be chosen in (q). Finally, we find from (b)

$$y_1 = \{a^2 - [\tfrac{1}{2}(h - r)]^2\}^{1/2}. \tag{s}$$

Thus, the coordinates of the equilibrium configuration are again completely determined. This case is shown in Fig. 9.4.2(c).

The Lagrange multipliers are readily found by substituting the equilibrium configuration coordinates into (g). They are

$$\nu = \tfrac{1}{2}m_1 g, \qquad \lambda = (\tfrac{1}{2}m_0 + m_1)(g/r). \tag{t}$$

The exceptional case for the first equilibrium position occurs when

$$r^2 - (h - 2a)^2 = 0,$$

and for the second equilibrium position when

$$a^2 - [\tfrac{1}{2}(h - r)]^2 = 0.$$

One sees from (m) that in the first exceptional case $y_0 = 0$, which means that the rod stands vertically in the x axis, and from (s) that in the second exceptional case $y_1 = 0$, which also means that the rod stands vertically in the x axis; this case is shown in Fig. 9.4.2(d). For that case, equations (g) become

$$-m_0 g + \lambda(h - 2a) - \mu a - \nu = 0,$$
$$-m_1 g + \mu a + 2\nu = 0,$$

and these are two equations in the three multipliers λ, μ, and ν. Thus, these multipliers cannot be determined uniquely. The reason is that this configuration is statically indeterminate, i.e., it is not possible to determine how much of the gravity force is resisted at the bottom of the circle, and how much by the horizontal line at the bottom of the rod.

Example 9.4.2. (Hamel, in part, p. 83). Under what forces is there no motion of the hatchet in a hatchet planimeter?

The hatchet of a hatchet planimeter behaves like a curved knife edge, as shown in Fig. 9.4.3(a). It can move in the direction of the knife edge, and this direction can change because the knife edge can rotate about the normal to the xy plane through its point of contact P. The hatchet or knife edge is a continuous solid, not a rigid system of a finite number of particles. Therefore, we shall model the hatchet by a "dumbbell" consisting of two masses m_0 and m_1, interconnected by a rigid massless rod of length a, as shown in Fig. 9.4.3(b). This dumbbell is subjected to the same constraint as the hatchet: It can slide only in the direction of the massless connection.

Let the coordinates of m_0 be (x_0, y_0), and let those of m_1 be (x_1, y_1). Further, let the x and y components of the force on m_0 be X_0, Y_0; those of the force on m_1 are X_1, Y_1. If the massless connection makes the angle θ with the x axis we have

$$\begin{aligned} x_1 &= x_0 + a\cos\theta, \\ y_1 &= y_0 + a\sin\theta, \end{aligned} \tag{a}$$

and these satisfy the constraint that the distance between the masses is a, or

$$(x_1 - x_0)^2 + (y_1 - y_0)^2 = a^2.$$

Hence, we use the coordinates x_0, y_0, and θ to describe the configuration. The condition that the displacement must be in the direction of the massless connection

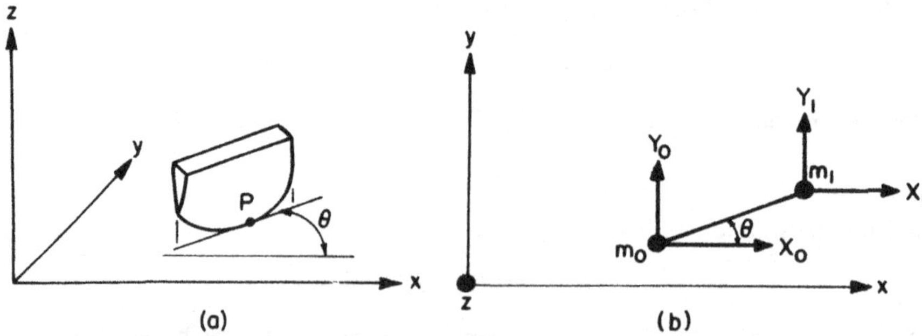

Fig. 9.4.3. Knife edge and its mathematical model of Example 9.4.2.

may now be written as

$$dx_0 \sin\theta - dy_0 \cos\theta = 0, \tag{b}$$

and this is a nonholonomic, catastatic constraint. Therefore, the virtual displacement components satisfy

$$\sin\theta\, \delta x_0 - \cos\theta\, \delta y_0 = 0, \tag{c}$$

and from (a),

$$\begin{aligned} \delta x_1 &= \delta x_0 - a \sin\theta\, \delta\theta, \\ \delta y_1 &= \delta y_0 + a \cos\theta\, \delta\theta. \end{aligned} \tag{d}$$

Inasmuch as we seek equilibrium, the acceleration terms in (9.4.12) vanish, and that equation is here

$$- X_0\, \delta x_0 - Y_0\, \delta y_0 - X_1\, \delta x_1 - Y_1\, \delta y_1 + \lambda(\sin\theta\, \delta x_0 - \cos\theta\, \delta y_0) = 0 \tag{e}$$

or, when (d) are substituted in (e) and the coefficients of δx_0, δy_0 and $\delta\theta$ are combined,

$$\begin{aligned} &(- X_0 - X_1 + \lambda \sin\theta)\, \delta x_0 + (- Y_0 - Y_1 - \lambda \cos\theta)\, \delta y_0 \\ &\quad + (X_1 a \sin\theta - Y_1 a \cos\theta)\, \delta\theta = 0. \end{aligned} \tag{f}$$

But δx_0, δy_0, and $\delta\theta$ in (f) are completely arbitrary. Therefore, the conditions for equilibrium are

$$\begin{aligned} - X_0 - X_1 + \lambda \sin\theta &= 0, \\ - Y_0 - Y_1 - \lambda \cos\theta &= 0, \\ X_1 a \sin\theta - Y_1 a \cos\theta &= 0. \end{aligned} \tag{g}$$

Eliminating λ between the first two gives

$$- \frac{X_0 + X_1}{Y_0 + Y_1} = \tan\theta, \tag{h}$$

and this equation states that the resultant of the forces (X_0, Y_0) and (X_1, Y_1) must be normal to the rigid connection. The last equation of (g) states that the moment of the force (X_1, Y_1) about the point (x_0, y_0) must vanish. These are the conditions on the forces (X_0, Y_0) and (X_1, Y_1) for which the system will not move.

We have already shown that the Stieltjes integral can be utilized to make the transition from rigid systems of a finite number of particles to continuous, rigid solids. We shall now show how (9.4.12) can be used to discuss the equilibrium of the knife edge regarded as a continuous solid. The general theory of this procedure will be discussed in a section on generalized coordinates.

We describe the configuration of the knife edge by the x, y coordinates of the contact point P, and the inclination θ of the knife edge to the positive x axis. Then, the constraint on the virtual displacements is

$$\cos\theta\, \delta y - \sin\theta\, \delta x = 0.$$

If the force components and the moment about P are denoted by X, Y, and M, we find in place of (e)

$$- X\, \delta x - Y\, \delta y - M\, \delta\theta + \lambda(\sin\theta\, \delta x - \cos\theta\, \delta y) = 0.$$

Proceeding as before, we find

$$-X + \lambda \,\text{xin}\, \theta = 0, \qquad -Y - \lambda \cos \theta = 0, \qquad M = 0.$$

On eliminating λ between the first two equations, we obtain

$$-X/Y = \tan \theta.$$

Hence, the conclusions are as before.

Example 9.4.3. The point of suspension of a simple plane pendulum is moved smoothly in the horizontal direction in a prescribed manner. Find the equations of motion.

Let the given horizontal motion of the point of suspension of the pendulum be the twice differentiable function $f(t)$, let its length be l, the mass of the bob m, and let the position of the bob be given by (x, y) with y positive vertically down, as shown in Fig. 9.4.4. The equation of constraint is

$$[x - f(t)]^2 + y^2 - l^2 = 0.$$

Evidently, this is a holonomic, rheonomic constraint. The possible infinitesimal displacements satisfy

$$[x - f(t)]\, dx + y\, dy - [x - f(t)] \dot{f}\, dt = 0.$$

This constraint is seen to be acatastatic. Hence, the virtual displacements satisfy

$$[x - f(t)]\, \delta x + y\, \delta y = 0,$$

The given forces are

$$X = 0, \qquad Y = mg.$$

We apply (9.4.12) and find

$$(m\ddot{x} - X)\, \delta x + (m\ddot{y} - mg)\, \delta y - m\lambda\{[x - f(t)]\, \delta x + y\, \delta y\} = 0,$$

where $m\lambda$ is the Lagrange multiplier.

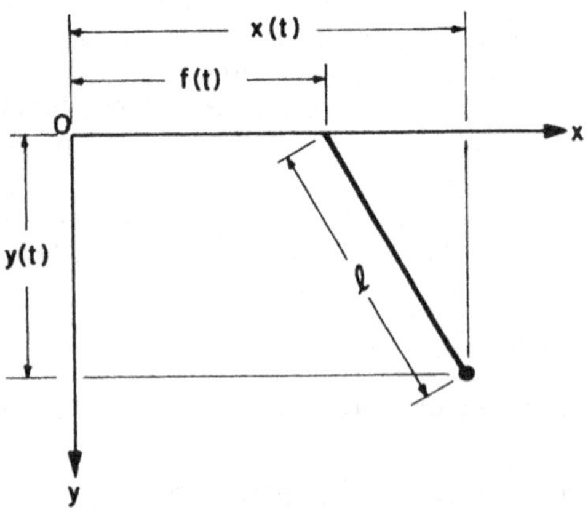

Fig. 9.4.4. Pendulum with moving suspension point of Example 9.4.3.

This last equation results in the two equations

$$\ddot{x} - \lambda[x - f(t)] = 0, \tag{a}$$

$$\ddot{y} - g - \lambda y = 0. \tag{b}$$

These equations, together with

$$[x - f(t)]^2 + y^2 - l^2 = 0, \tag{c}$$

determine $x(t)$, $y(t)$, and $\lambda(t)$. From (b) one finds

$$\lambda(t) = (\ddot{y} - g)/y. \tag{d}$$

It is typical of rheonomic constraints that the multipliers associated with them are not constants, but, are time dependent.

If one notices that the equations

$$\begin{aligned} x &= f(t) + l \sin \theta, \\ y &= l \cos \theta \end{aligned} \tag{e}$$

satisfy the constraint (c) for every value of θ, the problem can be greatly simplified. From (e)

$$\begin{aligned} \ddot{x} &= l\ddot{\theta} \cos \theta - l\dot{\theta}^2 \sin \theta + \ddot{f}, \\ \ddot{y} &= - l\ddot{\theta} \sin \theta - l\dot{\theta}^2 \cos \theta. \end{aligned} \tag{f}$$

These equations are readily used to combine (a) and (b) into the single equation

$$\ddot{\theta} = \frac{g}{l} \sin \theta = - \frac{\ddot{f}}{l} \cos \theta \tag{g}$$

and the multiplier becomes

$$\lambda = - \ddot{\theta} \tan \theta - \dot{\theta}^2 - g/(l \cos \theta). \tag{h}$$

The equation (g) could have been found directly; however, we wished to apply the preceding theory.

9.5. The Virtual Velocity

Formally, the virtual velocity $\delta\dot{u}$ may be introduced by writing

$$\frac{d}{dt}(\delta u) = \delta\left(\frac{du}{dt}\right) = \delta\dot{u}. \tag{9.5.1}$$

One may argue in defense of (9.5.1) that d/dt and δ symbolize two different operations and the sequence of these operations may be exchanged, or

$$\frac{d}{dt}\,\delta = \delta\,\frac{d}{dt}. \tag{9.5.2}$$

If certain conditions (which are derived below) are fulfilled, that argument is entirely correct; however, it poses some questions. If a virtual displacement is believed to "take place in zero time" or, as we prefer to put it, if time is not involved in a virtual displacement, it is entirely unclear what is meant by the quantity $d(\delta u)/dt$; there is no evident reason to suppose that this derivative exists or, if it exists, why it does not vanish identically. Moreover, one may of course write, in consequence of (9.5.1),

$$d\delta u = \delta\, du, \tag{9.5.3}$$

and that equation indicates that we are dealing with a differential of second order. Thus, if du and $\delta\, du$ (or if δu and $d\delta u$) occur in the same equation, one might suppose that the term of second order is negligible compared to that of first order. What then is the meaning of equations like (9.5.3)?

We establish now the conditions necessary and sufficient for (9.5.3) to be true, and interpret its meaning. For simplicity, consider a holonomic, catastatic constraint in the Pfaffian form, and let the constraint surface (or rather, an element of its tangent plane) contain two neighboring curves C_1 and C_2; one is a sequence of possible configurations, and the other is a different sequence of possible configurations neighboring on the first, as shown in Fig. 9.5.1. The points P and Q are simultaneous possible configurations at the time t, and the points $\bar{P}$ and $\bar{Q}$ are simultaneous possible configurations at a different time $\bar{t} = t + dt$. Thus, the curves connecting P with Q and $\bar{P}$ with $\bar{Q}$ are the loci of possible simultaneous states; we shall call them "isochrones." From the definitions of possible and virtual displacements, it is clear that the arc from P to $\bar{P}$ is a *possible* configuration

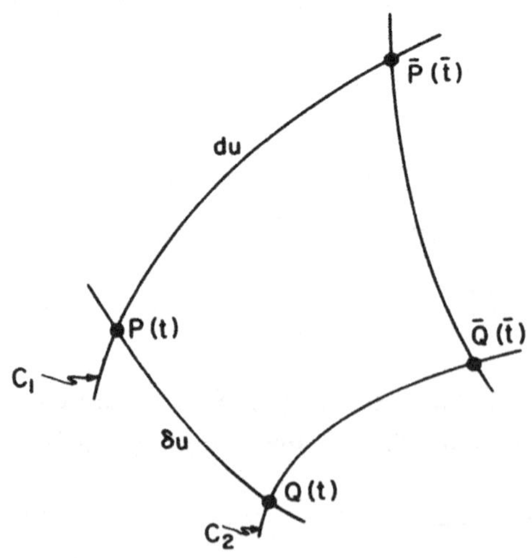

Fig. 9.5.1. Possible displacement du and virtual displacement δu.

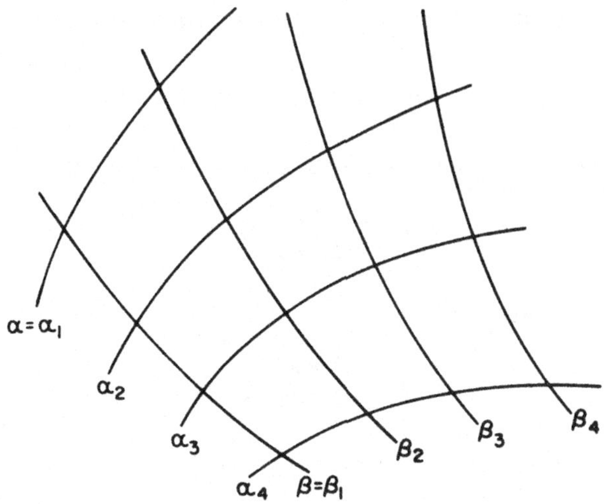

Fig. 9.5.2. Possible trajectories and isochrones.

change du, and that from P to Q is a *virtual* configuration change δu, i.e., the du lie along the possible configurations, and the δu along the isochrones. Now, in general, the virtual displacement at $\bar{t}$ (i.e., the arc from $\bar{P}$ to $\bar{Q}$) is different from that at t. Thus, this diagram makes it evident that δu is a function of time; it is therefore reasonable to consider the rate of change with time of that quantity.

Let us now generalize Fig. 9.5.1. Instead of only two curves C_1 and C_2, we construct a one-parameter family of such curves, the parameter being α, and instead of only two isochrones C_1 and C_2 we construct a one-parameter family of these curves also, the parameter being β. Thus, we consider the net of curves shown in Fig. 9.5.2.

Consider a fixed point (not necessarily in the plane of this net of curves) and let u be the position vector of any point of this net with respect to the fixed point. Then

$$u = u(\alpha, \beta), \tag{9.5.4}$$

and the arc between two neighboring points on a curve $\alpha =$ const is of length

$$du = \frac{\partial u}{\partial \beta}\, d\beta \tag{9.5.5}$$

while that along any curve $\beta =$ const is of length

$$\delta u = \frac{\partial u}{\partial \alpha}\, \delta\alpha. \tag{9.5.6}$$

These last two equations imply the operator equations

$$d = d\beta \frac{\partial}{\partial \beta}, \qquad \delta = \delta\alpha \frac{\partial}{\partial \alpha}. \tag{9.5.7}$$

It follows from (9.5.5) and (9.5.7) that

$$\delta\, du = \frac{\partial^2 u}{\partial \beta\, \partial \alpha}\, \delta\alpha\, d\beta, \tag{9.5.8}$$

and from (9.5.6) and (9.5.7) that

$$d\, \delta u = \frac{\partial^2 u}{\partial \alpha\, \partial \beta}\, d\beta\, \delta\alpha. \tag{9.5.9}$$

Now, the conditions necessary and sufficient for

$$\frac{\partial^2 u}{\partial \alpha\, \partial \beta} = \frac{\partial^2 u}{\partial \beta\, \partial \alpha}$$

to be an equality are that one of the second derivatives of u with respect to α and β exist and be continuous, and that both first derivatives of u with respect to α and β exist and be continuous. When these conditions are fulfilled (9.5.3) holds, and (9.5.1) follows directly.

It is interesting to give a geometrical interpretation of catastatic and acatastatic systems. In Fig. 9.5.2, the curves $\alpha = \text{const}$ are possible C trajectories, and elemental lengths along them are the du. The curves $\beta = \text{const}$ are the loci of possible, simultaneous configurations, and elemental lengths along them are the δu. Now, in catastatic systems the class of the du and δu is the same. This means that, in catastatic systems, the role of the curves $\alpha = \text{const}$ and $\beta = \text{const}$ may be interchanged, e.g., the curves $\beta = \text{const}$ also belong to the class of possible C trajectories, and the curves $\alpha = \text{const}$ also belong to the class of isochrones. In acatastatic systems, each family of one-parameter curves belongs to only one of these classes.

With the introduction of virtual displacements and virtual velocities it is now possible to define virtual changes in state. The "state of a system" has been defined in (3.3.2) as the point

$$(u, \dot{u}) = (u_1, u_2, \ldots, u_N; \dot{u}_1, \dot{u}_2, \ldots, \dot{u}_N)$$

in state space. We define a change of the system to a new state a "virtual change of state" if the new state is

$$(u + \delta u, \dot{u} + \delta \dot{u}) = (u_1 + \delta u_1, \ldots, u_N + \delta u_N; \dot{u}_1 + \delta \dot{u}_1, \ldots, \dot{u}_N + \delta \dot{u}_N),$$

where $\delta u = (\delta u_1, \delta u_2, \ldots, \delta u_N)$ is a virtual displacement, and $\delta\dot{u} = (\delta\dot{u}_1, \delta\dot{u}_2, \ldots, \delta\dot{u}_N)$ is a virtual velocity.

Moreover, if the displacements and velocities of a system satisfy the kinematical constraint equations (9.4.9), we say that the state of the system is a "possible state."

We shall now demonstrate the unexpected result that:

Virtual changes from a possible state do not, in general, lead to another possible state.

One might have imagined that this statement would be true in acatastatic systems because, in them, virtual and possible displacements do not satisfy the same equations. However, when virtual displacements do lead from one possible state to another, they do so whether the constraints are catastatic or not. In fact, when the system is nonholonomic, virtual displacements never lead from one possible state to another, and when the system is holonomic they always do. These conclusions are independent of whether the Pfaffian form of the constraints is catastatic or not.

To prove these statements we note that a possible state is one for which the displacements and velocities satisfy

$$\sum_{s=1}^{N} A_{rs}(u_1, u_2, \ldots, u_N, t)\dot{u}_s + A_r(u_1, u_2, \ldots, u_N, t) = 0 \qquad (r = 1, 2, \ldots, L). \tag{9.5.10}$$

Then, a virtual displacement leads to another possible state if and only if the new state satisfies

$$\sum_{s=1}^{N} A_{rs}(u_1 + \delta u_1, u_2 + \delta u_2, \ldots, u_N + \delta u_N, t)(\dot{u}_s + \delta\dot{u}_s) + A_r(u_1 + \delta u_1, u_2 + \delta u_2, \ldots, u_N + \delta u_N, t) = 0 \tag{9.5.11}$$

for every $r = 1, 2, \ldots, L$, where $\delta\dot{u}_s = d(\delta u_s)/dt$ for every $s = 1, 2, \ldots, N$ in accordance with (9.5.1).

We shall suppose that the A_{rs} and the A_r possess continuous first partial derivatives with respect to the u_s and to t. Then, we may write (9.5.11) as

$$\sum_{s=1}^{N} (A_{rs} + \delta A_{rs})(\dot{u}_s + \delta\dot{u}_s) + (A_r + \delta A_r) = 0 \qquad (r = 1, 2, \ldots, L), \tag{9.5.12}$$

where

$$\delta A_{rs} = \sum_{k=1}^{N} \frac{\partial A_{rs}}{\partial u_k}\,\delta u_k, \qquad \delta A_r = \sum_{k=1}^{N} \frac{\partial A_r}{\partial u_k}\,\delta u_k$$

and the A_{rs} and A_r are those of (9.5.10). Expanding (9.5.12) to first order in infinitesimal quantities and utilizing (9.5.10), we find that the new state will be a possible one if and only if

$$\sum_{s=1}^{N} (\delta A_{rs}\dot{u}_s + A_{rs}\,\delta\dot{u}_s) + \delta A_r = 0 \qquad (r = 1, 2, \ldots, L)$$

or, in expanded form, if

$$\sum_{s=1}^{N}\sum_{\alpha=1}^{N}\left(\frac{\partial A_{rs}}{\partial u_\alpha}\,\delta u_\alpha\dot{u}_s + A_{rs}\,\delta\dot{u}_s + \frac{\partial A_r}{\partial u_s}\,\delta u_s\right) = 0 \tag{9.5.13}$$

for every $r = 1, 2, \ldots, L$.

Now, the virtual displacements satisfy for all time

$$\sum_{s=1}^{N} A_{rs}\,\delta u_s = 0 \qquad (r = 1, 2, \ldots, L); \tag{9.5.14}$$

therefore one also has

$$\frac{d}{dt}\left[\sum_{s=1}^{N} A_{rs}\,\delta u_s\right] = 0 \qquad (r = 1, 2, \ldots, L) \tag{9.5.15}$$

or, in expanded form,

$$\sum_{s=1}^{N}\sum_{\alpha=1}^{N}\left[\left(\frac{\partial A_{rs}}{\partial u_\alpha}\,\dot{u}_\alpha + \frac{\partial A_{rs}}{\partial t}\right)\delta u_s + A_{rs}\,\delta\dot{u}_s\right] = 0 \tag{9.5.16}$$

for every $r = 1, 2, \ldots, L$.

If the indices s and α in the double sum in (9.5.16) are exchanged and the resulting equation is subtracted from (9.5.13), one finds

$$\sum_{s=1}^{N}\sum_{\alpha=1}^{N}\left(\frac{\partial A_{rs}}{\partial u_\alpha} - \frac{\partial A_{r\alpha}}{\partial u_s}\right)\dot{u}_s\,\delta u_\alpha - \sum_{s=1}^{N}\left(\frac{\partial A_{rs}}{\partial t} - \frac{\partial A_r}{\partial u_s}\right)\delta u_s = 0 \tag{9.5.17}$$

for all r. It is now convenient to rewrite (9.5.17) as

$$\sum_{s=1}^{N}\left\{\sum_{\alpha=1}^{N}\left(\frac{\partial A_{r\alpha}}{\partial u_s} - \frac{\partial A_{rs}}{\partial u_\alpha}\right)\dot{u}_\alpha + \left(\frac{\partial A_r}{\partial u_s} - \frac{\partial A_{rs}}{\partial t}\right)\right\}\delta u_s = 0$$

or, in Pfaffian form,

$$\sum_{s=1}^{N}\left\{\sum_{\alpha=1}^{N}\left(\frac{\partial A_{r\alpha}}{\partial u_s} - \frac{\partial A_{rs}}{\partial u_\alpha}\right)du_\alpha + \left(\frac{\partial A_r}{\partial u_s} - \frac{\partial A_{rs}}{\partial t}\right)dt\right\}\delta u_s = 0.$$

This equation takes on a more symmetrical form if one defines

$$t = u_{N+1}, \qquad A_r = A_{r,N+1};$$

it becomes, then,

$$\sum_{s=1}^{N} \sum_{\alpha=1}^{N+1} \left(\frac{\partial A_{r\alpha}}{\partial u_s} - \frac{\partial A_{rs}}{\partial u_\alpha} \right) du_\alpha \, \delta u_s = 0 \qquad (r = 1, 2, \ldots, L). \tag{9.5.18}$$

But this is precisely the Frobenius integrability condition given in (4.5.11) and (4.5.12) because the du_α and δu_s admitted here are those which satisfy, and hence are the solutions of, the algebraic equations

$$\sum_{s=1}^{N+1} A_{rs} \, du_s = 0, \qquad \sum_{s=1}^{N} A_{rs} \, \delta u_s = 0 \qquad (r = 1, 2, \ldots, L).$$

This proves the contention that virtual displacements do not lead from one possible state to another unless all constraints are holonomic.

9.6. The Variation

In the last section we saw that it is convenient to give the symbol δ an operational definition. This operational definition will now be made precise.

Consider a function of class C^1 of $m + n + 1$ independent arguments

$$f = f(x_1, x_2, \ldots, x_m; \dot{x}_1, \dot{x}_2, \ldots, \dot{x}_n; t). \tag{9.6.1}$$

Then, we define the operator d in the usual manner as

$$df = \sum_{r=1}^{m} \frac{\partial f}{\partial x_r} \, dx_r + \sum_{s=1}^{n} \frac{\partial f}{\partial \dot{x}_s} \, d\dot{x}_s + \frac{\partial f}{\partial t} \, dt. \tag{9.6.2}$$

Thus, the operator d is the familiar differential operator, and df is called the *differential* of f.

We define in an analogous manner

$$\delta f = \sum_{r=1}^{m} \frac{\partial f}{\partial x_r} \, \delta x_r + \sum_{s=1}^{n} \frac{\partial f}{\partial \dot{x}_s} \, \delta \dot{x}_s, \tag{9.6.3}$$

and we call δf the *variation* of f. This definition is in accord with our previous notions that the δ operation is restricted to *simultaneous* states. The verbalization of (9.6.3) is that, in the δ operation, the x_r $(r = 1, 2, \ldots, m)$ and the $\dot{x}_s$ $(s = 1, 2, \ldots, n)$ are "varied," but t is *not*.

Consider now a function

$$W = W(x_1, x_2, \ldots, x_n; \dot{x}_1, \dot{x}_2, \ldots, \dot{x}_n; t). \tag{9.6.4}$$

Then, the variation of W is

$$\delta W = \sum_{r=1}^{n} \frac{\partial W}{\partial x_r} \delta x_r + \sum_{r=1}^{n} \frac{\partial W}{\partial \dot{x}_r} \delta \dot{x}_r. \tag{9.6.5}$$

Let us suppose that the point of application of a force

$$F^r = F^r(x_1, x_2, \ldots, x_n; \dot{x}_1, \dot{x}_2, \ldots, \dot{x}_n; t) \tag{9.6.6}$$

moves through a displacement dx^r. Then, the work done by the force in this displacement is, *by definition*, the inner product

$$dW^r = F^r \cdot dx^r. \tag{9.6.7}$$

If the dx^r are increments of length along a trajectory (of the point of application of the force) which connects two distinct configurations C_1 and C_2 (e.g., the curve is a possible C trajectory from C_1 to C_2), the total work done is

$$W^r = \int_C F^r \cdot dx^r, \tag{9.6.8}$$

and this is a function of the form (9.6.4) because the force F^r depends in general on position, velocity, and time. Therefore, the *variation* of W contains not only terms linear in the δx^r, but terms linear in the $\delta \dot{x}^r$ as well, as shown in (9.6.5).

But, in analogy with (9.6.7), the *virtual work* is defined as

$$\delta W^r = F^r \cdot \delta x^r \tag{9.6.9}$$

and this quantity contains *no* terms in $\delta \dot{x}^r$, regardless of whether F^r depends on the $\dot{x}^r$ or not. We conclude:

The virtual work δW is not, in general, the variation of the work W.

Confusion may arise when it is not clearly realized that the virtual work cannot be found, in general, by constructing the variation of the work function. To avoid mistakes, it must be remembered that δW is merely a shorthand notation for the quantity $\sum_r F^r \cdot \delta x^r$; it is not, in general, the variation of the work.

9.7. Possible Velocities and Accelerations

In order to understand clearly the concepts to be introduced, it is useful to recall that *possible* displacements du satisfy

$$\sum_{s=1}^{N} A_{rs}\,du_s + A_r\,dt = 0 \qquad (r = 1, 2, \ldots, L) \tag{9.7.1}$$

and, under the same constraints, virtual displacements satisfy

$$\sum_{s=1}^{N} A_{rs}\,\delta u_s = 0 \qquad (r = 1, 2, \ldots, L). \tag{9.7.2}$$

The time rates of change of the virtual displacements, denoted by $\delta\dot{u} = (\delta\dot{u}_1, \delta\dot{u}_2, \ldots, \delta\dot{u}_N)$, are the *virtual* velocities.

Possible velocities $\dot{u} = (\dot{u}_1, \dot{u}_2, \ldots, \dot{u}_N)$, on the other hand, are defined as those satisfying the constraints

$$\sum_{s=1}^{N} A_{rs}\dot{u}_s + A_r = 0 \qquad (r = 1, 2, \ldots, L). \tag{9.7.3}$$

Thus, the $\dot{u}_s$ are the components of possible velocities at a configuration $u = (u_1, u_2, \ldots, u_N)$ and at a time t, these being the arguments of the A_{rs} and A_r in (9.7.3). Let us now consider another possible velocity $\dot{u} + \Delta\dot{u}$, where $\Delta\dot{u} = (\Delta\dot{u}_1, \Delta\dot{u}_2, \ldots, \Delta\dot{u}_N)$ *is not necessarily small.* Nevertheless, it satisfies by definition

$$\sum_{s=1}^{N} A_{rs}(\dot{u}_s + \Delta\dot{u}_s) + A_r = 0 \qquad (r = 1, 2, \ldots, L). \tag{9.7.4}$$

Now, if the possible velocities are both from the *same* configuration and time (or, in other words, from the same event), the difference between (9.7.3) and (9.7.4) is

$$\sum_{s=1}^{N} A_{rs}\,\Delta\dot{u}_s = 0 \qquad (r = 1, 2, \ldots, L). \tag{9.7.5}$$

On comparing (9.7.2) and (9.7.5), we have the result:

> *Possible velocity changes* (*which need not be small*) from the same event *satisfy the same constraints as virtual displacements.*

These possible velocity changes are not the virtual velocity changes defined earlier because the latter are always infinitesimals.

The idea which was just developed can be carried one step further. Differentiating (9.7.3) with respect to time (since it must hold for all t) one finds

$$\sum_{s=1}^{N}\left(A_{rs}\ddot{u}_s + \frac{dA_{rs}}{dt}\,\dot{u}_s\right) + \frac{dA_r}{dt} = 0 \qquad (r = 1, 2, \ldots, L), \tag{9.7.6}$$

where

$$\frac{d}{dt} = \sum_{\alpha} \dot{u}_\alpha \frac{\partial}{\partial u_\alpha} + \frac{\partial}{\partial t}.$$

We define as a *possible acceleration* $\ddot{u}$ the vector $\ddot{u} = (\ddot{u}_1, \ddot{u}_2, \ldots, \ddot{u}_N)$, whose components $\ddot{u}_s$ satisfy (9.7.6) when the $\dot{u}_s$ are possible velocity components.

Let us now consider some other possible acceleration $\ddot{u} + \Delta\ddot{u}$ with $\Delta\ddot{u} = (\Delta\ddot{u}_1, \Delta\ddot{u}_2, \ldots, \Delta\ddot{u}_N)$, not necessarily small. Then, by definition, it satisfies

$$\sum_{s=1}^{N}\left[A_{rs}(\ddot{u}_s + \Delta\ddot{u}_s) + \frac{dA_{rs}}{dt}\,\dot{u}_s\right] + \frac{dA_r}{dt} = 0 \quad (r = 1, 2, \ldots, L), \tag{9.7.7}$$

where the $\dot{u}_s$ are possible velocity components. If the two possible accelerations are from the *same* configuration, velocity, and time (or, in other words, from the same state–time) the difference between (9.7.7) and (9.7.6) is

$$\sum_{s=1}^{N} A_{rs}\,\Delta\ddot{u}_s = 0 \qquad (r = 1, 2, \ldots, L). \tag{9.7.8}$$

Expressed in words, this is:

> *Possible acceleration changes* (*which need not be small*) from the same state–time *satisfy the same constraints as virtual displacements.*

The most interesting application of possible large *velocity* changes is to impulsive motion because we saw that impulsive forces lead to velocity discontinuities, hence, to velocity changes that are large.

The most obvious application of possible large *acceleration* changes is to problems in which the acceleration is discontinuous, such as in the case of a marble rolling off a table. We remark here that one of the interesting applications of (9.7.8) is Gauss' principle of least constraint; however, in that principle, only small changes in possible accelerations are contemplated. (Gauss' principle is not discussed in this book.)

9.8. The Fundamental Equation

In (9.3.11) we wrote the fundamental equation as

$$\sum_{r=1}^{n} (m_r \ddot{x}^r - F^r) \cdot \delta x^r = 0,$$

and when the x_i^r were replaced by the u_s it took on the form

$$\sum_{s=1}^{N} (m_s \ddot{u}_s - F_s)\, \delta u_s = 0. \tag{9.8.1}$$

This equation states that, in a system of particles, the virtual work of the given forces balances the work done by the inertia forces. Pars calls this the *first form of the fundamental equation.*

In view of the equations (9.7.2) and (9.7.5) one may also write

$$\sum_{s=1}^{N} (m_s \ddot{u}_s - F_s)\, \Delta \dot{u}_s = 0, \tag{9.8.2}$$

where $\Delta \dot{u}_s$ is a possible (not necessarily small) velocity change. This is called by Pars (p. 40) the *second form of the fundamental equation*; actually, *it is a new principle.*

Finally, because of (9.7.8) we could also write

$$\sum_{s=1}^{N} (m_s \ddot{u}_s - F_s)\, \Delta \ddot{u}_s = 0, \tag{9.8.3}$$

where $\Delta \ddot{u}_s$ is a possible (not necessarily small) acceleration change. Pars calls (9.8.3) the *third form of the fundamental equation* (p. 41). We shall meet still other forms later on.

In the literature, the first form is often called "Lagrange's form of d'Alembert's principle." However, regardless of the name attached to (9.8.1), that equation has not only far-reaching consequences, but it may be regarded as occupying a central position in the theory of classical dynamics because all known principles of mechanics can be derived from it. For instance, Newton's second law is obviously an immediate consequence of the fundamental equation. However, it is not possible to derive the fundamental equation by invoking Newton's second law only.

9.9. The Nature of the Given Forces

The forces which remain in the fundamental equation (9.8.1) are the *given* forces (or rather their components)

$$F_s = F_s(u_1, u_2, \ldots, u_N; \dot{u}_1, \dot{u}_2, \ldots, \dot{u}_N; t). \tag{9.9.1}$$

Using the Newtonian viewpoint, one would separate these into internal and external forces. In Lagrangean mechanics, it is more useful to separate the given forces into *potential* and *nonpotential* forces. We begin with the definition:

If there exists a scalar function $U^p = U^p(u_1, u_2, \ldots, u_N)$ of class C^1 such that a given force

$$F^p = (F_1{}^p, F_2{}^p, \ldots, F_N{}^p)$$

satisfies the relation

$$F^p = \text{grad } U^p$$

we call U^p a potential *function and F^p a* potential *force.*

Every given force F which does not satisfy this definition is a *nonpotential* force.

Potential forces are sometimes called "conservative" because in certain circumstances, to be described later, when all given forces acting on the particles of a system are potential, the total energy of the system is "conserved," i.e., it remains constant in time. However, the circumstances required for energy conservation are not always met, even when all forces are potential, and they are almost never met when some forces are nonpotential; yet, the concept of the potential force is useful even in these instances. Therefore, we shall use the term "potential" rather than "conservative" to describe them.

We also define:

The negative of a potential function

$$V^p = -U^p$$

is called a potential energy.

The force components of a potential force are, thus,

$$F_s^p = -\frac{\partial}{\partial u_s}[V^p(u_1, u_2, \ldots, u_N)]. \tag{9.9.2}$$

One sees then that potential forces, as defined here, are functions of the configuration only, while forces in general are functions of displacements, velocities, and time (see Section 2.5).

Then, if F^r in (9.6.9) is a potential force F^p, one has

$$\sum_{s=1}^{N} F_s^p \,\delta u_s = -\sum_{s=1}^{N} \frac{\partial V^p}{\partial u_s}\,\delta u_s, \tag{9.9.3}$$

where δu_s is a component of the virtual displacement

$$\delta u = (\delta u_1, \delta u_2, \ldots, \delta u_N).$$

The left-hand side of (9.9.3) is, by definition, the virtual work δW^p done by the potential force F^p in a virtual displacement δu, and it is in general not zero because the F^p are given forces, not constraint forces. The right-hand side is, by definition, the negative of the variation of the potential energy and, physically, it is the negative of the change δV^p of the potential energy in a virtual displacement. Therefore, (9.9.3) may be written as

$$\delta W^p = -\delta V^p. \tag{9.9.4}$$

The most significant result to be deduced from (9.9.4) is obtained when that equation is integrated. Suppose C_1 and C_2 are two distinct points in configuration space satisfying the constraints, i.e., they both lie in the surface defined by the constraints. Then one can, of course, find an infinity of different sequences of configurations, starting with the configuration C_1 such that all lead to the configuration C_2, and such that the path element, in going from one configuration to a neighboring one of the same sequence, is a virtual displacement. If we wish to integrate (9.9.4) along such a sequence of virtual displacements, we find for the right-hand side

$$\mathop{\mathrm{S}}_{C_1}^{C_2} \delta V^p = V^p(C_2) - V^p(C_1). \tag{9.9.5}$$

The symbol S is to the operator δ what the integral symbol $\int$ is to the differential operator d. Thus, it is defined by the operator equation

$$\mathrm{S}\,\delta = I, \tag{9.9.6}$$

where I is the identity operator, i.e.,

$$Iu = u. \tag{9.9.7}$$

One sees, therefore, that:

The total virtual work done by potential forces in a sequence of virtual displacements leading from a configuration C_1 to another configuration C_2 depends on the end points C_1 and C_2 only, but not on the path.

In the special case when the path is closed, it follows from (9.9.5) that

$$\mathfrak{S}_{C_1}^{C_1} \delta W^p = \oint \delta W^p = 0. \tag{9.9.8}$$

9.10. Given Forces Which Are Functions of Constraint Forces

Up to this point we have only considered forces which are functions of position, velocity, and/or time (and possibly of certain constants), and all were either given, or constraint† forces. One consequence of these considerations has been that the constraint forces are absent from the fundamental equation.

However, given forces which are functions of the constraint forces are occasionally encountered in dynamical systems. The best known example of such a force is the Coulomb friction force. Let a particle slide on an imperfectly rough horizontal surface. Then, the Coulomb friction force is

$$F_f = -\mu N v / | v |,$$

where μ is the coefficient of sliding friction, N is the normal force exerted by the surface on the particle, and $v/| v |$ is the unit vector in the direction of the particle velocity relative to the surface.

Evidently, in this example, the gravitational force as well as the normal force N is a constraint force because neither does work in a virtual displacement. However, there are two important differences between them: One of these is that the normal force disappears when the constraint is removed, while the gravitational force does not; it acts whether the constraint is present or absent. The presence of the constraint merely transforms the gravitational force from the class of given to that of constraint forces.

† The constraint forces are functions of position and time only.

The other is that the gravitational force is known; it is one of the given elements when the problem is posed. The normal force is not known at the outset; if we wish to determine its magnitude, we must invoke some principle of mechanics. It becomes evident, therefore, that we shall have to consider two distinct subclasses of constraint forces: the class F', which acts whether the constraints are present or absent, and the class F'', which acts only in the presence of the constraints. We shall write for the resultant of the latter

$$F'' = -\sum_{s=1}^{N}\sum_{r=1}^{L} \lambda_r A_{rs}\hat{e}_s \tag{9.10.1}$$

when the constraints are given by

$$\sum_{s=1}^{N} A_{rs}\dot{u}_s + A_r = 0 \qquad (r = 1, 2, \ldots, L). \tag{9.10.2}$$

We shall also have to consider two distinct classes of given forces: the class F^{I}, which is independent of the constraints, and the class F^{II}, which is a function of the constraints. Among the latter we only admit forces which vanish in the absence of constraints; hence we have

$$\begin{aligned} F^{\mathrm{II}} &= F^{\mathrm{II}}(F''), \\ F^{\mathrm{II}}(0) &= 0. \end{aligned} \tag{9.10.3}$$

Then, the $\hat{e}_s$ component of the totality of the given forces is

$$F_s = F_s^{\mathrm{I}} + F_s^{\mathrm{II}}\Big(\sum_{\alpha=1}^{N}\sum_{r=1}^{L} \lambda_r A_{r\alpha}\Big), \tag{9.10.4}$$

and the $\hat{e}_s$ component of the totality of the constraint forces is

$$F_s' + F_s'' = F_s' + \sum_{r=1}^{L} \lambda_r A_{rs}. \tag{9.10.5}$$

It follows that Newton's second law becomes here

$$m_s\ddot{u}_s - F_s^{\mathrm{I}} - F_s^{\mathrm{II}}\Big(\sum_{r=1}^{L}\sum_{\alpha=1}^{N} \lambda_r A_{r\alpha}\Big) - F_s' + \sum_{r=1}^{L} \lambda_r A_{rs} = 0$$
$$(s = 1, 2, \ldots, N). \tag{9.10.6}$$

These N equations together with the L equations (9.10.2) are sufficient to determine the N variables u_s and the L parameters λ_r. If we define an *arbitrary* displacement by

$$\delta u = (\delta u_1, \delta u_2, \ldots, \delta u_N)$$

(not necessarily a virtual displacement), multiply each of the equations (9.10.6) by the corresponding component of δu, and then add them, we find, similar to (9.4.12),

$$\sum_{s=1}^{N}\left[m_s\ddot{u}_s - F_s^{\mathrm{I}} - F_s^{\mathrm{II}}\left(\sum_{r=1}^{L}\sum_{\alpha=1}^{N}\lambda_r A_{r\alpha}\right) - F_s' - \sum_{r=1}^{L}\lambda_r A_{rs}\right]\delta u_s = 0, \qquad (9.10.7)$$

but the left-hand sum is zero because each relation of (9.10.6) equals zero, not because the constraint forces do no virtual work.

As they must, either (9.10.6) or (9.10.7) satisfy the requirement that, if the constraints are removed, the unconstrained problem emerges. We may remove the constraint by setting all A_{rs} and A_r equal to zero. Then, (9.10.7) becomes

$$\sum_{s=1}^{N}(m_s\ddot{u}_s - F_s^{\mathrm{I}} - F_s')\,\delta u_s = 0. \qquad (9.10.8)$$

The force component F_s' is now no longer a constraint force because there are no constraints. But it does not cease to act; instead it now belongs to the class of given forces.

The system of equations defining the dynamical system comprises (9.10.2) and (9.10.6). Now, (9.10.6) is simply a statement of Newton's second law because it states that the acceleration is proportional to the sum of all forces. One may well wonder whether the dynamical representation given here does not, therefore, belong to the domain of Newtonian rather than Lagrangean mechanics. It is our opinion that the representation is Lagrangean, not Newtonian. Newtonian mechanics of systems of more than one particle necessarily divides the forces into those external and internal to the system because one of its essential axioms is the third law, which utilizes a special property of internal forces. In fact the third law may be regarded as defining internal forces. Thus, any representation of the dynamics of systems of particles which does not utilize Newton's third law cannot be "Newtonian," in our view. The dynamical representation given here has retained the Lagrangean classification of forces into given and constraint forces; hence, we regard it as Lagrangean. But, when dealing with forces which are functions of constraints we must also introduce the notion that there are forces which are always present but which may belong to one or the other of the Lagrangean classes depending on the circumstances.

While the above treatment of given forces that depend on constraint forces is Lagrangean in its approach, Lagrangean mechanics is not a convenient vehicle for dealing with them. For one, it is necessary to complicate the simple division of forces into two classes by creating two subclasses for each. For another, the elegance of the fundamental equation in the beautiful and simple form (9.3.12) is lost. This loss becomes particularly evident, if the problem is stated in the form

$$\sum_{s=1}^{N} \left[m_s \ddot{u}_s - F_s^{\mathrm{I}} - F_s^{\mathrm{II}}\left(\sum_{r=1}^{L} \sum_{\alpha=1}^{N} \lambda_r A_{r\alpha} \right) \right] \delta u_s = 0, \tag{9.10.9}$$

where the δu_s must satisfy

$$\sum_{s=1}^{N} A_{rs}\, \delta u_s = 0 \qquad (r = 1, 2, \ldots, L) \tag{9.10.10}$$

when the constraints are

$$\sum_{s=1}^{N} A_{rs} \dot{u}_s + A_r = 0. \tag{9.10.11}$$

Let us utilize (9.10.10) to solve for L of the δu_s in terms of the $N - L$ remaining ones and substitute the result in (9.10.9). This is a process called "embedding" of constraints, and it is treated in greater detail in Chapter 14. The $N - L$ remaining δu_s are now arbitrary and, hence, their coefficients in (9.10.9) are arbitrary, and each must be equal to zero. This yields $N - L$ equations, and together with the L equations (9.10.11) we have only N equations to solve for $N + L$ unknowns, which is not a properly posed problem. Therefore, when given forces occur which are functions of constraint forces, the latter must be adjoined by the multiplier rule; embedding of constraints is, then, not a possible technique for solving the problem.

Nevertheless, problems of the class discussed here can be dealt with by the methods described above, as we shall now demonstrate on a simple example.

Example 9.10.1. A heavy particle of unit mass slides under Coulomb friction on an imperfectly rough, horizontal surface. The magnitude of the friction force is μ times that of the constraint force, and its direction is opposite that of the particle velocity relative to the surface. Discuss the motion.

Let the xy-axes of a right-handed x, y, z-system of cartesian coordinates be fixed in the surface, and let the z-axis be positive in the up-direction. Let the unit vectors along the axes be $\hat{\imath}$, $\hat{\jmath}$ and $\hat{k}$, respectively.

The single constraint is

$$f = z - c = 0, \tag{a}$$

where c is a constant, and the constraint force is

$$F'' = \lambda \operatorname{grad} f = \lambda \hat{k}.$$

Hence, the magnitude of F'' is

$$|F''| = \lambda. \tag{b}$$

The particle velocity relative to the surface is

$$\dot{r} = \dot{x}\hat{i} + \dot{y}\hat{j} + \dot{z}\hat{k},$$

and the unit vector in the direction of the velocity is

$$\frac{\dot{r}}{|\dot{r}|} = \frac{\dot{x}i + \dot{y}\hat{j}}{(\dot{x}^2 + \dot{y}^2)^{1/2}}. \tag{c}$$

Therefore, the Coulomb friction force is

$$F^{\text{II}} = -\mu\lambda \frac{\dot{x}\hat{i} + \dot{y}\hat{j}}{(\dot{x}^2 + \dot{y}^2)^{1/2}}. \tag{d}$$

The force F' is here

$$F' = -g\hat{k}, \tag{e}$$

and this is a force which is a constraint force in the presence of the constraint, and a given force when the constraint is absent. If we introduce an *arbitrary* displacement

$$\delta r = \delta x\, \hat{i} + \delta y\, \hat{j} + \delta z\, \hat{k} \tag{f}$$

we find that

$$\left[\ddot{x}\hat{i} + \ddot{y}\hat{j} + \ddot{z}\hat{k} - g\hat{k} + \mu\lambda \frac{\dot{x}\hat{i} + \dot{y}\hat{j}}{(\dot{x}^2 + \dot{y}^2)^{1/2}} + \lambda\hat{k}\right] \cdot (\delta x\, \hat{i} + \delta y\, \hat{j} + \delta z\, \hat{k}) = 0$$

and in view of the arbitrariness of δr,

$$\begin{aligned} \ddot{x} + \frac{\mu\lambda\dot{x}}{(\dot{x}^2 + \dot{y}^2)^{1/2}} &= 0, \\ \ddot{y} + \frac{\mu\lambda\dot{y}}{(\dot{x}^2 + \dot{y}^2)^{1/2}} &= 0, \\ \ddot{z} - g + \lambda &= 0. \end{aligned} \tag{g}$$

These equations together with (a) determine the motion.

Introducing (a) in the third equation of (g) gives

$$\lambda = g \tag{h}$$

and the substitution of (h) in the first two equations of (g) results in

$$\begin{aligned} \frac{du}{dt} &= -\frac{\mu\lambda u}{(u^2 + v^2)^{1/2}}, \\ \frac{dv}{dt} &= -\frac{\mu\lambda v}{(u^2 + v^2)^{1/2}}, \end{aligned} \tag{i}$$

where $u = \dot{x}$, $v = \dot{y}$. From (i), one has

$$\frac{dv}{du} = \frac{v}{u},$$

which integrates to

$$v = Cu, \tag{j}$$

where C is a constant. Equation (j) states that the trajectory is a straight line of inclination α to the x axis, where $\tan \alpha = C$.

The substitution of (j) into (i) gives

$$\ddot{x} = -\frac{\mu g}{(1 + C^2)^{1/2}}, \qquad \ddot{y} = -\frac{\mu g C}{(1 + C^2)^{1/2}}. \tag{h}$$

Hence, the particle moves under constant deceleration until it stops.

It should be noted that, when the constraint is removed ($\lambda = 0$), equations (g) state, as they should, that the horizontal component of the velocity is zero, and the vertical acceleration is the acceleration of gravity.

9.11. Problems

9.1. A heavy, uniform, smooth ladder of length l stands on a horizontal floor and leans against a wall of height $h < l$ as shown. Let the coordinates of its lower extremity be (x_1, y_1) and those of its upper (x_2, y_2) with y vertical, positive in the up direction. What are the equations of constraint on finite, possible, and virtual displacements for the cases $y_2 > h$ and $h > y_2 > 0$?

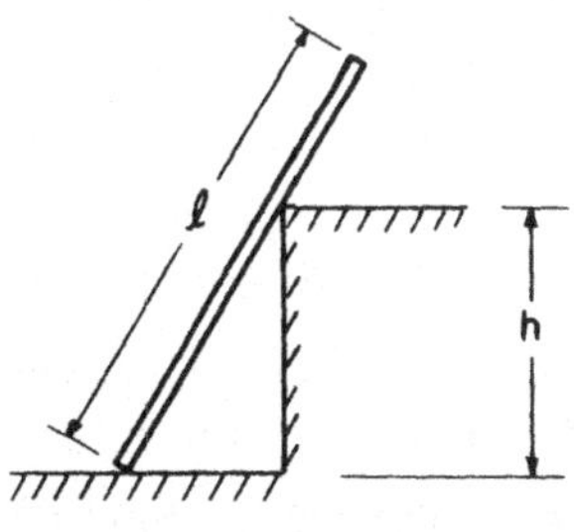

9.2. A particle of mass m is attached to one end of a massless inextensible string of length l. The particle and string are placed on a smooth horizontal table so that the string is straight. At the time t_0, the free end of the string is set in motion with uniform velocity v_0 in the plane of the table and normal to the string; this velocity is maintained constant for all $t > t_0$. Give the equations of constraint on finite, possible, and virtual displacements of the particle position in Cartesian coordinates, write down the fundamental equation and the equations of motion, and discuss the C, E, and S trajectories.

9.3. The smooth ends of a rigid, heavy, uniform rod of length $2a$ are always in contact with the parabola $y = x^2$, which lies in a vertical plane, open in the up direction. Find the equilibrium positions; in particular, show that there is only one such position if $2a \leq 1$, but there are three for $2a > 1$. (Hint: Use the angle θ subtended by the rod and the positive x axis to describe the configuration.)

9.4. Show that, when the virtual displacements satisfy the knife edge constraint $\sin\theta\,\delta x = \cos\theta\,\delta y$, virtual displacements from a possible state do not lead to another possible state.

9.5. Show that, when the virtual displacements satisfy $\sin\theta_0\,\delta x = \cos\theta_0\,\delta y$ where θ_0 is a nonzero constant, virtual displacements from a possible state do lead to another possible state.

9.6. Below are given the Cartesian components X, Y, Z of certain forces. Which of these are potential?

(a)
$$X = 4a_1x^3 + 3a_2x^2y + 2a_3xy^2 + a_4y^3,$$
$$Y = a_2x^3 + 2a_3x^2y + 3a_4xy^2 + 4a_5y^3,$$
$$Z = 0,$$

where the a_i are constants ($i = 1, 2, ..., 5$).

(b)
$$X = \cos y + z\sec^2 x,$$
$$Y = x\cos y + \sin z,$$
$$Z = \tan x.$$

(c)
$$X = \sin y + z\sec^2 x,$$
$$Y = x\cos y + \sin z,$$
$$Z = y\cos z + \tan x.$$

(d)
$$X = f(x), \qquad Y = g(y), \qquad Z = h(z).$$

9.7. Let

$$X = f_1(x, y) + g_1(x, z),$$
$$Y = f_2(x, y) + h_1(y, z),$$
$$Z = g_2(x, z) + h_2(y, z),$$

where $f_{1,2}$, $g_{1,2}$, and $h_{1,2}$ are analytic functions. What are the necessary and sufficient conditions for $F = (X, Y, Z)$ to be a potential force?

9.8. A particle of mass m is subjected to a force whose Cartesian components are

$$X = -\frac{x^2 + y^2 - z^2 - a^2}{(x^2 + y^2)^{3/2}}\,x,$$

$$Y = -\frac{x^2 + y^2 - z^2 - a^2}{(x^2 + y^2)^{3/2}}\,y,$$

$$Z = -\frac{2z}{(x^2 + y^2)^{1/2}}.$$

(a) Show that the force is potential and discuss the equipotential curve $V = \text{const.}$

(b) Suppose the particle is constrained to move on a smooth sphere centered at the origin of the Cartesian coordinate system. Write the fundamental equation and the equations of motion. Find the equilibrium positions.

9.9. A particle of mass m is constrained to move on the curve defined by

$$x = \cos\theta, \qquad y = \sin\theta, \qquad z = 2.$$

It is subjected to a force whose Cartesian components are

$$X = \frac{z^2 - y^2}{(x + y)^2}, \qquad Y = \frac{2y}{x + z}, \qquad Z = \frac{x^2 - y^2}{(x + z)^2}.$$

(a) Calculate the work done by this force when the particle moves on the arc corresponding to $0 \leq \theta \leq \pi/2$.

(b) Answer the same question when the curve is defined by

$$x = 2\sin^2\theta, \qquad y = \cos 2\theta, \qquad z = 2\cos^2\theta.$$

9.10. An unconstrained particle is acted upon by a force whose Cartesian components are

$$X = \frac{x}{x^2 + y^2 + z^2}, \qquad Y = \frac{y}{x^2 + y^2 + z^2}, \qquad Z = \frac{z}{x^2 + y^2 + z^2}.$$

Find the equilibrium positions.

9.11. Answer the same question as in Problem 9.10 when the spherical components of the force are

(a) $R = \dfrac{-2a\cos\theta}{r^3}, \qquad \Theta = \dfrac{-a\sin\theta}{r^2}, \qquad \Phi = 0,$

(b) $R = -e^{-kr}\left(\dfrac{1}{r^2} + \dfrac{1}{r}\right), \qquad \Theta = 0, \quad \Phi = 0.$

10

Hamilton's Principle

10.1. The Kinetic Energy

If a particle P_r of mass m_r moves with velocity $\dot{x}^r$ relative to an inertial reference frame, the quantity

$$T^r = \tfrac{1}{2} m_r \dot{x}^r \cdot \dot{x}^r = \tfrac{1}{2} m_r (\dot{x}^r)^2 \tag{10.1.1}$$

is called its kinetic energy. Evidently, T^r is a scalar quantity; thus, in a system of particles in which m_r moves with the velocity $\dot{x}^r$ ($r = 1, 2, \ldots, n$), the total kinetic energy of the particles in the system is

$$T = \sum_{r=1}^{n} T^r = \frac{1}{2} \sum_{r=1}^{n} m_r (\dot{x}^r)^2. \tag{10.1.2}$$

The double of this quantity was formerly called the "living force" (French: *force vive*; German: *lebendige Kraft*), and in modern French texts, T is still sometimes referred to as the "*demi force vive*").

Equations (10.1.1) and (10.1.2) are completely general provided the velocity exists (i.e., they hold at all times for which the velocity is defined). If we proceed from the x_i^r ($r = 1, 2, \ldots, n$; $i = 1, 2, 3$) to the u_s ($s = 1, 2, \ldots, N$), the expression for the kinetic energy of a system of particles becomes

$$T = \frac{1}{2} \sum_{s=1}^{N} m_s \dot{u}_s^2. \tag{10.1.3}$$

In certain circumstances to be described, the kinetic energy becomes an important dynamic property of the motion of a system of particles.

10.2. Kinetic Energy in Catastatic Systems

In catastatic systems, *but only in catastatic systems*, we may write the fundamental equation as

$$\sum_{s=1}^{N} (m_s \ddot{u}_s - F_s)\, du_s = 0, \tag{10.2.1}$$

where the du_s are the components of *possible* displacements

$$du = (du_1, du_2, \ldots, du_N);$$

(10.2.1) follows from (9.3.12) because the class of virtual and possible displacements coincides in catastatic systems. It may also be written as

$$\sum_{s=1}^{N} m_s \ddot{u}_s \dot{u}_s = \sum_{s=1}^{N} F_s \dot{u}_s, \tag{10.2.2}$$

where the $\dot{u}_s$ are the components of possible velocities $\dot{u} = (\dot{u}_1, \dot{u}_2, \ldots, \dot{u}_N)$, i.e., the $\dot{u}_s$ satisfy the equations

$$\sum_{s=1}^{N} A_{rs} \dot{u}_s = 0 \qquad (r = 1, 2, \ldots, L). \tag{10.2.3}$$

It is readily ascertained that the left-hand side of (10.2.2) is the time rate of change of the kinetic energy because

$$\frac{dT}{dt} = \frac{d}{dt}\left(\frac{1}{2} \sum_{s=1}^{N} m_s \dot{u}_s^2\right) = \sum_{s=1}^{N} m_s \ddot{u}_s \dot{u}_s.$$

Here, the $\ddot{u}_s$ are the time rates of change of the components of the possible velocities. Thus, they belong to the class of possible acceleration components. We may write (10.2.2) as

$$\frac{dT}{dt} = \sum_{s=1}^{N} F_s \dot{u}_s. \tag{10.2.4}$$

The product $F \cdot \dot{u} = \sum_{s=1}^{N} F_s \dot{u}_s$ is called the power. Equation (10.2.4) states:

The time rate of change of the kinetic energy of a catastatic system equals the power of the given forces under possible velocities.

10.3. The Energy Relations in Catastatic Systems

Under certain conditions listed below, (10.2.4) may be integrated; one finds then

$$T = \int \sum_{s=1}^{N} F_s \dot{u}_s \, dt + h = \int \sum_{s=1}^{N} F_s \, du_s + h, \tag{10.3.1}$$

where h is a constant of integration. The conditions under which this integration may be carried out are:

(i) The velocity components $\dot{u}_s$ must be defined for every value of the configuration, i.e., there may be no velocity discontinuities during the interval of integration. This condition is insured if the system is strictly Newtonian (SN).

(ii) N is a constant, i.e., no particles may be acquired or lost by the system during the interval of integration. This stipulation implies that *mass is conserved.*

Let us now suppose that some of the given forces are potential forces F^p, and some are nonpotential forces F^n. For generality, we assume that every particle of the system is subjected to given forces of each kind. Thus, (10.3.1) becomes

$$T = \int \sum_{s=1}^{N} \sum_{p=1,2,\ldots} F_s^p \, du_s + \int \sum_{s=1}^{N} \sum_{n=1,2,\ldots} F_s^n \, du_s + h. \tag{10.3.2}$$

But, we saw in (9.9.2) that every

$$F_s^p = -\frac{\partial V^p}{\partial u_s} \qquad (s = 1, 2, \ldots, N; \ p = 1, 2, \ldots). \tag{10.3.3}$$

Thus, if we define the total potential energy of the system as

$$V = \sum_{p=1,2,\ldots} V^p, \tag{10.3.4}$$

we find

$$T + V = \int \sum_{s=1}^{N} \sum_{n=1,2,\ldots} F_s^n \, du_s + h. \tag{10.3.5}$$

When only potential forces act, i.e., $F_s^n \equiv 0$ for all $s = 1, 2, \ldots, N$ and for all $n = 1, 2, \ldots$, (10.3.5) reduces to one of the most celebrated equations of classical dynamics. That equation is known as the *energy integral*; it is

$$T + V = h. \tag{10.3.6}$$

It states:

> *In catastatic systems of particles in which the total mass is conserved the total energy is a constant of the motion provided all given forces are potential forces.*

Systems which possess a first integral of the form (10.3.6) are called *conservative*, or *closed* systems.

It is clear that the constant h is a constant of the *motion*, not of the *system*. For suppose that two experiments are conducted on the same system. In both, the system is given the same configuration at the time $t = t_0$. Thus, the initial potential energy is the same in both. In the first experiment, the system is released from rest, or the initial kinetic energy is zero. In the second experiment, the system is given a nonvanishing velocity at $t = t_0$; thus, the initial kinetic energy is positive. Then, the total energy is different in the two motions at $t = t_0$ and, therefore, for all time. The constant h is called the *energy level* of the motion.

The energy integral has an interesting geometric interpretation. By definition T is nonnegative. Thus, if the system is closed $T = h - V \geq 0$, or $V \leq h$. The surface

$$V(u_1, u_2, \ldots, u_N) = h \tag{10.3.7}$$

is called the *maximum equipotential surface*. It divides the configuration space into domains where $V < h$ and those where $V > h$. Therefore, the C trajectories of the motion of closed systems must remain in the domain where $V < h$, except that they may intercept, but not pierce, the surface $V = h$. Expressed differently, C trajectories of closed systems satisfy the nonholonomic configuration constraint

$$V(u_1, u_2, \ldots, u_N) \leq h. \tag{10.3.8}$$

The maximum equipotential surface is the locus of rest points of the motion of closed systems because, when $V = h$, $T = 0$ in view of the energy integral, and $T = 0$ implies $\dot{u}_s = 0$ for all $s = 1, 2, \ldots, N$.

The general energy integral for nonclosed, catastatic systems is (10.3.5). In it, we shall put

$$\sum_{n=1,2,\ldots} F_s^n = F_s. \tag{10.3.9}$$

[This represents actually a slight change in nomenclature because, up to this point, we have used the symbol F_s to denote all given forces, both potential and nonpotential, while the F_s in (10.3.9) are nonpotential. Thus, we shall introduce the following rule: when the symbols F_s and V occur in the same equations, F_s refers to nonpotential forces only.]

The equation corresponding to the energy integral then becomes

$$T + V - \int \sum_{s=1}^{N} F_s \, du_s = h. \tag{10.3.10}$$

This important equation states:

In catastatic systems of constant total mass, the total energy diminished by the work done by the nonpotential forces is a constant of the motion.

The companion to this equation is

$$\frac{d}{dt}(T + V) = \sum_{s=1}^{N} F_s \dot{u}_s, \tag{10.3.11}$$

which may be obtained either by differentiating (10.3.10), or directly from (10.2.4) if one separates the given forces in that equation into potential and nonpotential ones.

All relations given here which involve energies are more general than is often supposed. Since they apply to all catastatic systems, they are in fact valid for constrained systems in which the constraints may be functions of time. Pars given an interesting example of a nonholonomic system of this type.

Example 10.3.1. (Pars, p. 32). Let a particle move under the action of gravity and let it be steered in such a way that the slope of its trajectory varies as the time. The particle is projected horizontally with initial velocity u. Choose the y axis vertical, positive in the downward direction, and let the x axis be horizontal, positive in the direction of the initial velocity u. The steering control is such that $dy/dx = t$ or $dy - t\,dx = 0$; this is a nonholonomic constraint. Then, the fundamental equation is

$$m\ddot{x}\,\delta x + (m\ddot{y} - mg)\,\delta y + m\lambda\,(dy - t\,\delta x) = 0, \tag{a}$$

where $m\lambda$ is a Lagrange multiplier. From (a) one finds the two equations

$$\begin{aligned} \ddot{x} - \lambda t &= 0, \\ \ddot{y} - g + \lambda &= 0. \end{aligned} \tag{b}$$

These equations, together with the catastatic time-dependent constraint

$$\dot{y} - t\dot{x} = 0, \tag{c}$$

are sufficient to find $x(t)$, $y(t)$, and $\lambda(t)$. Eliminating λ from (b), one has

$$\ddot{x} = -\ddot{y}t + gt \tag{d}$$

and, on differentiating (c), one finds

$$\ddot{y} - \dot{x} - t\ddot{x} = 0. \tag{e}$$

When $\ddot{y}$ is eliminated between (d) and (e) there results the following equation in x:

$$\ddot{x}(1 + t^2) + \dot{x}t - gt = 0. \tag{f}$$

The first and second integrals of (f) satisfying the initial conditions $x(0) = 0$, $\dot{x}(0) = u$ are

$$\begin{aligned} \dot{x} &= g + (u - g)/(1 + t^2)^{1/2}, \\ x &= gt + (u - g)\sinh^{-1} t. \end{aligned} \tag{g}$$

In view of (c),

$$\begin{aligned} \dot{y} &= t\dot{x} = gt + (u - g)t/(1 + t^2)^{1/2}, \\ y &= \tfrac{1}{2}gt^2 + (u - g)[(1 + t^2)^{1/2} - 1]. \end{aligned} \tag{h}$$

These satisfy $y(0) = \dot{y}(0) = 0$.

Then, from the first equations of (g) and (h) the double of the kinetic energy per unit mass is

$$\dot{x}^2 + \dot{y}^2 = g^2(1 + t^2) + 2g(u - g)(1 + t^2)^{1/2} + (u - g)^2. \tag{i}$$

Moreover, from the second equation of (h), the double of the potential energy per unit mass with respect to the origin is

$$-2gy = -g^2t^2 - 2g(u - g)[(1 + t^2)^{1/2} - 1], \tag{j}$$

and the sum of the double of kinetic and potential energies becomes

$$\dot{x}^2 + \dot{y}^2 - 2gy = u^2. \tag{k}$$

Thus, the total energy is a constant, equal to the initial energy. This last result (k) is the same as would have been found if there had been no steering. However, under the identical initial conditions, but without steering, the trajectory equations would have been

$$x = ut, \qquad y = \tfrac{1}{2}gt^2.$$

One also sees that λ is not a constant since, from the first equation of (b) and with the aid of the first equation of (g),

$$\lambda(t) = \ddot{x}/t = -(u - g)/(1 + t^2)^{3/2}.$$

The difference between a steered and an unsteered trajectory having the same energy integral is shown in Fig. 10.3.1.

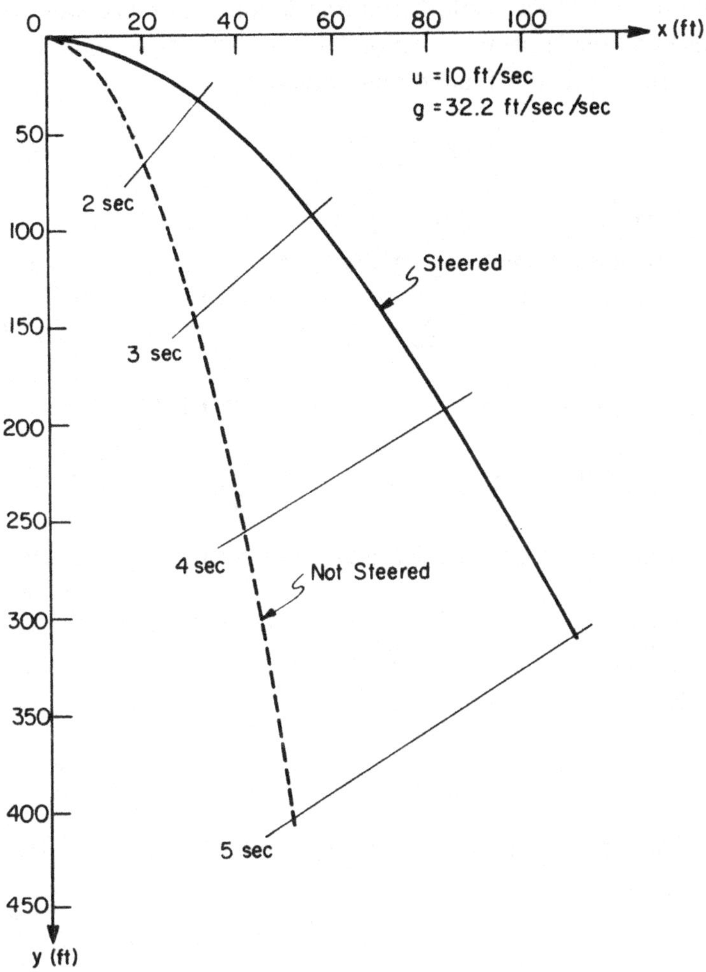

Fig. 10.3.1. The steered trajectory of Example 10.3.1. and the "unsteered" trajectory under the same initial conditions. Both have the same total energy.

10.4. The Central Principle

The relations established in the preceding two sections are valid for catastatic systems only, for only in them can one deduce (10.2.1) from the fundamental equation. It was with the aid of (10.2.1) that we found relations for the time rate of the kinetic energy or, when integrating with respect to time, for the kinetic energy. In this section, we establish a relation which was called by Hamel (p. 233, after Heun) "Zentralgleichung," and which is valid for acatastatic systems as well. We translate Hamel's term as "central principle" because the literal translation "central equation" sounds so much like "fundamental equation" as to invite confusion.

We begin with the fundamental equation

$$\sum_{r=1}^{n} (m_r \ddot{x}^r - F^r) \cdot \delta x^r = 0, \tag{10.4.1}$$

where the F^r are given forces.

Our attention is centered on the term

$$\sum_{r=1}^{n} m_r \ddot{x}^r \cdot \delta x^r. \tag{10.4.2}$$

Consider the time derivative of $\sum_{r=1}^{n} m_r \dot{x}^r \cdot \delta x^r$; it is

$$\frac{d}{dt}\left(\sum_{r=1}^{n} m_r \dot{x}^r \cdot \delta x^r\right) = \sum_{r=1}^{n} m_r \ddot{x}^r \cdot \delta x^r + \sum_{r=1}^{n} m_r \dot{x}^r \cdot \frac{d}{dt}(\delta x^r).$$

If we exchange the operations d/dt and δ (see Section 9.5) we find

$$\frac{d}{dt}\left(\sum_{r=1}^{n} m_r \dot{x}^r \cdot \delta x^r\right) = \sum_{r=1}^{n} m_r \ddot{x}^r \cdot \delta x^r + \sum_{r=1}^{n} m_r \dot{x}^r \cdot \delta \dot{x}^r,$$

so that the quantity of interest is

$$\sum_{r=1}^{n} m_r \ddot{x}^r \cdot \delta x^r = \frac{d}{dt}\left(\sum_{r=1}^{n} m_r \dot{x}^r \cdot \delta x^r\right) - \sum_{r=1}^{n} m_r \dot{x}^r \cdot \delta \dot{x}^r.$$

But, the last sum in this equation can be further transformed. The kinetic energy (10.1.2) is

$$T = \frac{1}{2} \sum_{r=1}^{n} m_r (\dot{x}^r)^2.$$

Therefore, the variation of T is

$$\delta T = \sum_{r=1}^{n} m_r \dot{x}^r \cdot \delta \dot{x}^r, \tag{10.4.3}$$

and (10.4.2) becomes

$$\sum_{r=1}^{n} m_r \ddot{x}^r \cdot \delta x^r = \frac{d}{dt} \sum_{r=1}^{n} m_r \dot{x}^r \cdot \delta x^r - \delta T. \tag{10.4.4}$$

This is the equation called by Hamel the *central principle*. It states:

The virtual work done by the inertia forces is equal to the time rate of change of the work done by the momentum, diminished by the virtual change in kinetic energy.

It is difficult to give (10.4.4) a readily comprehensible physical interpretation. Hamel's reason for regarding this equation as having great importance is that it establishes a relation between an invariant (under a Galilean transformation) of second order and one of first order.

To us, its importance is twofold: First, it is valid for catastatic and acatastatic systems, and second, it is in fact very nearly the differential form of Hamilton's principle.

10.5. The Principle of Hamilton

Hamilton's principle is one of the best known integral principles of mechanics. We derive it from (10.4.1) and (10.4.4). From the first of these, one has

$$\sum_{r=1}^{n} m_r \ddot{x}^r \cdot \delta x^r = \sum_{r=1}^{n} F^r \cdot \delta x^r = \delta W. \tag{10.5.1}$$

It follows that (10.4.4) may be written as

$$\frac{d}{dt} \sum_{r=1}^{n} m_r \dot{x}^r \cdot \delta x^r = \delta T + \delta W.$$

On integrating this last relation with respect to time over the interval from t_0 to t_1 one finds

$$\left[\sum_{r=1}^{n} m_r \dot{x}^r \cdot \delta x^r\right]_{t_0}^{t_1} = \int_{t_0}^{t_1} (\delta T + \delta W)\, dt. \tag{10.5.2}$$

This equation is Hamilton's principle in its most general form.

Usually, this equation is written for virtual displacements satisfying

$$\delta x^r(t_0) = \delta x^r(t_1) = 0 \qquad (r = 1, 2, \ldots, n). \tag{10.5.3}$$

The meaning of (10.5.3) is this: Let the actual motion have the time sequence of configurations (i.e., the C trajectory) $x^r(t)$ $(r = 1, 2, \ldots, n)$, with end-configurations $x^r(t_0)$ and $x^r(t_1)$, respectively. Let neighboring, simultaneous configurations be $\bar{x}^r(t) = x^r(t) + \delta x^r(t)$. Then, at t_0 and t_1, $\bar{x}^r = x^r$, or all paths of the C trajectories $\bar{x}^r$ and x^r coincide at the end points. During the open time interval (t_0, t_1), simultaneous configurations are such that the δx^r xatisfy the constraint equations which result from putting all $A_r = 0$. Under (10.5.3), Hamilton's principle becomes

$$\int_{t_0}^{t_1} (\delta T + \delta W)\, dt = 0. \tag{10.5.4}$$

In (10.5.4), the term δT is the virtual change in kinetic energy which results from virtual displacements, and δW is the work done by the given forces in this displacement. Therefore, Hamilton's principle states:

The time integral over any interval of the sum of virtual kinetic energy change and virtual work vanishes when the virtual displacements are made from configurations of the actual motion and when the end configurations are given.

Essentially, the use of Hamilton's principle is this: If one computes the expressions for δT and δW from any *arbitrary* configuration sequence and then sets the time integral of their sum equal to zero, one imposes thereby the condition that the virtual displacements were made from the *actual sequence of configurations*; hence, one has a condition on this actual C trajectory which may permit its determination.

When all given forces are potential, we saw that

$$\delta W = -\delta V. \tag{10.5.5}$$

However, it will be recalled that δW is *not*, in general, the variation of W while δV *is* the variation of V, as is evident from (9.9.3) and (10.3.4). Moreover, δT is the variation of T as seen from (10.4.3). Therefore,

$$\delta T - \delta V = \delta(T - V),$$

or, in closed systems it is always true that

$$\int_{t_0}^{t_1} \delta(T - V)\, dt = 0. \tag{10.5.6}$$

This form of Hamilton's principle is frequently written in terms of the so-called *Lagrangean function*

$$L = T - V. \tag{10.5.7}$$

Thus, we may write instead of (10.5.6)

$$\int_{t_0}^{t_1} \delta L\, dt = 0 \tag{10.5.8}$$

when the system is closed. The meaning of this principle is this:

The time integral of the variation of the Lagrangean function vanishes along the actual *T trajectory connecting two points C_0 and C_1 in state–time space.*

We point out that, in holonomic systems, one may write

$$\int_{t_0}^{t_1} \delta L \, dt = \delta \int_{t_0}^{t_1} L \, dt, \tag{10.5.9}$$

but in nonholonomic systems

$$\int_{t_0}^{t_1} \delta L \, dt \neq \delta \int_{t_0}^{t_1} L \, dt. \tag{10.5.10}$$

In these equations, the expression

$$\delta \int_{t_0}^{t_1} L \, dt = 0$$

defines the problem in the calculus of variations of finding stationary values of the integral

$$\int_{t_0}^{t_1} L \, dt.$$

The meaning of equations (10.5.9) and (10.5.10) is that, in holonomic systems, Hamilton's principle formulates a problem in the calculus of variations, but in nonholonomic systems it does not. Thus:

Hamilton's principle is not, in general, a variational principle.

This question will be more fully discussed (in Section 13.5) after the introduction of generalized coordinates. Here, we merely state that the only difference between Hamilton's principle and variational problems lies in the treatment of the side conditions, and in dynamical problems the side conditions are the equations of constraint. Briefly, the side conditions to be satisfied in variational problems are the kinematical constraints, i.e., all states must be possible, while in Hamilton's principle the side conditions to be satisfied are those which define the virtual displacements. Now, we showed in Section 9.5 that virtual displacements from possible states lead to possible states in holonomic problems, but not in nonholonomic ones. It is for this reason that (10.5.9) is true only for conservative, holonomic systems. For them, Hamilton's principle may be written in the frequently seen form

$$\delta \int_{t_0}^{t_1} L \, dt = 0. \tag{10.5.11}$$

The meaning of (10.5.11) is very different from that of (10.5.8). We note from the definition of the d and δ operators that $df = 0$ defines the stationary value of f when t is varied, and $\delta f = 0$ defines the stationary value of f when t is not varied. Therefore, the meaning of (10.5.11) is:

The time integral of the Lagrangean function is stationary along the actual path relative to all other possible comparison paths having the same endpoints and differing from the actual one by virtual displacements lying in an open, small neighborhood of the actual path.

Example 10.5.1. A particle, subjected to the gravitational force only, is constrained to move in a smooth surface. What is Hamilton's principle in this case?

Let z be vertical, positive up. The particle has kinetic energy

$$T = \tfrac{1}{2}m(\dot{x}^2 + \dot{y}^2 + \dot{z}^2).$$

The gravitational force is potential, the potential energy being

$$V = mgz.$$

The constraint of the surface

$$z = f(x, y)$$

is holonomic. Thus (10.5.11) is applicable.

Since

$$\dot{z} = \frac{\partial f}{\partial x}\dot{x} + \frac{\partial f}{\partial y}\dot{y},$$

Hamilton's principle is

$$\delta \int_{t_0}^{t_1} \frac{m}{2}\left\{\left[\dot{x}^2 + \dot{y}^2 + \left(\frac{\partial f}{\partial x}\dot{x} + \frac{\partial f}{\partial y}\dot{y}\right)^2\right] - 2gf(x, y)\right\} dt = 0.$$

Example 10.5.2. Let a particle of unit mass be acted on by a potential force, derivable from $V = V(x, y)$, and let it be constrained in such a way that the slope of its trajectory is proportional to the time. (This is a slight generalization of Example 10.3.1. where we had $V = gy$.)

The constraint is

$$t\dot{x} - \dot{y} = 0,$$

which is not holonomic. Hence, the applicable form of Hamilton's principle is (10.5.8). The kinetic energy is

$$T = \tfrac{1}{2}(\dot{x}^2 + \dot{y}^2),$$

so that

$$\delta T = \dot{x}\,\delta\dot{x} + \dot{y}\,\delta\dot{y},$$

and

$$\delta V = \frac{\partial V}{\partial x}\,\delta x + \frac{\partial V}{\partial y}\,\delta y$$

$$= \left(\frac{\partial V}{\partial x} + t\,\frac{\partial V}{\partial y}\right)\delta x.$$

Thus,

$$\delta L = \dot{x}\,\delta\dot{x} + \dot{y}\,\delta\dot{y} - \left(\frac{\partial V}{\partial x} + t\,\frac{\partial V}{\partial y}\right)\delta x,$$

where the constraint equation has been used. Hence, Hamilton's integral is

$$\int_{t_0}^{t_1}\left[\dot{x}\,\delta\dot{x} + \dot{y}\,\delta\dot{y} - \left(\frac{\partial V}{\partial x} + t\,\frac{\partial V}{\partial y}\right)\delta x\right]dt = 0.$$

The quantities δy, $\delta\dot{y}$, and $\dot{y}$ may be eliminated by means of

$$\dot{y} = t\dot{x}, \qquad \delta y = t\,\delta x, \qquad \delta\dot{y} = \delta x + t\,\delta\dot{x},$$

where the third results from differentiating the second. Substitution in Hamilton's integral gives

$$\int_{t_0}^{t_1}\left[(1 + t^2)\,\dot{x}\,\delta\dot{x} + \left(t\dot{x} - \frac{\partial V}{\partial x} - t\,\frac{\partial V}{\partial y}\right)\delta x\right]dt = 0;$$

this is the required answer.

It is important to note that the following procedure is *in*correct and leads to wrong results: Substitute $\dot{y}$ in the kinetic energy to give

$$T^{\dagger} = \tfrac{1}{2}(1 + t^2)\dot{x}^2.$$

Hence,

$$\delta T^{\dagger} = (1 + t^2)\dot{x}\,\delta\dot{x}.$$

Substitute in

$$\int_{t_0}^{t_1}(\delta T^{\dagger} - \delta V)\,dt = 0,$$

resulting in

$$\int_{t_0}^{t_1}\left[(1 + t^2)\,\dot{x}\,\delta\dot{x} - \left(\frac{\partial V}{\partial x} + t\,\frac{\partial V}{\partial y}\right)\delta x\right]dt = 0.$$

It is obvious that this last equation cannot be the correct answer because it differs from the earlier one found by correct methods. The error arose because it was implicitely assumed that

$$\delta T^{\dagger} = \delta T,$$

but this is not in general true.

Example 10.5.3. Let a particle having a single degree of freedom be subjected to a force which depends linearly on the velocity, to another force which depends linearly on the displacement, and to a time-dependent force $F(t)$. What is Hamilton's principle in this case?

Let

$$F_1 = -c\dot{x}, \qquad F_2 = -kx, \qquad F_3 = F(t),$$

where x is the degree of freedom. The only force which is potential is

$$-kx = d(-kx^2/2)\,dx.$$

Hence $V = kx^2/2$. The kinetic energy is $T = m\dot{x}^2/2$. The work done by F_1 and F_3 in a virtual displacement δx is $[-c\dot{x} + F(t)]\,\delta x$. Thus, Hamilton's principle is

$$\int_{t_0}^{t_1} \{m\dot{x}\,\delta\dot{x} - [kx + c\dot{x} - F(t)]\,\delta x\}\,dt = 0.$$

Note, that we could introduce a slight generalization by considering a time-dependent potential function. Consider the same problem with $c = 0$. Then, we may define a potential energy

$$V = \tfrac{1}{2}kx^2 - xF(t).$$

In consequence, we would have

$$\delta V = kx\,\delta x - F(t)\,\delta x,$$

and the applicable form for Hamilton's principle is (10.5.6). One finds

$$\int_{t_1}^{t_2} \{m\dot{x}\,\delta\dot{x} - [kx - F(t)]\,\delta x\}\,dt = 0$$

identical with the previous answer if one sets $c = 0$. One could also have used (10.5.11). For $c \neq 0$, one could *not* have written

$$V = \tfrac{1}{2}kx^2 + cx\dot{x} - xF(t)$$

because the variation of this quantity is

$$\delta V = kx\,\delta x + c\dot{x}\,\delta x + cx\,\delta\dot{x} - F(t)\,\delta x.$$

Thus, the answer would have been incorrect.

10.6. Noncontemporaneous Variations

The principle of Hamilton is stated in order of decreasing generality in (10.5.2), (10.5.4), (10.5.8), and (10.5.11). In all but the first of these, the end configurations are supposed given and fixed for all trajectories admitted to the competition. Moreover, Hamilton's principle contemplates that the departure time t_0 and the arrival time t_1, and hence the transit time $t_1 - t_0$ are also fixed. However, cases arise where the transit time cannot be the same for varied neighboring trajectories passing through the same end configurations. This is most easily demonstrated by considering the force-free motion of a holonomic, scleronomic system. Evidently, we may regard this system as conservative with

$$V \equiv 0. \tag{10.6.1}$$

For that system, Hamilton's principle (10.5.11) reduces to

$$\delta \int_{t_0}^{t_1} T\,dt = 0. \tag{10.6.2}$$

But, since energy is conserved throughout the motion and $V \equiv 0$, T is a constant, so that (10.6.2) becomes

$$\delta(t_1 - t_0) = 0. \tag{10.6.3}$$

This equation shows that the transit time must be varied in this case, if varied paths under constant energy are admitted. We are thus led to consider "noncontemporaneous variations," i.e., variations in which time is varied as well as the state variables. We denote such a variation by the operator symbol δ_t.

To carry out the δ_t variation, we introduce an "auxiliary time" τ, and we suppose that t is a function of τ. To be more precise, let t be a once differentiable, monotonically increasing function of τ on the interval $0 \leq \tau \leq 1$, or

$$t = t(\tau), \qquad \frac{dt}{d\tau} = t' > 0, \tag{10.6.4}$$

and with boundary values

$$t(0) = t_0, \qquad t(1) = t_1. \tag{10.6.5}$$

Then, the noncontemporaneous variation of a function $F(u_1, \ldots, u_n; \dot{u}_1, \ldots, \dot{u}_n; t)$ is

$$\delta_t F = \sum_{s=1}^{n} \frac{\partial F}{\partial u}\,\delta_t u_s + \sum_{s=1}^{n} \frac{\partial F}{\partial \dot{u}}\,\delta_t\!\left(\frac{u_s'}{t'}\right) + \frac{\partial F}{\partial t}\,\delta_t t, \tag{10.6.6}$$

where the prime denotes differentiation with respect to τ. But,

$$\begin{aligned}
\delta_t\left(\frac{u_s'}{t'}\right) &= \frac{t'\,\delta_t u_s' - u_s'\,\delta_t t'}{t'^2} \\
&= \frac{\delta_t(du_s/d\tau)}{dt/d\tau} - \frac{du_s/d\tau}{dt/d\tau}\cdot\frac{\delta_t(dt/d\tau)}{dt/d\tau} \\
&= \frac{d(\delta_t u_s)/d\tau}{dt/d\tau} - \dot{u}_s\frac{d(\delta_t t)/d\tau}{dt/d\tau} \\
&= \frac{d}{dt}(\delta_t u_s) - \dot{u}_s\frac{d}{dt}(\delta_t t) \\
&= \frac{d}{dt}(\delta u_s) - \dot{u}_s\frac{d}{dt}(\delta_t t) \\
&= \delta\dot{u}_s - \dot{u}_s\frac{d}{dt}(\delta_t t),
\end{aligned} \tag{10.6.7}$$

where we have utilized

$$\delta_t u_s = \delta u_s \tag{10.6.8}$$

because time is not involved in the variation of a configuration variable; hence, the δ and δ_t variations are identical in that case. The substitution of (10.6.7) and (10.6.8) in (10.6.6) together with the definition of the δ operator results in

$$\delta_t F = \delta F - \sum_{s=1}^{N}\dot{u}_s\frac{\partial F}{\partial\dot{u}_s}\frac{d}{dt}(\delta_t t) + \frac{\partial F}{\partial t}\delta_t t. \tag{10.6.9}$$

This equation shows that the δ and δ_t variations become identical when t is not varied.

10.7. Lagrange's Principle of Least Action

The variational principle (10.6.2) which holds for force-free motion is due to Leibnitz. It is frequently attributed to Maupertuis, who, without knowledge of Leibnitz's work, announced in 1747 his "principle of least action" in rather vague terms. He defended it with a teleological argument concerning the frugality of nature in dispensing "action" but he was unable to explain why it was the action rather than some other dynamical quantity that nature wished to minimize.

Both Euler and Lagrange considered the generalization of Leibnitz's principle to the holonomic case possessing an energy integral, and in which $V \neq 0$. All trajectories pass through the same, fixed end configurations, and the energy level is the same for all varied trajectories. This problem leads to Lagrange's principle of least action.

It should be clearly understood that the type of variation contemplated in Hamilton's principle is contemporaneous while that in Lagrange's principle of least action is not. Both lead to virtual displacements within the meaning of definition (9.2.4); nevertheless, the variations leading to them are different. In Hamilton's principle, virtual displacements from a possible configuration are considered at equal times, i.e., the variation is made under fixed *time*. In Lagrange's principle of least action, virtual displacements from a possible configuration are considered at equal energy levels, i.e., the variation is made under fixed *energy level*, not time. A formalization of these observations is that, if

$$T + V = h,$$

then a contemporaneous variation of this equation gives

$$\delta T + \delta V = \delta h$$

where $\delta h \neq 0$ in general, while the noncontemporaneous variation in Lagrange's principle of least action gives

$$\delta_t T + \delta_t V = 0.$$

Here, we derive Lagrange's principle from that of Hamilton in the applicable form (10.5.8). The substitution of T in place of F in (10.6.9) gives

$$\delta_t T = \delta T - 2T \frac{d}{dt} (\delta_t t) \tag{10.7.1}$$

because

$$\sum_{s=1}^{N} \frac{\partial T}{\partial \dot{u}_s} \dot{u}_s = \sum_{s=1}^{N} m_s \dot{u}_s^2 = 2T,$$

and in the scleronomic problem considered here,

$$\frac{\partial T}{\partial t} = 0.$$

Hence

$$\delta T = \delta_t T + 2T \frac{d}{dt} (\delta_t t). \tag{10.7.2}$$

Moreover,

$$\delta V = \delta_t V \tag{10.7.3}$$

because V is a function of the configuration variables only, and not time-dependent. Then, the substitution of (10.7.2) and (10.7.3) in (10.5.8) gives

$$\int_{t_0}^{t_1}\left[\delta_t T + 2T\frac{d}{dt}(\delta_t t) - \delta_t V\right] dt = 0. \tag{10.7.4}$$

Since the energy level is the same for all trajectories and is

$$T + V = h = \text{const},$$

we find

$$\delta_t V = -\delta_t T. \tag{10.7.5}$$

This last relation gives in place of (10.7.4)

$$\int_{t_0}^{t_1} 2[\delta_t T\,dt + T\,d(\delta_t t)] = 0. \tag{10.7.6}$$

But from (10.6.4) and (10.6.5) we have

$$dt = t'\,d\tau, \qquad d(\delta_t t') = \frac{d}{d\tau}(\delta_t t)\,d\tau = \delta_t(t')\,d\tau.$$

These relations transform (10.7.6) into

$$\int_0^1 2[t'\,\delta_t T + T\,\delta_t t']\,d\tau = \int_0^1 2\,\delta_t(Tt')\,d\tau = 0.$$

Finally, if we exchange the variational and integral operators, which we may do in holonomic systems [see (10.5.11)], we obtain

$$\delta_t\int_0^1 2Tt'\,d\tau = \delta_t\int_{t_0}^{t_1} 2T\,dt = 0. \tag{10.7.7}$$

This is Lagrange's principle of least action, which is also frequently written as

$$\delta_t\int_{t_0}^{t_1} T\,dt = 0. \tag{10.7.8}$$

The meaning of (10.7.8) is this:

In holonomic systems possessing an energy integral in which the energy level is fixed in moving from a prescribed initial to a prescribed terminal configuration, the time integral of the kinetic energy is stationary under noncontemporaneous variations.

It should be noted that the transit times are different for different trajectories.

10.8. Jacobi's Principle of Least Action

Jacobi's principle of least action is readily obtained from (10.7.7); however, its meaning is quite different from Lagrange's principle.

If we write $T = (T^2)^{1/2}$ and utilize $T = h - V$, we find in place of (10.7.7)

$$\delta_t \int_{t_0}^{t_1} 2[T(h - V)]^{1/2}\, dt = 0$$

or

$$\delta_t \int_{t_0}^{t_1} \left[2 \sum_{r=1}^{N} m_r \dot{u}_r^2 (h - V)\right]^{1/2} dt = 0. \tag{10.8.1}$$

The solution to the dynamical problem may be represented as a C trajectory in N-dimensional configuration space, and the arc length s along the C trajectory, measured from some fixed point on the trajectory, may be used as a parameter to describe the configuration; thus, one writes

$$u_r = u_r(s) \qquad (r = 1, 2, \ldots, N).$$

Introducing the parameters s in (10.8.1), that equation becomes

$$\delta \int_{s_0}^{s_1} \left[2 \sum_{r=1}^{N} m_r \dot{u}_r^2 (h - V)\right]^{1/2} \frac{ds}{\dot{s}} = 0, \tag{10.8.2}$$

where s_0 and s_1 are the values of s at the initial and terminal configurations, and the operator δ_t has been replaced by δ because the limits of the integral in (10.8.2) are fixed, and time is not being varied. We now introduce the transformations

$$q_r = (m_r)^{1/2} u_r \qquad (r = 1, 2, \ldots, N); \tag{10.8.3}$$

this is merely a scale change. Then, (10.8.2) becomes

$$\delta \int_{s_0}^{s_1} \left[2 \sum_{r=1}^{N} \dot{q}_r^2 (h - V)\right]^{1/2} \frac{ds}{\dot{s}} = 0. \tag{10.8.4}$$

But, the line element ds and the velocity $\dot{s}$ are given in terms of the q_r by

$$ds = \left(\sum_{r=1}^{N} dq_r^2\right)^{1/2}, \qquad \dot{s}^2 = \sum_{r=1}^{N} \dot{q}_r^2, \tag{10.8.5}$$

and the substitution of these in (10.8.4) gives that equation its final form:

$$\delta \int_{s_0}^{s_1} \{2[h - V(q_1, q_2, \ldots, q_N)]\}^{1/2}\, ds = 0 \tag{10.8.6}$$

or

$$\delta \int_{q_0}^{q_1} \left\{2[h - V(q_1, q_2, \ldots, q_N)] \sum_{r=1}^{N} dq_r^2\right\}^{1/2} = 0, \tag{10.8.7}$$

where

$$\begin{aligned} q^0 &= (q_1^0, q_2^0, \ldots, q_N^0), \\ q^1 &= (q_1^1, q_2^1, \ldots, q_N^1) \end{aligned} \tag{10.8.8}$$

are the initial and terminal configurations, respectively. In contradistinction to Hamilton's principle or to Lagrange's principle of least action, the Jacobi principle (10.8.7) or (10.8.8) is *geometrical*, not dynamical, i.e., the independent variables are the configuration coordinates, not time.

If we define the action integral as

$$A = \int_{s_0}^{s_1} [2(h - V)]^{1/2}\, ds, \tag{10.8.9}$$

we may verbalize Jacobi's principle of least action as follows:

In holonomic systems possessing an energy integral in which the energy level is fixed in moving from a prescribed initial to a prescribed terminal configuration, the action integral along the actual C trajectory connecting initial and terminal configurations is stationary relative to all other trajectories connecting the same end configurations and which differ from the actual trajectory by virtual displacements lying in an open neighborhood of the actual C trajectory.

This last provision is a general requirement in variational problems; however, it is stressed here because in the case of Jacobi's principle one can sometimes find trajectories for which Jacobi's action integral is a global minimum, but where these trajectories are inadmissible because they do not possess an open neighborhood. We shall now demonstrate this fact.

Consider the surface

$$V(q_1, q_2, \ldots, q_N) = h, \tag{10.8.10}$$

and suppose that the terminal configurations q^0 and q' lie in it. Since $T + V = h$, the surface (10.8.10) is the locus of points on which the kinetic energy vanishes, i.e., the locus of rest points. It separates the q space into open domains in which the potential energy is less than the total energy and where it is larger. But, the potential energy of conservative systems can never exceed the total energy; hence, C trajectories of these systems can exist only in the domain

$$V(q_1, q_2, \ldots, q_N) \leq h. \tag{10.8.11}$$

It follows that trajectories lying partly or entirely *in* the surface $V = h$ do not possess everywhere an open neighborhood in which comparison trajectories can lie; hence, they are inadmissible. Now, the action integral is nonnegative by definition, and one sees from (10.8.9) that its value is zero for trajectories lying entirely in the surface $V = h$. It is also intuitively clear that a trajectory consisting only of rest points (i.e., no motion) is an absurdity.

An interesting interpretation of Jacobi's principle of least action results from writing

$$[2(h - V)]^{1/2}\, ds = d\sigma. \tag{10.8.12}$$

Then, (10.8.9) becomes

$$\delta \int_{\sigma_0}^{\sigma_1} d\sigma = 0, \tag{10.8.13}$$

where $\sigma_0 = \sigma(s_0)$ and $\sigma_1 = \sigma(s_1)$. If we regard $d\sigma$ as a distance function defining a metric space, the Jacobi principle becomes a problem of finding the stationary σ distance between the points σ_0 and σ_1 in that space, i.e., geodesics in a Riemann space.

10.9. Problems

10.1. A weight of mass $4m$ is attached to a massless, inextensible string which passes over a frictionless, massless pulley, as shown on page 182. The other end of this string is attached to the center of a frictionless, homogeneous pulley of mass m. A second massless inextensible string having masses m and $2m$ attached to its extremities passes over the pulley of mass m. Gravity is the only force acting on this system.

(a) Give the kinetic energy for this system;
(b) give the energy integral, if one exists;
(c) write down Hamilton's principle.

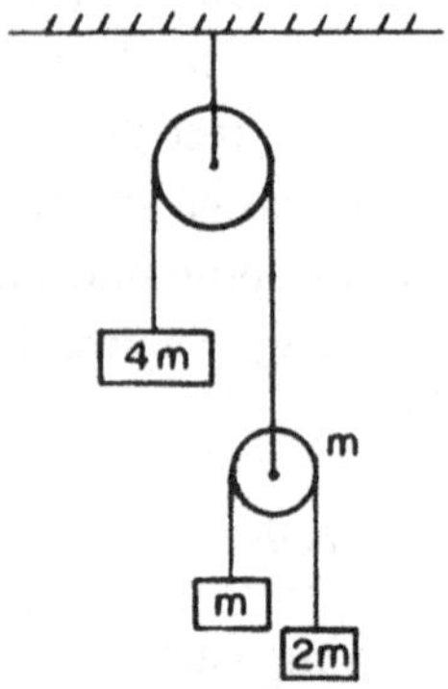

10.2. A homogeneous disk of mass M, constrained to remain in a vertical plane, rolls without sliding on a horizontal line as shown. A massless horizontal, linear spring of rate k is attached to the center of the disk and to a fixed point. If the free length of the spring is l, and the disk radius is R,

(a) Give the kinetic energy for this system;
(b) give the energy integral, if one exists;
(c) write down Hamilton's principle.

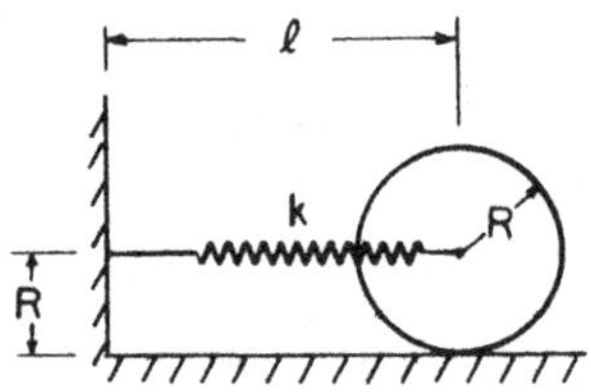

10.3. Give the same answers as in Problem 10.2 when the configuration is changed so that the line is inclined by the angle α to the horizontal, as shown.

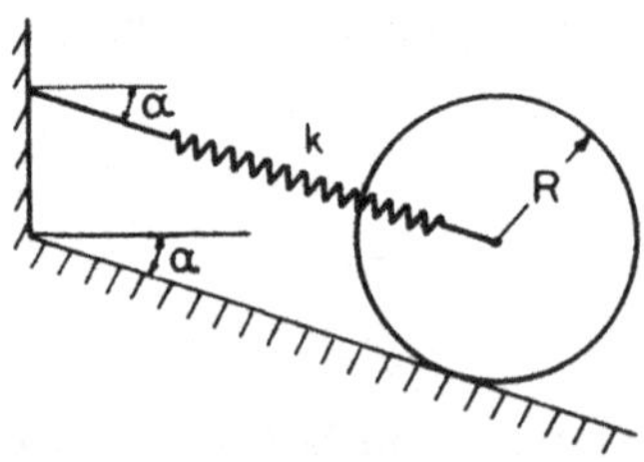

10.4. Five masses m_1, m_2, m_3, m_4, and m_5 are connected by massless, inextensible strings passing over massless, frictionless pulleys as shown. The vertical distance between the center of the top pulley and a fixed datum is given by the prescribed, smooth function $f(t)$.

(a) Give the kinetic energy for this system;
(b) give the energy integral, if one exists;
(c) write down Hamilton's principle.

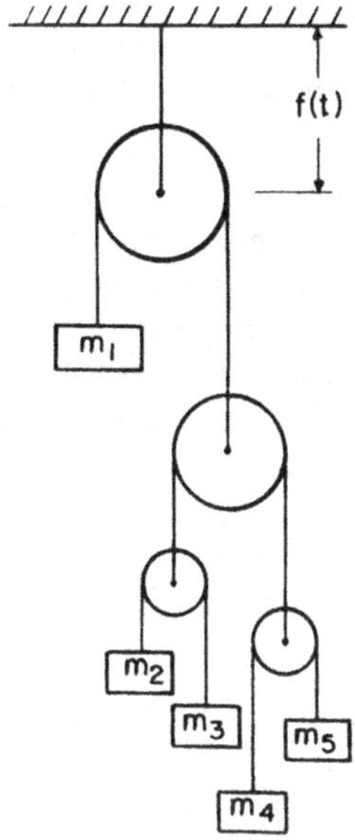

10.5. Three particles of mass m_1, m_2, and m_3, respectively, are constrained to move so that they lie for all time on a straight line passing through a fixed point. For the force-free problem in Cartesian coordinates:

(a) Give the kinetic energy;
(b) give the energy integral, if one exists;
(c) write down Hamilton's principle.

10.6. Three particles of mass m_1, m_2, and m_3, respectively, are constrained to move in a vertical plane under their mutual mass attraction and under the force of gravity (assumed constant).

(a) Give the energy integral, if one exists;
(b) write down Hamilton's principle;
(c) write down the fundamental equation.

10.7. What is Hamilton's principle for the motion of the knife edge under no forces? (See Example 9.4.2.)

10.8. A heavy, homogeneous inextensible string of given length remains for all time in a vertical plane. It lies in part on a smooth, horizontal table, and in part, it hangs vertically down over the table edge. What is Hamilton's principle?

10.9. A particle of mass m moves in the x, y plane under a force which is derivable from a potential energy. The particle velocity is directed for all time toward a point P which moves along the x axis so that its distance from the origin is given by the prescribed function $\xi(t)$.

(a) How many degrees of freedom does the particle have?
(b) What is Hamilton's principle?
(c) Give the energy integral, if one exists.

10.10. One point of a rigid body is constrained to move on a smooth, warping space curve defined by $f(x_0, y_0, z_0, t) = 0$. If the forces and moments acting on the body are potential, give Hamilton's principle. Does an energy integral exist? If so, write it down. If none exists, explain why.

10.11. Two rigid bodies moving in 3-space are constrained by a massless rigid rod which connects a given point of one of the bodies with a given point on the other. The extremities of the rod are smoothly hinged to the bodies so that it can turn freely in any direction relative to the bodies. One of the bodies rolls on a given, perfectly rough surface. How many degrees of freedom does the system have?

11

Generalized Coordinates

11.1. Introductory Remarks

In our treatment of systems of n particles we have, in general, considered the configuration to be fixed by $3n = N$ Cartesian coordinates in configuration space. If the system is constrained by equality constraints we conclude that these constraints define either surfaces or elements of tangent planes to surfaces, and the point defining the configuration in the configuration space (or the point defining the state in state space) must lie in these surfaces. To mention a concrete example, a simple spherical pendulum consists of a particle that moves on the surface of a sphere. Therefore, three Cartesian coordinates define its position, but that position must be a point in the spherical surface.

We have not examined in a systematic way whether or not a constraint can be utilized to reduce the number of coordinates needed to define the configuration uniquely. Yet, it is well known that such is the case. For instance, the position of the bob of a plane, simple pendulum is defined by two numbers x and y, but a single angle is sufficient to specify its configuration. Similarly, three Cartesian coordinates define the position of the bob of a spherical pendulum, but two angles are sufficient to do this. The configuration of a rigid system of any number n of particles is defined by $3n$ Cartesian coordinates, but we saw that six numbers are sufficient to define the configuration uniquely.

It is of great advantage to reduce the number of configuration coordinates to their minimum, because a dynamical system (SN) is governed by as many second-order differential equations as there are configuration coordinates. The study of generalized coordinates is a systematic study of the least number of coordinates needed to specify uniquely the configuration of constrained systems of particles. Obviously, the case of rigid bodies is included; we treat an example of a rigid body in Example 11.3.3.

11.2. The Theory of Generalized Coordinates

Consider a system of $N/3$ particles in the configuration space having Cartesian coordinates $u_1, u_2, \ldots, u_N$. Let it be constrained by L *independent* equations of constraint:

$$\sum_{s=1}^{N} A_{rs}\, du_s + A_r\, dt = 0 \qquad (r = 1, 2, \ldots, L). \tag{11.2.1}$$

We saw in Section 4.3 that all admissible equality constraints may be written in the form (11.2.1) whether they are holonomic or not.

We also defined the notion of "degrees of freedom" by saying that the number of degrees of freedom of a system of $N/3$ particles, constrained by L equality constraints, is equal to $N - L$. Let us suppose that L' of the constraints are holonomic ($L' \leq L$), i.e., they are of, or can be put into, the form

$$\sum_{s=1}^{N} \frac{\partial f_i}{\partial u_s}\, du_s + \frac{\partial f_i}{\partial t}\, dt = 0 \qquad (i = 1, 2, \ldots, L'); \tag{11.2.2}$$

the remaining $L - L'$ constraints are nonholonomic. These are the constraints

$$\sum_{s=1}^{N} A_{js}\, du_s + A_j\, dt = 0 \qquad (j = L' + 1, L' + 2, \ldots, L). \tag{11.2.3}$$

We shall now show that the L' holonomic constraints may be utilized to reduce to $n = N - L'$ the number of configuration coordinates[†] which are needed to specify uniquely the configuration of a system of $N/3$ particles.

We define "coordinates" as follows:

Any set of real numbers $\{q_1, q_2, \ldots, q_n\}$ which can be used to describe the configuration of a system is called a set of coordinates.

[†] Here, n is *not* the number of particles. Henceforth, we use n to denote the quantity $N - L'$.

Then, we can define further:

> *Any set of coordinates* $\{q_1, q_2, \ldots, q_n\}$ *is called a set of* generalized coordinates *of a system if and only if the number n of its members is necessary and sufficient to define the configuration of the system uniquely.*

Thus, in colloquial language, any smallest set of coordinates is a set of generalized coordinates.

Consider the holonomic equations of constraint (11.2.2). Each is a perfect differential:

$$df_i(u_1, u_2, \ldots, u_N, t) = 0 \qquad (i = 1, 2, \ldots, L'). \tag{11.2.4}$$

Therefore, each may be integrated to give

$$f_i(u_1, u_2, \ldots, u_N, t) = \alpha_i, \tag{11.2.5}$$

where the α_i are constants that are determined from the initial conditions $u_1(t_0) = u_1{}^0$, $u_2(t_0) = u_2{}^0$, $\ldots$, $u_N(t_0) = u_N{}^0$.

Let us now introduce transformations

$$q_s = \varphi_s(u_1, u_2, \ldots, u_N, t) \qquad (s = 1, 2, \ldots, N), \tag{11.2.6}$$

where the φ_s are *single-valued* functions of their arguments. Moreover, we select the first L' of these functions to be the f_i defined in (11.2.5). The remaining $N - L'$ functions are *linearly independent*, *arbitrary* functions which are at least of class C^1 in their arguments.

We may regard the φ_s as mapping functions which map the point $(u_1, u_2, \ldots, u_N)$ in the u space into a point $(q_1, q_2, \ldots, q_N)$ in the q space; this mapping is done under fixed t. Obviously, it is unique because of the single-valuedness of all φ_s.

We suppose moreover that there exists a domain D in the u space on which the Jacobian

$$J = \begin{vmatrix} \dfrac{\partial \varphi_1}{\partial u_1} & \cdots & \dfrac{\partial \varphi_1}{\partial u_N} \\ \vdots & & \vdots \\ \dfrac{\partial \varphi_N}{\partial u_1} & \cdots & \dfrac{\partial \varphi_N}{\partial u_N} \end{vmatrix} = \frac{\partial(\varphi_1, \varphi_2, \ldots, \varphi_N)}{\partial(u_1, u_2, \ldots, u_N)}$$

is not zero for any bounded t. Then, it is known from the implicit function theorem that the mapping is locally one-to-one, i.e., it is not only true that to every point u in D there belongs one and only one point q in a corresponding domain Δ of the q space, but to every point q in Δ there belongs one and only one point u in D. This means that there exists an inverse mapping from Δ into D defined by the transformations

$$u_s = u_s(q_1, q_2, \ldots, q_N, t) \qquad (s = 1, 2, \ldots, N). \tag{11.2.7}$$

Combining (11.2.5) and (11.2.6), we find that the first L' of the q_s are

$$q_i = f_i(u_1, u_2, \ldots, u_N, t) = \alpha_i \qquad (i = 1, 2, \ldots, L'), \tag{11.2.8}$$

where the α_i are constants that are fixed once for all time by the surfaces $f_i = \alpha_i$. Hence, the inverse mapping (11.2.7) is

$$u_s = u_s(\alpha_1, \alpha_2, \ldots, \alpha_{L'}, q_{L'+1}, q_{L'+2}, \ldots, q_N, t) \qquad (s = 1, 2, \ldots, N). \tag{11.2.9}$$

In words, the N quantities u_s are determined uniquely by the L' constants, and by the $N - L'$ variables q_j $(j = L' + 1, L' + 2, \ldots, N)$. Hence, only $N - L' = n$ coordinates q_k $(k = 1, 2, \ldots, n)$ are required to fix uniquely the N coordinates u_s when these satisfy L' holonomic constraints. Moreover, the q_k $(k = 1, 2, \ldots, n)$ are now no longer subject to the holonomic constraints.

From (11.2.9) the differential displacements are

$$du_s = \sum_{k=1}^{n} \frac{\partial u_s}{\partial q_k} dq_k + \frac{\partial u_s}{\partial t} dt \qquad (s = 1, 2, \ldots, N). \tag{11.2.10}$$

The virtual displacements δu_s are formed by using the δ operator (see Section 9.6) on (11.2.9) rather than the d operator. Hence, since time is not involved in the δ operation, one sees from (11.2.10) that

$$\delta u_s = \sum_{k=1}^{n} \frac{\partial u_s}{\partial q_k} \delta q_k \qquad (s = 1, 2, \ldots, N). \tag{11.2.11}$$

It is evident that the nonholonomic constraints (11.2.3) cannot be utilized to reduce the number of q_k to less than n because, not being integrable, they do not admit constants of integration. Thus we have:

The number of generalized coordinates q_k $(k = 1, 2, \ldots, n)$ of a system of $N/3$ particles, subject to L' independent, holonomic constraints is precisely $n = N - L'$.

In consequence of this theorem and of the definition of "degrees of freedom" we note that, in general, the number of generalized coordinates exceeds that of the degrees of freedom. In fact, the number of generalized coordinates is equal to the number of degrees of freedom if, and only if, all constraints are holonomic, or if the system is unconstrained.

It is now simple to find an expression for possible infinitesimal displacements dq_k. The possible du_s satisfy the constraints

$$\sum_{s=1}^{N} A_{rs}\, du_s + A_r\, dt = 0 \qquad (r = 1, 2, \ldots, L). \tag{11.2.12}$$

Substituting (11.2.10) in that equation, we find

$$\sum_{s=1}^{N} \left\{ A_{rs} \left[\sum_{k=1}^{n} \frac{\partial u_s}{\partial q_k}\, dq_k + \frac{\partial u_s}{\partial t}\, dt \right] \right\} + A_r\, dt = 0$$

or, exchanging the sequence of summing,

$$\sum_{k=1}^{n} \left(\sum_{s=1}^{N} A_{rs} \frac{\partial u_s}{\partial q_k} \right) dq_k + \left(\sum_{s=1}^{N} A_{rs} \frac{\partial u_s}{\partial t} + A_r \right) dt = 0.$$

If we now introduce the notation

$$B_{rk} = \sum_{s=1}^{N} A_{rs} \frac{\partial u_s}{\partial q_k}, \qquad B_r = \sum_{s=1}^{N} A_{rs} \frac{\partial u_s}{\partial t} + A_r, \tag{11.2.13}$$

the nonholonomic constraint equations become

$$\sum_{s=1}^{n} B_{rs}\, dq_s + B_r\, dt = 0 \qquad (r = 1, 2, \ldots, l), \tag{11.2.14}$$

where $l = L - L'$. This may also be written as

$$\sum_{s=1}^{n} B_{rs} \dot{q}_s + B_r = 0 \qquad (r = 1, 2, \ldots, l). \tag{11.2.15}$$

Then, the possible displacements satisfy (11.2.14), the possible velocities satisfy (11.2.15), and the virtual displacements satisfy

$$\sum_{s=1}^{n} B_{rs}\, \delta q_s = 0 \qquad (r = 1, 2, \ldots, l). \tag{11.2.16}$$

These formulas for the q_s resemble the corresponding ones for Cartesian coordinates. However, one evident difference is that, when generalized coordinates are used, all constraint equations (11.2.14) are nonholonomic; when Cartesian coordinates u_s are used, this is not necessarily so.

One may be tempted to conclude that the two formulations are always identical when the system is holonomic; nevertheless, in general there remain significant differences. When the system is holonomic, the minimum number of *Cartesian* coordinates is, in fact, one set of generalized coordinates, but it is not the only one. When one chooses in a holonomic system a set of generalized coordinates $q_s \neq u_s$, (11.2.10) and (11.2.11) cause a significant change in the form of the kinetic energy (discussed in Section 12.1).

11.3. The Nature of Generalized Coordinates

The essential nature of generalized coordinates is that they *are* "general," i.e., they are undefined except as a class of C^1 functions of the Cartesian coordinates having an inverse given in (11.2.9). In particular, their dimension is not specified. While Cartesian coordinates always have the dimension of length, this need not be the case with generalized coordinates. For instance, in the set

$$x = q_1 \cos q_2, \qquad y = q_1 \sin q_2,$$

q_1 has the dimension of length, and q_2 is usually nondimensional. Similarly, in

$$x = 2lq_1q_2, \qquad y = l(q_1^2 - q_2^2), \qquad z = q_3$$

where l has the dimensions of length, q_1 and q_2 are dimensionless, but q_3 has the dimensions of length. Thus, the notion of generalized coordinates obliterates the distinction between angles and lengths or nondimensional coordinates.

It is evident from the theory of generalized coordinates that L' of the q_s are the constants defined by the holonomic constraints, and the remaining ones are the generalized coordinates; they are so chosen that the holonomic constraints are satisfied by the inverse functions (11.2.9). In many cases, the choice of suitable generalized coordinates becomes evident from the problem under consideration. This is best illustrated by some examples.

Example 11.3.1. We consider first the well-known case of the plane, simple pendulum. The configuration is specified by the Cartesian coordinates x, y of the pendulum bob, and this bob moves on a circle of radius l, the length of the pendulum. Let the circle be centered on the origin of the x, y plane. Then, x and y must satisfy the holonomic constraint

$$x^2 + y^2 - l^2 = 0, \tag{a}$$

and this constraint shows that one must always have

$$|x|, \ |y| \leq l.$$

As two Cartesian coordinates satisfy one holonomic constraint, there is a single generalized coordinate q_1, the other being replaced by l^2. We write $q_1 = \theta$, and chose for the inverse functions (11.2.9) the pair

$$\begin{aligned} x &= f_1(l^2, \theta) = l \cos \theta, \\ y &= f_2(l^2, \theta) = l \sin \theta. \end{aligned} \tag{b}$$

These satisfy the constraint (a) for every $|x|$, $|y| \leq l$. The Jacobian

$$\frac{\partial(f_1, f_2)}{\partial(l, \theta)} = \begin{vmatrix} \dfrac{\partial f_1}{\partial l} & \dfrac{\partial f_1}{\partial \theta} \\ \dfrac{\partial f_2}{\partial l} & \dfrac{\partial f_2}{\partial \theta} \end{vmatrix} = l \neq 0,$$

which is the sufficient condition for one-to-one mapping from l, θ to x, y.

We could also choose $q_1 = x$, the other q being equal to l^2, and we choose for (11.2.9) the set of functions

$$\begin{aligned} x &= f_1(l^2, x) = x \\ y &= f_2(l^2, x) = +(l^2 - x^2)^{1/2}. \end{aligned} \tag{c}$$

These satisfy the constraint (a) for every value of x in $-l \leq x \leq l$, and for every y in $0 \leq y \leq l$. Thus, x is a suitable generalized coordinate on this domain of the x, y plane. If we chose, instead of (c),

$$\begin{aligned} x &= f_1(l^2, x) = x, \\ y &= f_2(l^2, x) = -(l^2 - x^2)^{1/2}, \end{aligned} \tag{d}$$

the domain of x would be $-l \leq x \leq l$ as before, but for y we would have $-l \leq y \leq 0$.

Another possible choice of generalized coordinate is $q_1 = \varphi$, with a choice of (11.2.9) as

$$\left.\begin{aligned} x &= l \cosh \varphi, \\ y &= il \sinh \varphi, \end{aligned}\right\} \qquad i = (-1)^{1/2}. \tag{e}$$

In fact, once a generalized coordinate q_1 and a pair of inverse functions (11.2.9) have been selected which satisfy the constraint on the domain $|x|, |y| \le l$, any arbitrary, once differentiable, monotonic function $\bar{q}_1 = f(q_1)$ furnishes another generalized coordinate. The question as to which is the best choice of generalized coordinate is determined from the form which the differential equation of motion takes on for any given coordinate. Consideration of this problem of the *choice* of generalized coordinates will be delayed until the Lagrangean equations of motion have been introduced.

Example 11.3.2. As a generalization of the problem just discussed, consider a system of $N/3$ particles whose Cartesian coordinates are $u_1, u_2, \ldots, u_N$. Let the position vector $u = (u_1, u_2, \ldots, u_N)$ of a configuration in the configuration space be subject to the holonomic constraint

$$\sum_{s=1}^{N} u_s^2 = l^2. \tag{a}$$

In words, the configuration of the system is a point in the surface of an N-dimensional sphere of radius l, centered on the origin of the configuration space. Since there are N Cartesian coordinates and one holonomic constraint, this system has $N - 1$ generalized coordinates q_s $(s = 1, 2, \ldots, n = N - 1)$. Let $q_s = \theta_s$, and choose for the functions (11.2.9) the set

$$\begin{aligned} u_s &= l \sin \theta_{N+1-s} \prod_{j=1}^{N-s} \cos \theta_j \qquad (s = 1, 2, \ldots, N), \\ \theta_N &= \pi/2. \end{aligned} \tag{b}$$

These are spherical coordinates in N-space, and they satisfy the constraint (a) for all $|u_s| \le l$. They reduce to the familiar spherical coordinates of 3-space for $N = 3$, and to the polar coordinates of the previous example when $N = 2$.

The two examples treated so far are of systems subject to holonomic constraints only. A more instructive example is one in which the system is subject to holonomic constraints as well as to nonholonomic ones. In this case, one seeks generalized coordinates which satisfy the holonomic constraints, and then one must find the form which the nonholonomic constraints take.

Example 11.3.3. Consider a sphere of radius r which rolls without sliding inside a rough sphere of radius $R + r$. Let the origin of the X, Y, Z space coincide with the center of the fixed sphere of radius $R + r$. Then, the center of the rolling sphere has the position (X, Y, Z) and that position satisfies the holonomic constraint

$$X^2 + Y^2 + Z^2 = R^2. \tag{a}$$

Let α be the latitude of the center of the rolling sphere below the equator, and let β be the longitude, measured from the X axis, as shown in Fig. 11.3.1. These are two

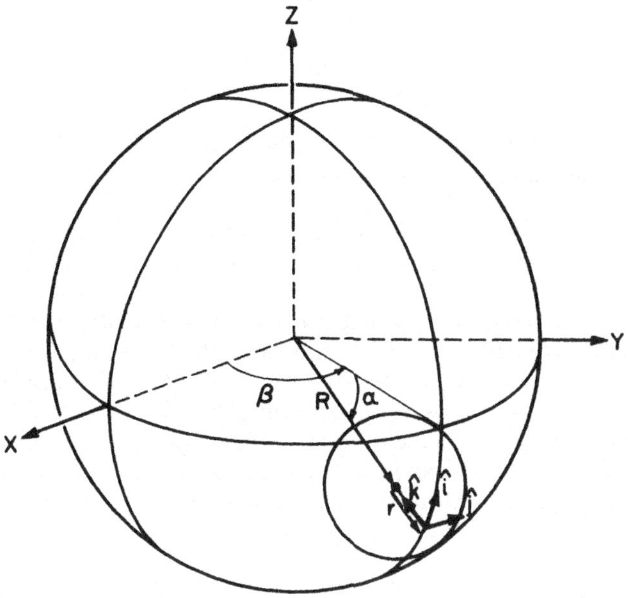

Fig. 11.3.1. Sphere rolling inside fixed sphere of Example 11.3.3.

generalized coordinates for the center of the rolling sphere, and the functions

$$\begin{aligned} X &= R \cos \alpha \cos \beta, \\ Y &= R \cos \alpha \sin \beta, \\ Z &= - R \sin \alpha \end{aligned} \tag{b}$$

satisfy (a) for every value of α and β, and for every $|X|$, $|Y|$, $|Z| \leq R$.

Next, we write the equation of pure rolling. Let v be the velocity of the point on the rolling sphere which is in contact with the fixed sphere. Then, the condition of pure rolling (or of no slipping) is $v = 0$.

If we denote by v_m the velocity of the center of the rolling sphere, we have

$$v = v_m + \omega \times r, \tag{c}$$

where r is the position vector of the contact point relative to the center of the rolling sphere, and ω is the angular velocity of the rolling sphere.

We now construct a Cartesian triad of unit vectors $\hat{i}$, $\hat{j}$, $\hat{k}$ with origin at the contact point such that $\hat{i}$ points north, $\hat{j}$ points east, and $\hat{k}$ points toward the center of the fixed sphere. Then,

$$\begin{aligned} r &= - r\hat{k}, \\ \omega &= \omega_x \hat{i} + \omega_y \hat{j} + \omega_z \hat{k}, \\ v_m &= - R\dot{\alpha}\hat{i} + R\dot{\beta}\hat{j}. \end{aligned} \tag{d}$$

Substituting (d) in (c) and setting $v = 0$, the conditions of pure rolling are found to be

$$\begin{aligned} R\dot{\alpha} + r\omega_y &= 0, \\ R\dot{\beta} + r\omega_x &= 0. \end{aligned} \tag{e}$$

We express the components of the vector ω in terms of the Euler angles [see (6.8.14)]. Then, the nonholonomic constraints become

$$\begin{aligned} R\dot{\alpha} + r(\dot{\varphi}\sin\theta\cos\psi - \dot{\theta}\sin\psi) &= 0, \\ R\dot{\beta} + r(\dot{\varphi}\sin\theta\sin\psi + \dot{\theta}\cos\psi) &= 0. \end{aligned} \tag{f}$$

We note that this system has five generalized coordinates α, β, θ, φ, and ψ, but it has only three degrees of freedom because the five generalized coordinates must satisfy the two nonholonomic constraints (f). We may visualize the three degrees of freedom as follows: Two of them define the position of the contact point in the fixed sphere, and the third is a rotation of the rolling sphere about a line connecting the contact point with the center of the fixed sphere, i.e., the third is a rotation of the sphere about the contact point.

11.4. The δ Operator for Generalized Coordinates

In Section 9.5 we saw that

$$d\,\delta u = \delta\,du, \tag{11.4.1}$$

where u is the vector $(u_1, u_2, \ldots, u_N)$. A similar result holds for the vector $q = (q_1, q_2, \ldots, q_n)$. With a change in subscript notation, (11.2.10) becomes

$$du_r = \sum_{\sigma=1}^{n} \frac{\partial u_r}{\partial q_\sigma}\,dq_\sigma + \frac{\partial u_r}{\partial t}\,dt \qquad (r = 1, 2, \ldots, N), \tag{11.4.2}$$

and (11.2.11) becomes

$$\delta u_r = \sum_{\sigma=1}^{n} \frac{\partial u_r}{\partial q_\sigma}\,\delta q_\sigma \qquad (r = 1, 2, \ldots, N). \tag{11.4.3}$$

If we write (11.4.1) for the rth component and substitute (11.4.2) and (11.4.3) in it, we find

$$\begin{aligned} 0 = d\,\delta u_r - \delta\,du_r &= \sum_\sigma \frac{\partial u_r}{\partial q_\sigma}(d\,\delta q_\sigma - \delta\,dq_\sigma) + \sum_\sigma d\left(\frac{\partial u_r}{\partial q_\sigma}\right)\delta q_\sigma \\ &\quad - \sum_\sigma \delta\left(\frac{\partial u_r}{\partial q_\sigma}\right)dq_\sigma - \delta\left(\frac{\partial u_r}{\partial t}\right)dt. \end{aligned} \tag{11.4.4}$$

Now, it is easy to show that the sum of the last three terms on the right-hand side of (11.4.4) is zero. These terms are

$$\sum_{\sigma} d\left(\frac{\partial u_r}{\partial q_\sigma}\right) \delta q_\sigma - \sum_{\sigma} \delta\left(\frac{\partial u_r}{\partial q_\sigma}\right) dq_\sigma - \delta\left(\frac{\partial u_r}{\partial t}\right) dt$$

$$= \sum_{\alpha,\beta} \frac{\partial^2 u_r}{\partial q_\alpha \, \partial q_\beta} dq_\alpha \, \delta q_\beta + \sum_{\sigma} \frac{\partial^2 u_r}{\partial t \, \partial q_\sigma} dt \, \delta q_\sigma$$

$$- \sum_{\alpha,\beta} \frac{\partial^2 u_r}{\partial q_\alpha \, \partial q_\beta} \delta q_\alpha \, dq_\beta - \sum_{\sigma} \frac{\partial^2 u_r}{\partial q_\sigma \, \partial t} \delta q_\sigma \, dt.$$

The second and fourth terms on the right-hand side are identical and, hence, cancel so that there remains

$$\sum_{\alpha,\beta} \frac{\partial^2 u_r}{\partial q_\beta \, \partial q_\alpha} dq_\alpha \, \delta q_\beta - \sum_{\alpha,\beta} \frac{\partial^2 u_r}{\partial q_\alpha \, \partial q_\beta} \delta q_\alpha \, dq_\beta .$$

In the second sum, we exchange the indices α and β, in which case the two sums become identical, because we showed that (11.4.1) itself is only correct if

$$\frac{\partial^2 u_r}{\partial q_\alpha \, \partial q_\beta} = \frac{\partial^2 u_r}{\partial q_\beta \, \partial q_\alpha} .$$

It follows from (11.4.4) that

$$\sum_{\sigma=1}^{n} \frac{\partial u_r}{\partial q_\sigma} (d \, \delta q_\sigma - \delta \, dq_\sigma) = 0, \tag{11.4.5}$$

and this must hold for all q and for any transformations $u = u(q)$. Hence, (11.4.5) implies

$$d \, \delta q_\sigma - \delta \, dq_\sigma = 0 \qquad (\sigma = 1, 2, \ldots, n). \tag{11.4.6}$$

It follows that

$$d \, \delta q = \delta \, dq, \tag{11.4.7}$$

where q is the vector $(q_1, q_2, \ldots, q_n)$.

11.5. Exceptional Cases

In Section 11.2, it was stated that, in the theory of generalized coordinates, the transformation from some coordinate system to generalized coordinates must be one-to-one.

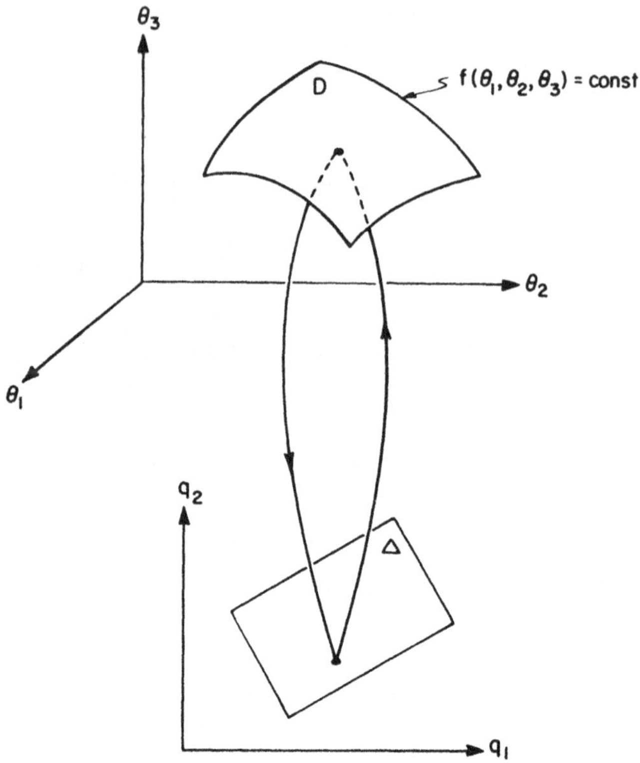

Fig. 11.5.1. One-to-one mapping from $\{\theta_1, \theta_2, \theta_3\}$ into $\{q_1, q_2\}$.

Suppose we have described the configuration of a system by the three coordinates θ_1, θ_2, and θ_3, but two generalized coordinates q_1 and q_2 are sufficient to define the configuration. Then, if the system is holonomic, there exists some surface $f(\theta_1, \theta_2, \theta_3) = \text{const}$, as shown in Fig. 11.5.1, which constitutes a constraint between the θ_i. If the transformation between the θ_i and the q_j is one-to-one on some domain D of the $\theta_1, \theta_2, \theta_3$ space [evidently, this domain belongs to the surface $f(\theta_1, \theta_2, \theta_3) = \text{const}$], then there exists a corresponding domain Δ of the q_1, q_2 space such that every point in D maps into a unique point in Δ, and vice versa. This is illustrated in Fig. 11.5.1.

We want to show now by means of an example that the theory described above fails when the mapping is not one-to-one, and that it is not always obvious on inspection that the theory is not applicable; in fact, Problem 4.14 is a case in point.

Example 11.5.1. Consider a chain of three bars of lengths l_1, l_2, l_3, respectively, constrained to move in a plane. One end of this chain is attached to the origin O of a q_1, q_2 system of Cartesian coordinates, as shown in Fig. 11.5.2; the other end carries a particle P.

Let the angles which the bars make with the q_2 axis be θ_1, θ_2 and θ_3, respectively. It is evident that the position of P is uniquely described either by its Cartesian coordinates q_1 and q_2, or by the angles θ_1, θ_2, θ_3. Since two coordinates are sufficient to define the configuration of P, there must exist a constraint equation

$$f(\theta_1, \theta_2, \theta_3) = \text{const}$$

between the θ_i. Find that constraint.

No such constraint exists. In fact, the mapping between the θ_i and the q_j is not one-to-one, and the theory of transformation to generalized coordinates is not applicable. To see this consider the mapping $(\theta_1, \theta_2, \theta_3) \to (q_1, q_2)$, given by

$$q_1 = \sum_{i=1}^{3} l_i \sin \theta_i, \qquad q_2 = \sum_{i=1}^{3} l_i \cos \theta_i. \tag{a}$$

Let us now examine the mapping of a point (q_{10}, q_{20}) into the θ_1, θ_2, θ_3 space. From (11.5.1) we have

$$F(q_{10}, \theta_1, \theta_2, \theta_3) = q_{10} - \sum_{i=1}^{3} l_i \sin \theta_i = 0, \tag{b}$$

and

$$G(q_{20}, \theta_1, \theta_2, \theta_3) = q_{20} - \sum_{i=1}^{3} l_i \cos \theta_i = 0. \tag{c}$$

Evidently, (b) defines a surface in θ_1, θ_2, θ_3 space with q_{10} as a parameter, and to each value of q_{10} there belongs one such surface. A similar statement holds for (c) with

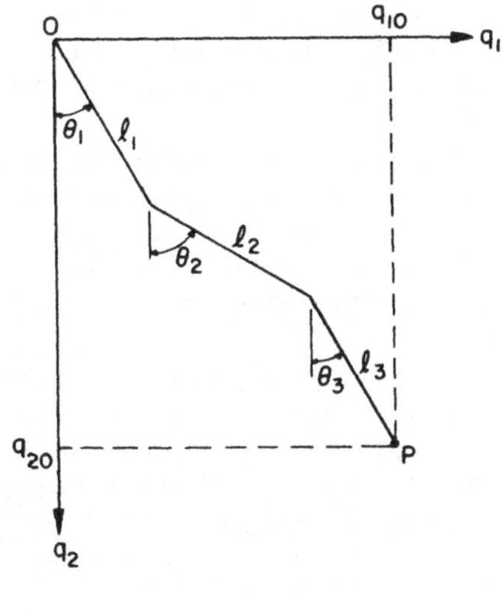

Fig. 11.5.2. Three linked bars of Example 11.5.1.

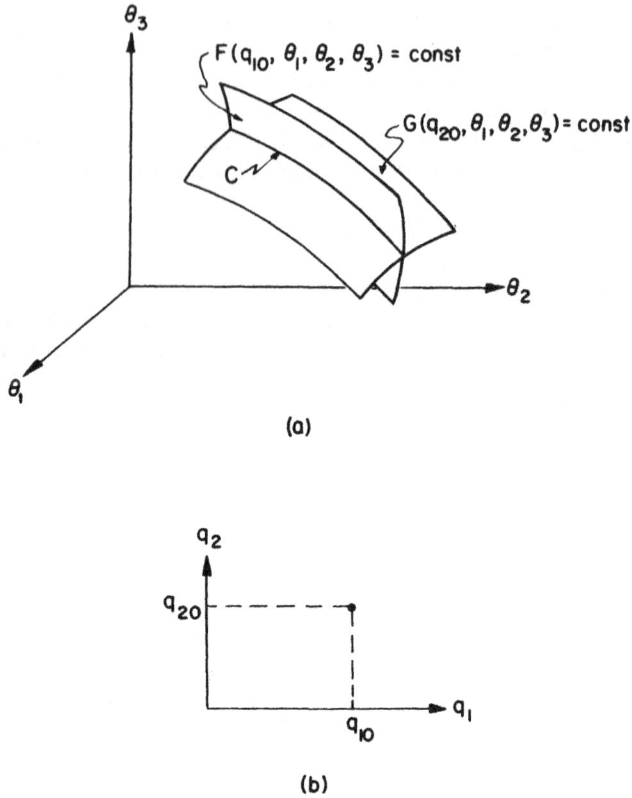

Fig. 11.5.3. The mapping from $\{\theta_1, \theta_2, \theta_3\}$ into $\{q_1, q_2\}$ of Example 11.5.1. It is not one-to-one.

q_{20} as a parameter. As both equations (b) and (c) must be satisfied these surfaces must intersect. Therefore, the point (q_{10}, q_{20}) maps in the $\theta_1, \theta_2, \theta_3$ space into a curve C formed by the intersection of the surfaces defined by (b) and (c) as shown in Fig. 11.5.3(a). Conversely, every point of the curve C maps into the point (q_{10}, q_{20}).

It is easy to show that the transformation matrix for the infinitesimal displacements $(d\theta_1, d\theta_2, d\theta_3) \to (dq_1, dq_2)$ has, in general, maximum rank unless $\theta_1 = \theta_2 = \theta_3$. Moreover, the intersection of the surfaces (b) and (c) is, in general, a curve of finite length, and it reduces to a point when the transformation matrix does not have maximum rank. The proof of these statements is left as an exercise for the reader.

It is also physically clear that the three angles θ_1, θ_2, and θ_3 define the location of P uniquely, but a given location of P does not define the three angles uniquely. For, consider the point P fixed; then, since the other end of the chain of links is fixed at O, the three links may be given many different orientations without violating the constraint $|\overline{OP}| \leq \sum_{i=1}^{3} |l_i|$ (unless the equal sign holds, in which case $\theta_1 = \theta_2 = \theta_3$, and the links form the straight line between O and P).

11.6. Problems

11.1. A particle moves on the surface of a three-dimensional sphere.

(a) Choose suitable generalized coordinates for the motion.

(b) What are the equations (11.2.7) for this case?

(c) Examine the Jacobian.

11.2. A particle moves on the surface of a right circular cylinder whose radius expands according to the law $r = f(t)$ while its axis remains stationary. Answer the same questions as in Problem 11.1.

11.3. A sphere of radius r rolls without sliding on the outside of a sphere of radius R. Answer the same questions as in Problem 11.1.

11.4. A rod of length $2l$ moves in the x, y plane on the inside of a smooth circle of radius r with $r > l$ so that its ends are always in contact with the circle. Answer the same questions as in Problem 11.1.

11.5. A rod of length $2l$ moves so that its ends are always in contact with an ellipse having major and minor axes $2a$ and $2b$, respectively. Answer the same questions as in Problem 11.1 when $l < b$.

11.6. Choose suitable generalized coordinates for the simple plane pendulum. What are the equations (11.2.7) for your choice?

11.7. Choose parabolic coordinates to express the motion of an unconstrained particle in the plane.

(a) What are the equations (11.2.7) for this case?

(b) Examine the Jacobian of the transformation.

11.8. A bar of fixed length l moves in the plane so that each of its endpoints is for all time in contact with one of two concentric circles of radii r_1 and r_2, respectively, and $l > r_2 - r_1 > 0$.

(a) Show that the bar has one degree of freedom.

(b) It is clear that to each position of one of the endpoints there correspond two possible positions of the other. This appears to contradict (a) above. Explain this contradiction.

(c) Choose suitable generalized coordinates and construct equations (11.2.7) for your choice.

11.9. A bar of fixed length l moves in the plane so that each of its endpoints is for all time in contact with one of two equal nonintersecting circles which are "side by side," i.e., the center of neither lies within the finite area surrounded by the other, and $D > l > d$, where D and d are the maximum and minimum distances between the circles.

(a) Answer the same questions as in Problem 11.8.

(b) Give the boundary of the configuration space.

11.10. A centrifugal governor has the configuration shown. If unconstrained, six coordinates would be required to define the configurations of the flyballs.

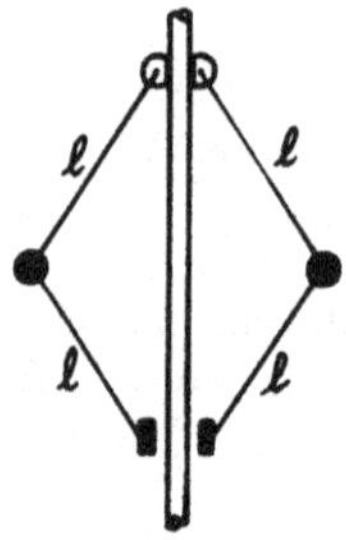

How many constraints must the Cartesian coordinates satisfy? What are they? Choose suitable generalized coordinates to describe the position of the flyballs. Construct equations (11.2.7) for this problem.

12

The Fundamental Equation in Generalized Coordinates

12.1. The Kinetic Energy

For a system of $N/3$ particles, the kinetic energy is given by

$$T = \frac{1}{2} \sum_{r=1}^{N} m_r \dot{u}_r^2, \tag{12.1.1}$$

and we have from (11.4.2)

$$\dot{u}_r = \sum_{s=1}^{n} \frac{\partial u_r}{\partial q_s} \dot{q}_s + \frac{\partial u_r}{\partial t} \qquad (r = 1, 2, \ldots, N). \tag{12.1.2}$$

It follows that the kinetic energy in terms of generalized coordinates is

$$T = \frac{1}{2} \sum_{r=1}^{N} m_r \left[\sum_{s=1}^{n} \frac{\partial u_r}{\partial q_s} \dot{q}_s + \frac{\partial u_r}{\partial t} \right]^2.$$

But, we may write

$$\begin{aligned} \left[\sum_{s=1}^{n} \frac{\partial u_r}{\partial q_s} \dot{q}_s + \frac{\partial u_r}{\partial t} \right]^2 &= \left[\sum_{\alpha=1}^{n} \frac{\partial u_r}{\partial q_\alpha} \dot{q}_\alpha + \frac{\partial u_r}{\partial t} \right] \left[\sum_{\beta=1}^{n} \frac{\partial u_r}{\partial q_\beta} \dot{q}_\beta + \frac{\partial u_r}{\partial t} \right] \\ &= \sum_{\alpha=1}^{n} \sum_{\beta=1}^{n} \left(\frac{\partial u_r}{\partial q_\alpha} \frac{\partial u_r}{\partial q_\beta} \right) \dot{q}_\alpha \dot{q}_\beta \\ &\quad + 2 \sum_{\alpha=1}^{n} \frac{\partial u_r}{\partial q_\alpha} \frac{\partial u_r}{\partial t} \dot{q}_\alpha + \left(\frac{\partial u_r}{\partial t} \right)^2. \end{aligned}$$

Therefore, the kinetic energy is

$$T = \frac{1}{2} \sum_{\alpha=1}^{n} \sum_{\beta=1}^{n} a_{\alpha\beta}\dot{q}_\alpha\dot{q}_\beta + \sum_{\alpha=1}^{n} b_\alpha\dot{q}_\alpha + c, \tag{12.1.3}$$

where

$$\begin{aligned} a_{\alpha\beta} &= \sum_{r=1}^{N} m_r \frac{\partial u_r}{\partial q_\alpha} \frac{\partial u_r}{\partial q_\beta}, \\ b_\alpha &= \sum_{r=1}^{N} m_r \frac{\partial u_r}{\partial q_\alpha} \frac{\partial u_r}{\partial t}, \\ c &= \frac{1}{2} \sum_{r=1}^{N} m_r \left(\frac{\partial u_r}{\partial t}\right)^2. \end{aligned} \tag{12.1.4}$$

One sees that

$$a_{\beta\alpha} = \sum_{r=1}^{N} m_r \frac{\partial u_r}{\partial q_\beta} \frac{\partial u_r}{\partial q_\alpha},$$

and comparing this with the first equation of (12.1.4) one finds the important relation

$$a_{\alpha\beta} = a_{\beta\alpha}. \tag{12.1.5}$$

If the u_r did not depend explicitly on time, i.e., $\partial u_r/\partial t$ were absent from (12.1.2), one would have

$$T = \frac{1}{2} \sum_{\alpha=1}^{n} \sum_{\beta=1}^{n} a_{\alpha\beta}\dot{q}_\alpha\dot{q}_\beta, \tag{12.1.6}$$

and this must be positive-definite because it is equal to (12.1.1). It follows that the first term in (12.1.3) is a positive quadratic form in the velocities. Moreover, the third term in that equation is necessarily nonnegative because of the definition of c in (12.1.4), and the sum of all three terms in (12.1.3) is also necessarily nonnegative because it equals (12.1.1). Therefore, while $\sum_\alpha b_\alpha\dot{q}_\alpha$ may be negative for some values of q_s, $\dot{q}_s$, and t, for $s = 1, 2, \ldots, n$, it can never render T itself negative.

Evidently, the general form of the kinetic energy in generalized coordinates is much more complex than that in Cartesian coordinates. This is the price we must pay for using non-Cartesian generalized coordinates. We pay it for the benefit that comes from reducing the number of configuration coordinates to the least possible and the most suitable to the problem, and for the great generality that is achieved by a formulation that holds for the entire class of generalized coordinates without having to specify in advance which particular set is to be used.

Equation (12.1.6) is the kinetic energy of catastatic systems; it resembles (12.1.1), i.e., the kinetic energy in terms of Cartesian coordinates. However, appearances are deceiving here. Not only does (12.1.6) contain, in general, all mixed products of the velocities, rather than their squares only as in (12.1.1) but, more important, the coefficients $\alpha_{\alpha\beta}$ are, in general, functions of the q_s and t for $s = 1, 2, \ldots, n$, while the m_r of (12.1.1) are constants.

Some simple examples will illustrate these points.

Example 12.1.1. The kinetic energy of an unconstrained particle of mass m in terms of the Cartesian coordinates x, y, z is given by $T = \frac{1}{2}m(\dot{x}^2 + \dot{y}^2 + \dot{z}^2)$. Let us change to the spherical coordinates r, θ, φ by using

$$x = r \sin\theta \sin\varphi,$$
$$y = r \sin\theta \cos\varphi,$$
$$z = r \cos\theta.$$

Then, differentiating these and adding their squares, we find

$$T = \tfrac{1}{2}m(\dot{r}^2 + r^2\dot{\theta}^2 + r^2 \sin^2\theta\dot{\varphi}^2).$$

In this example, only squares of the velocity components occur, but the coefficient of the second is a function of r, and that of the third is a function of r and θ.

Example 12.1.2. As a second example we choose an acatastatic system. Consider a plane, simple pendulum whose point of suspension is moved in the x direction with the prescribed motion $f(t)$, as shown in Fig. 12.1.1 (see Example 9.4.3).

The generalized coordinate chosen is the angle which the pendulum makes with the y axis. We have

$$x = f(t) + l \sin\theta, \qquad y = l \cos\theta;$$

the kinetic energy in Cartesian coordinates is $T = \frac{1}{2}m(\dot{x}^2 + \dot{y}^2)$, and the holonomic constraint is $[x - f(t)]^2 + y^2 = l^2$. Here

$$T = \tfrac{1}{2}m\{l^2\dot{\theta}^2 + 2l\dot{\theta}\cos\theta\dot{f}(t) + [\dot{f}(t)]^2\}.$$

In this example, the quantities (12.1.4) are

$$a_{11} = \tfrac{1}{2}ml^2, \qquad b_1 = ml\cos\theta\dot{f}(t), \qquad c = \tfrac{1}{2}m[\dot{f}(t)]^2.$$

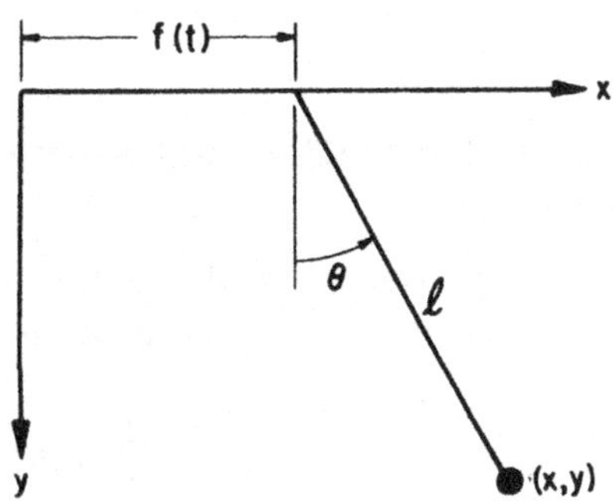

Fig. 12.1.1. Pendulum with moving suspension point of Example 12.1.2.

12.2. Two Equalities

Starting with the equations

$$u_r = u_r(q_1, q_2, \ldots, q_n, t) \qquad (r = 1, 2, \ldots, N), \tag{12.2.1}$$

and

$$\dot{u}_r = \sum_{s=1}^{n} \frac{\partial u_r}{\partial q_s} \dot{q}_s + \frac{\partial u_r}{\partial t} \tag{12.2.2}$$

we wish to demonstrate, for later use, the two equalities

$$\frac{\partial \dot{u}_r}{\partial \dot{q}_s} = \frac{\partial u_r}{\partial q_s} \tag{12.2.3}$$

and

$$\frac{\partial \dot{u}_r}{\partial q_s} = \frac{d}{dt}\left(\frac{\partial u_r}{\partial q_s}\right). \tag{12.2.4}$$

The first of these follows immediately from differentiating (12.2.2) with respect to $\dot{q}_s$. To demonstrate the second, we substitute (12.2.2) into the left-hand side of (12.2.4), or

$$\frac{\partial \dot{u}_r}{\partial q_s} = \frac{\partial}{\partial q_s}\left(\sum_{\alpha=1}^{n} \frac{\partial u_r}{\partial q_\alpha} \dot{q}_\alpha + \frac{\partial u_r}{\partial t}\right).$$

We find

$$\begin{aligned} \frac{\partial \dot{u}_r}{\partial q_s} &= \sum_{\alpha=1}^{n} \frac{\partial^2 u_r}{\partial q_\alpha \, \partial q_s} \dot{q}_\alpha + \frac{\partial^2 u_r}{\partial t \, \partial q_s} \\ &= \sum_{\alpha=1}^{n} \frac{\partial^2 u_r}{\partial q_s \, \partial q_\alpha} \dot{q}_\alpha + \frac{\partial^2 u_r}{\partial q_s \, \partial t}, \end{aligned}$$

and this is precisely the formula for $d(\partial u_r/\partial q_s)/dt$, which was to be shown.

12.3. The Fundamental Equation

The fundamental equation in Cartesian coordinates has been derived in Section 9.3. It is (with a change of index notation)

$$\sum_{r=1}^{N} (m_r \ddot{u}_r - F_r) \, \delta u_r = 0. \tag{12.3.1}$$

Substituting (11.2.11) for δu_r in this equation we have

$$\sum_{s=1}^{n} \left(\sum_{r=1}^{N} (m_r \ddot{u}_r - F_r) \frac{\partial u_r}{\partial q_s} \right) \delta q_s = 0. \tag{12.3.2}$$

Consider the time derivative

$$\frac{d}{dt}\left(\dot{u}_r \frac{\partial u_r}{\partial q_s} \right) = \ddot{u}_r \frac{\partial u_r}{\partial q_s} + \dot{u}_r \frac{d}{dt}\left(\frac{\partial u_r}{\partial q_s} \right).$$

If we substitute (12.2.3) and (12.2.4) in this last equation and solve it for $\ddot{u}_r \, \partial u_r / \partial q_s$ we find

$$\ddot{u}_r \frac{\partial u_r}{\partial q_s} = \frac{d}{dt}\left(\dot{u}_r \frac{\partial \dot{u}_r}{\partial \dot{q}_s} \right) - \dot{u}_r \frac{\partial \dot{u}_r}{\partial q_s}. \tag{12.3.3}$$

The substitution of (12.3.3) in (12.3.2) results in

$$\sum_{s=1}^{n} \left\{ \sum_{r=1}^{N} m_r \left[\frac{d}{dt}\left(\dot{u}_r \frac{\partial \dot{u}_r}{\partial \dot{q}_s} \right) - \dot{u}_r \frac{\partial \dot{u}_r}{\partial q_s} \right] - F_r \frac{\partial u_r}{\partial q_s} \right\} \delta q_s = 0. \tag{12.3.4}$$

But, from (12.1.1) we see that

$$\begin{aligned} \sum_{r=1}^{N} m_r \dot{u}_r \frac{\partial \dot{u}_r}{\partial \dot{q}_s} &= \frac{\partial T}{\partial \dot{q}_s}, \\ \sum_{r=1}^{N} m_r \dot{u}_r \frac{\partial \dot{u}_r}{\partial q_s} &= \frac{\partial T}{\partial q_s}. \end{aligned} \tag{12.3.5}$$

Therefore, (12.3.4) is

$$\sum_{s=1}^{n} \left\{ \left[\frac{d}{dt}\left(\frac{\partial T}{\partial \dot{q}_s} \right) - \frac{\partial T}{\partial q_s} \right] - \sum_{r=1}^{N} F_r \frac{\partial u_r}{\partial q_s} \right\} \delta q_s = 0. \tag{12.3.6}$$

Finally, we define the *generalized force*

$$Q = (Q_1, Q_2, \ldots, Q_n), \tag{12.3.7}$$

where the sth component is

$$Q_s = \sum_{r=1}^{N} F_r \frac{\partial u_r}{\partial q_s}. \tag{12.3.8}$$

This gives the *fundamental equation* the celebrated form, due to Lagrange,

$$\sum_{s=1}^{n} \left[\frac{d}{dt}\left(\frac{\partial T}{\partial \dot{q}_s} \right) - \frac{\partial T}{\partial q_s} - Q_s \right] \delta q_s = 0. \tag{12.3.9}$$

In it, the quantity $Q_s\,\delta q_s$ is the work done by the generalized force component Q_s in the δq_s component of a generalized virtual displacement δq.

The generalized form (12.3.9) has more far-reaching importance than the earlier forms (9.3.11) and (9.3.12), for the following reasons:

(a) We have already noted that, in a formal sense, generalized coordinates which are angles (for instance) are indistinguishable from others having the dimension of length. The only requirements imposed by (12.3.9) with respect to dimensions are that:

(i) for a given $s = k$, the terms $d(\partial T/\partial \dot{q}_k)/dt$, and Q_k must have the same dimension, and

(ii) the terms $Q_s\,\delta q_s$ must have the same dimension for all $s = 1, 2, \ldots, n$.

For instance, if Q_1 is a force and δq_1 has the dimension of length, their product has the same dimension as $Q_2\,\delta q_2$ if Q_2 is a moment and δq_2 is nondimensional. This property was already mentioned in the second part of Example 9.4.2, where it was pointed out that a general discussion would be given later on.

(b) The form (12.3.9) of the fundamental equation greatly enhances its utility over the earlier formulations, in which the coordinates were Cartesian (either in 3-space or in N-space). This becomes particularly evident in the case of rigid body dynamics. We saw earlier what benefits are gained from describing the configuration of an unconstrained rigid body in terms of displacement components along body-fixed principal axes, and of three rotational components about them. These benefits are now accessible to the Lagrangean formulation.

(c) Even when nonrigid systems of constrained particles are considered, the advantages of (12.3.9) over (9.3.12) are rewarding. For instance, the motion of a particle on a fixed circle of radius R is much more easily described in terms of an angle than in terms of Cartesian coordinates satisfying the constraint $x^2 + y^2 = R^2$.

12.4. Generalized Potential Forces

We saw in Section 9.9 that the component F_s^p of a potential force is derivable from a potential energy V^p as

$$F_s^p = -\frac{\partial V^p}{\partial u_s}. \tag{12.4.1}$$

We show now that this property is preserved if one proceeds to generalized coordinates, i.e., the generalized force component Q_r^p of a potential force is related to the potential energy V^p, given in generalized coordinates, by

$$Q_r^p = -\frac{\partial V^p}{\partial q_r}. \tag{12.4.2}$$

Inasmuch as the u_s are functions of the q_s in accordance with (11.2.8), one has

$$-\frac{\partial V^p}{\partial q_r} = \sum_{s=1}^{N} -\frac{\partial V^p}{\partial u_s}\frac{\partial u_s}{\partial q_r} = \sum_{s=1}^{N} F_s^p \frac{\partial u_s}{\partial q_r} = Q_r^p, \tag{12.4.3}$$

which was to be shown. In (12.4.3) we have utilized (12.4.1) to obtain the second equality, and the definition (12.3.8) of the generalized force to obtain the third.

12.5. Velocity-Dependent Potentials

The Lagrangean form (12.3.9) of the fundamental equation suggests a generalization of the potential function to the form derived below.

Let us suppose that there exist some potential forces Q_s^p whose resultant is

$$\sum_p Q_s^p = -\sum_p \frac{\partial V^p}{\partial q_s} = -\frac{\partial V}{\partial q_s} \qquad (s = 1, 2, \ldots, n). \tag{12.5.1}$$

Then, if we agree to denote by Q_s all nonpotential forces, (12.3.9) may be written as

$$\sum_{s=1}^{n} \left[\frac{d}{dt}\left(\frac{\partial T}{\partial \dot{q}_s}\right) - \frac{\partial L}{\partial q_s} - Q_s\right] \delta q_s = 0,$$

where L is the Lagrangean function defined in (10.5.7).

Heretofore, we have assumed that the potential energy V is a function of the coordinates and possibly of time (see Example 10.5.3) but *not* of the velocities. In consequence of that assumption

$$\frac{\partial L}{\partial \dot{q}_s} = \frac{\partial T}{\partial \dot{q}_s} \qquad (s = 1, 2, \ldots, n),$$

so that the fundamental equation may be written as

$$\sum_{s=1}^{n} \left[\frac{d}{dt}\left(\frac{\partial L}{\partial \dot{q}_s}\right) - \frac{\partial L}{\partial q_s} - Q_s\right] \delta q_s = 0 \tag{12.5.2}$$

and the virtual work done by the potential forces is

$$-\sum_{s=1}^{n} (\partial V/\partial q_s)\, \delta q_s.$$

The generalization to be introduced consists in admitting velocity-dependent potentials which are such that the virtual work done the force derivable from them is

$$\delta W = \sum_{s=1}^{n} \left[\frac{d}{dt}\left(\frac{\partial V}{\partial \dot{q}_s}\right) - \frac{\partial V}{\partial q_s}\right] \delta q_s. \tag{12.5.3}$$

The substitution of (12.5.3) in (12.3.9) gives (12.5.2) again. Hence, forces derivable from

$$\sum_p Q_s^p = \sum_p \left[\frac{d}{dt}\left(\frac{\partial V^p}{\partial \dot{q}_s}\right) - \frac{\partial V^p}{\partial q_s}\right] = \frac{d}{dt}\left(\frac{\partial V}{\partial \dot{q}_s}\right) - \frac{\partial V}{\partial q_s} \tag{12.5.4}$$

preserve the form (12.5.2) of the fundamental equation. Consequently, potential energies $V(q_1, q_2, \ldots, q_n, \dot{q}_1, \dot{q}_2, \ldots, \dot{q}_N; t)$ which give rise to forces calculated by means of (12.5.4) are admitted.

When the first term on the right-hand side of (12.5.4) is differentiated out, there results

$$\frac{d}{dt}\left(\frac{\partial V}{\partial \dot{q}_s}\right) = \sum_{\alpha=1}^{n} \frac{\partial^2 V}{\partial \dot{q}_s\, \partial q_\alpha}\, \dot{q}_\alpha + \sum_{\beta=1}^{n} \frac{\partial^2 V}{\partial \dot{q}_s\, \partial \dot{q}_\beta}\, \ddot{q}_\beta. \tag{12.5.5}$$

But it was shown in Example 2.5.1 that forces acting on particles may not be functions of any particle accelerations. This requires that

$$\frac{\partial^2 V}{\partial \dot{q}_\alpha\, \partial \dot{q}_\beta} = 0$$

for all α and β, or *velocity-dependent potentials may at most be linear functions of the velocities*, i.e., velocity-dependent potentials of systems accessible to Newtonian mechanics must be of the form

$$V = \sum_{s=1}^{n} a_s \dot{q}_s + V_0, \tag{12.5.6}$$

where the a_s and V_0 may be functions of the q_i $(i = 1, 2, \ldots, n)$ and possibly of time. This observation is due to Pars (p. 82).

An application of a potential depending linearly on the velocity occurs in the motion of an electrically charged particle in a magnetic field.

12.6. Problems

12.1. A system of n particles is so constrained that the kth particle having position (x_k, y_k, z_k) moves on the surface of a fixed sphere of radius r_k, centered at the point (ξ_k, η_k, ζ_k). What is the total kinetic energy of the system in generalized coordinates?

12.2. How does the answer to Problem 12.1 change if each sphere expands according to the law $r_k = r_k(t)$, where $r_k(t)$ is a given, smooth function?

12.3. How does the answer to Problem 12.1 change if the radius r_k is fixed, but each sphere moves according to the law $\xi_k = f_{1k}(t)$, $\eta_k = f_{2k}(t)$, $\zeta_k = f_{3k}(t)$, where the $f_{ik}(t)$ are given, smooth functions?

12.4. A particle of mass m moves on a smooth surface of revolution about the z axis.

(a) What is the kinetic energy in cylindrical coordinates?

(b) Specialize this result for the cases in which the surface is a cone and a sphere.

12.5. A body rotates about a fixed axis. Show that if the mass moment of inertia is being reduced while the angular momentum remains unchanged, the kinetic energy increases.

12.6. A heavy particle of mass m is constrained to move on a circle of radius r which lies in the vertical plane, as shown. It is attached to a linear spring of rate k, which is anchored at a point on the x axis a distance a from the origin of the x, y system, and $a > r$. The free length of the spring is $a - r$. Using the angle θ as generalized coordinate, utilize (12.3.8) to calculate the generalized forces raising from the gravitational and the spring force. Does the answer change if $a < r$ and, if so, how?

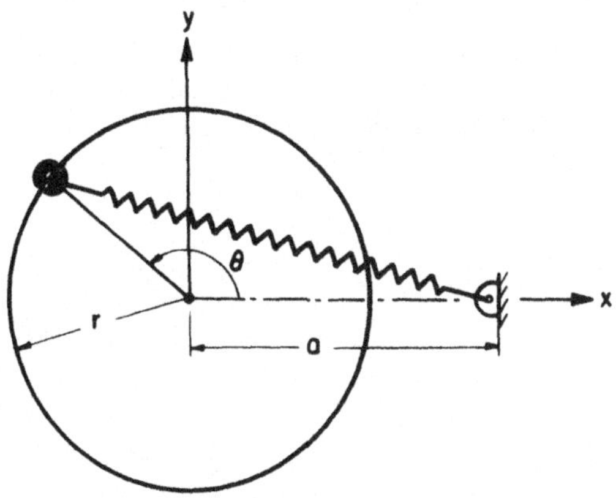

12.7. A heavy particle of mass m is attached to one extremity of a linear, massless spring of rate k, and of free length l. The other extremity of the spring is free to rotate about a fixed point. This system is, therefore, an elastic, spherical pendulum. Using spherical coordinates, calculate the generalized forces acting on the particle.

12.8. The Cartesian components of a force are

$$X = 2ax(y + z),$$
$$Y = 2ay(x + z),$$
$$Z = 2az(x + y).$$

Calculate the generalized forces for cylindrical and spherical coordinates.

12.9. A heavy monkey M climbs up a massless, inextensible rope which passes over two smooth, fixed pegs as shown. The other end of the rope carries a weight W. Give the kinetic energy in terms of generalized coordinates. (This problem is known as "The monkey on the counterpoise.")

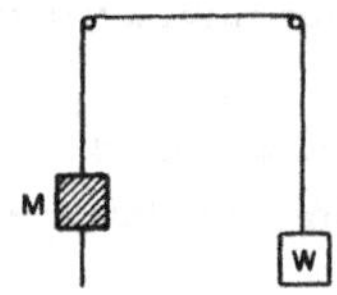

12.10. A particle is constrained to move in a plane. What is its kinetic energy in elliptic coordinates ξ, η, where $x = l \cosh \xi \cos \eta$, $y = l \sinh \xi \sin \eta$. Is this transformation one-to-one?

12.11. Answer the same question as in Problem 12.10 for biaxial coordinates ξ, η, where

$$x = \frac{l \sinh \xi}{\cosh \xi + \cos \eta}, \qquad y = \frac{l \sin \eta}{\cosh \xi + \cos \eta}.$$

12.12. Answer the same question as in Problem 12.10 for parabolic coordinates ξ, η, where

$$x = l(\xi^2 - \eta^2), \qquad y = 2l\xi\eta.$$

12.13. How do the answers in Problem 12.10 to 12.12 change when the plane translates according to the law $f(t)$ and remains always parallel to itself, where $f(t)$ is a once-differentiable function of time.

13

Lagrange's Equations

13.1. The Dynamical Problem

The fundamental equation in generalized coordinates has been found in (12.3.9) as

$$\sum_{s=1}^{n} \left(\frac{d}{dt} \frac{\partial T}{\partial \dot{q}_s} - \frac{\partial T}{\partial q_s} - Q_s \right) \delta q_s = 0, \tag{13.1.1}$$

where the kinetic energy T and the generalized forces Q_s are, in general, functions of all the q_s and $\dot{q}_s$ ($s = 1, 2, \ldots, n$) and of t.

The actual velocities $\dot{q}_s$ belong to the class of possible ones, and the latter satisfy the *linearly independent* system of constraint equations

$$\sum_{s=1}^{n} B_{rs}\dot{q}_s + B_r = 0 \qquad (r = 1, 2, \ldots, l). \tag{13.1.2}$$

Therefore, among all the q_s, $\dot{q}_s = dq_s/dt$ which satisfy (13.1.2) we seek those satisfying (13.1.1) as well. The δq_s in (13.1.1) are the virtual, generalized displacements which satisfy the set of equations

$$\sum_{s=1}^{n} B_{rs}\,\delta q_s = 0 \qquad (r = 1, 2, \ldots, l). \tag{13.1.3}$$

From the linear independence of (13.1.2) it follows that the equations (13.1.3) are linearly independent as well.

13.2. The Multiplier Rule

Let us temporarily denote by R_s the quantity

$$R_s = \frac{d}{dt}\frac{\partial T}{\partial \dot{q}_s} - \frac{\partial T}{\partial q_s} - Q_s \qquad (s = 1, 2, \ldots, n). \qquad (13.2.1)$$

Then, if we regard R as the n-dimensional vector

$$R = (R_1, R_2, \ldots, R_n) \qquad (13.2.2)$$

and δq as the n-dimensional vector

$$\delta q = (\delta q_1, \delta q_2, \ldots, \delta q_n), \qquad (13.2.3)$$

we may write the fundamental equation (13.1.1) as

$$R \cdot \delta q = 0. \qquad (13.2.4)$$

This equation states that the inner product of R and δq vanishes, i.e., R and δq are orthogonal vectors.

Let us now denote by $\tilde{B}_r$ the vectors

$$\tilde{B}_r = (B_{r1}, B_{r2}, \ldots, B_{rn}) \qquad (r = 1, 2, \ldots, l). \qquad (13.2.5)$$

Then, it follows from the linear independence of the constraint equations that the $\tilde{B}_r$ are linearly independent vectors. The mathematical meaning of "linear independence" is that one cannot find multipliers λ_r $(r = 1, 2, \ldots, l)$ not all zero such that the equation (see also Sections 2.3 and 4.2)

$$\sum_{r=1}^{l} \lambda_r \tilde{B}_r = 0 \qquad (13.2.6)$$

can be satisfied.

Now, the equations (13.1.3) may be written in the form

$$\tilde{B}_r \cdot \delta q = 0 \qquad (r = 1, 2, \ldots, l), \qquad (13.2.7)$$

and these state that δq must also be orthogonal to every vector $\tilde{B}_r$ $(r = 1, 2, \ldots, l)$.

The multiplier rule was given in Section 9.4. In terms of generalized coordinates it states that, if the q_s and $\dot{q}_s$ satisfy the fundamental equation as well as the constraints, then one may write

$$\left(R + \sum_{r=1}^{l} \lambda_r \tilde{B}_r\right) \cdot \delta q = 0 \qquad (13.2.8)$$

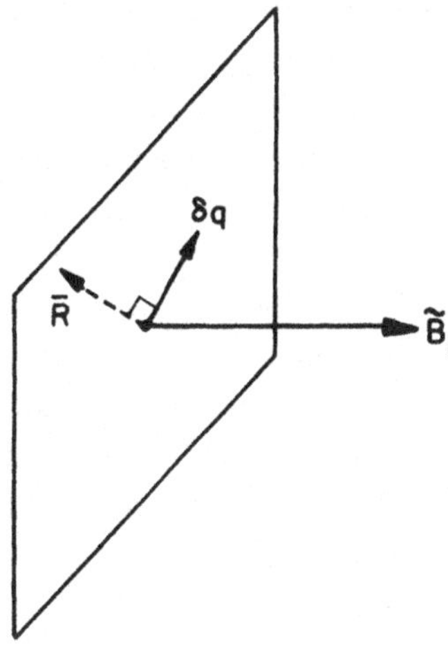

Fig. 13.2.1. Schematic diagram illustrating the Lagrange multiplier rule.

and *in that equation the δq are completely arbitrary*, i.e., they are no longer subject to the constraints (13.1.3).

To prove this rule we shall use geometrical arguments. We begin by supposing temporarily that $n = 3$ and $l = 1$. In other terms, R and δq are 3-vectors, and there is only a single constraint, or there is only a single $\tilde{B}_1 \neq 0$. In that case, one cannot find a multiplier $\lambda_1 \neq 0$ such that $\lambda_1 \tilde{B}_1 = 0$. Now, (13.2.7) states that δq must lie in a plane normal to $\tilde{B}_1$, as shown in Fig. 13.2.1. Expressed differently, among all possible 3-vectors δq, the constraint implies that only those are admitted which lie in a plane normal to $\tilde{B}_1$. However, beyond this requirement, the δq are arbitrary; in particular, the direction of δq is *not* prescribed. Now, (13.2.4) requires that R be *always* orthogonal to δq. We claim that, in consequence, R must have the same direction as $\tilde{B}_1$. For, suppose that R had a component $\bar{R}$ in the plane normal to $\tilde{B}_1$, as shown by the dotted vector in Fig. 13.2.1. Then, $\bar{R}$ would have to be normal to δq, as shown, in order to satisfy (13.2.4). But the direction of δq in the plane normal to $\tilde{B}_1$ is arbitrary. Thus, we could then rotate δq so as to destroy its orthogonality to $\bar{R}$, and this would violate (13.2.4). Hence, R must have the same direction as $\tilde{B}_1$. In that case, R may be expressed as a multiple of $\tilde{B}_1$, or one may write

$$R = -\lambda_1 \tilde{B}_1. \tag{13.2.9}$$

If we now form the scalar product of (13.2.4) with any *completely arbitrary* 3-vector δq we find

$$(R + \lambda_1 \tilde{B}_1) \cdot \delta q = 0 \tag{13.2.10}$$

and in this equation the components of δq are no longer subject to the constraints (13.1.3).

This result is readily extended to any n and to any $l < n$. From the linear independence of the vectors $\tilde{B}_r$ $(r = 1, 2, \ldots, l)$ it follows that one can find an l-dimensional subspace of the n-space for which the vectors $\tilde{B}_1, \tilde{B}_2, \ldots, \tilde{B}_l$ are base vectors; we denote this l-dimensional subspace by $\mathscr{E}_l$. Then, (13.2.7) states that δq must lie in the space $\mathscr{E}_l{}^\perp$ which is orthogonal to $\mathscr{E}_l$; however, except for that restriction, the direction of δq is not prescribed. We now show that R cannot have a component $\bar{R}$ belonging to $\mathscr{E}_l{}^\perp$ for, if it did, that component would have to be orthogonal to δq. But then, one could rotate δq in $\mathscr{E}_l{}^\perp$ so as to destroy this orthogonality, thereby violating (13.2.4). It follows that R belongs to $\mathscr{E}_l$ and must be expressible as a linear combination of the base vectors $\tilde{B}_r$, or

$$R = -\sum_{r=1}^{l} \lambda_r \tilde{B}_r. \tag{13.2.11}$$

Then, forming the scalar product of (13.2.10) with any completely arbitrary n-vector δq, one finds (13.2.8).

13.3. Derivation from the Fundamental Equation

If one writes (13.2.8) in component form and restores to R_s its meaning as defined in (13.2.1) one has

$$\sum_{s=1}^{n} \left[\frac{d}{dt} \frac{\partial T}{\partial \dot{q}_s} - \frac{\partial T}{\partial q_s} - Q_s + \sum_{r=1}^{l} \lambda_r B_{rs} \right] \delta q_s = 0 \tag{13.3.1}$$

and in (13.3.1), the δq_s are completely free. In a purely formal manner, one may therefore construct (13.3.1) from the fundamental equation and the constraint equations (13.1.3) by multiplying each $B_{rs}\, \delta q_s$ by a multiplier λ_r, and by adding their sum to the fundamental equation. When this is done, the δq_s are completely free.

From (13.3.1), one may deduce directly the *Fundamental Theorem of Lagrange's Mechanics*:

For (13.3.1) *to be satisfied, it is necessary and sufficient that*

$$\frac{d}{dt} \frac{\partial T}{\partial \dot{q}_s} - \frac{\partial T}{\partial q_s} - Q_s + \sum_{r=1}^{l} \lambda_r B_{rs} = 0 \tag{13.3.2}$$

for every $s = 1, 2, \ldots, n$.

The proof that the condition is necessary is very simple. Suppose one, several, or all of the quantities in the brackets of (13.3.1) are different from zero. Then, since the δq_s are arbitrary, we may always choose for each such nonzero term a δq_s having the same sign as the nonzero term. Then the sum (13.3.1) is necessarily positive, which contradicts the requirement that it be zero. This proves the necessity; the sufficiency is obvious.

The equations

$$\frac{d}{dt}\frac{\partial T}{\partial \dot{q}_s} - \frac{\partial T}{\partial q_s} - Q_s + \sum_{r=1}^{l} \lambda_r B_{rs} = 0 \qquad (s = 1, 2, \ldots, n) \tag{13.3.3}$$

are called *Lagrange's equations of motion*. They are n equations in $n + l$ unknowns, the unknowns being the q_s $(s = 1, 2, \ldots, n)$ and the λ_r $(r = 1, 2, \ldots, l)$. The n equations (13.3.3) together with the l equations of constraint (13.1.2) furnish precisely the right number of equations to solve for these unknowns. The quantities λ_r $(r = 1, 2, \ldots, l)$, are called *Lagrangean multipliers*.

[Lagrange's equations (13.3.3) may also be applied when the number of coordinates used exceeds the minimum number, i.e., when the coordinates are not a set of generalized ones or, which is the same thing, when the problem is subject to holonomic as well as nonholonomic constraints.

Suppose the number of coordinates v_σ $(\sigma = 1, 2, \ldots, \bar{n})$ exceeds the minimum number n by l', and that there are l constraints. Then, there are l' holonomic constraints

$$f_\alpha(v_1, v_2, \ldots, v_{\bar{n}}, t) = 0 \qquad (\alpha = 1, 2, \ldots, l')$$

whose Pfaffian form is

$$\sum_{n=1}^{\bar{n}} B_{\alpha\sigma}\, dv_\sigma + B_\alpha\, dt = 0 \qquad (\alpha = 1, 2, \ldots, l'),$$

and there are $l - l'$ nonholonomic constraints

$$\sum_{n=1}^{\bar{n}} B_{\beta\sigma}\, dv_\sigma + B_\beta\, dt = 0 \qquad (\beta = l' + 1, l' + 2, \ldots, l).$$

The method of the Lagrange multipliers used to adjoin the constraints to the problem does not depend on the integrability of the constraint equations. Therefore, Lagrange's equations are now

$$\frac{d}{dt}\frac{\partial T}{\partial \dot{v}_\sigma} - \frac{\partial T}{\partial v_\sigma} - Q_\sigma + \sum_{\alpha=1}^{l'} \lambda_\alpha B_{\alpha\sigma} + \sum_{\beta=l'+1}^{l} \lambda_\beta B_{\beta\sigma} = 0 \quad (\sigma = 1, 2, \ldots, \bar{n}).$$

This can, of course, be written as

$$\frac{d}{dt}\frac{\partial T}{\partial \dot{v}_\sigma} - \frac{\partial T}{\partial v_\sigma} - Q_\sigma + \sum_{\gamma=1}^{l} \lambda_\gamma B_{\gamma\sigma} = 0 \qquad (\sigma = 1, 2, \ldots, \bar{n}),$$

which is identical in form to (13.3.3). However, it is rarely advisable to use Lagrange's equations with more coordinates than necessary because it results in an excessive number of equations of motion.]

13.4. Derivation from the Central Principle

The central principle was derived in (10.4.4). It is the relation

$$\sum_{r=1}^{N/3} m_r \ddot{x}^r \cdot \delta x^r = \frac{d}{dt}\left(\sum_{r=1}^{N/3} m_r \dot{x}^r \cdot \delta x^r\right) - \delta T. \tag{13.4.1}$$

Written in terms of the u_s, it becomes

$$\sum_{s=1}^{N} m_s \ddot{u}_s\, \delta u_s = \frac{d}{dt}\left(\sum_{s=1}^{N} m_s \dot{u}_s\, \delta u_s\right) - \delta T. \tag{13.4.2}$$

We shall derive Lagrange's equations here with the aid of the central principle. For simplicity, we assume that the system is holonomic; if it is nonholonomic, one merely adds the appropriate term $\sum_{r=1}^{l} \lambda_r B_{rs}$ to the equations of motion.

We fix our attention on the first term of the right-hand side of (13.4.2). Inasmuch as

$$\delta u_s = \sum_{r=1}^{n} \frac{\partial u_s}{\partial q_r}\, \delta q_r,$$

one has

$$\begin{aligned}\sum_{s=1}^{N} m_s \dot{u}_s\, \delta u_s &= \sum_{s=1}^{N} m_s \dot{u}_s \sum_{r=1}^{n} \frac{\partial u_s}{\partial q_r}\, \delta q_r \\ &= \sum_{r=1}^{n}\sum_{s=1}^{N}\left(m_s \dot{u}_s \frac{\partial u_s}{\partial q_r}\right) \delta q_r \\ &= \sum_{r=1}^{n}\sum_{s=1}^{N}\left(m_s \dot{u}_s \frac{\partial \dot{u}_s}{\partial \dot{q}_r}\right) \delta q_r,\end{aligned}$$

where the last equality results from (12.2.3). Then, by (12.3.5), we obtain

$$\sum_{s=1}^{N} m_s \dot{u}_s\, \delta u_s = \sum_{r=1}^{n} \frac{\partial T}{\partial \dot{q}_r}\, \delta q_r. \tag{13.4.3}$$

Also, from the definition of the δ operation,

$$\delta T = \sum_{r=1}^{n} \frac{\partial T}{\partial q_r} \delta q_r + \sum_{r=1}^{n} \frac{\partial T}{\partial \dot{q}_r} \delta \dot{q}_r. \tag{13.4.4}$$

When (13.4.3) and (13.4.4) are substituted in (13.4.2) one finds

$$\sum_{s=1}^{N} m_s \ddot{u}_s \, \delta u_s = \frac{d}{dt}\left(\sum_{r=1}^{n} \frac{\partial T}{\partial \dot{q}_r} \delta q_r\right) - \sum_{r=1}^{n} \frac{\partial T}{\partial q_r} \delta q_r - \sum_{r=1}^{n} \frac{\partial T}{\partial \dot{q}_r} \delta \dot{q}_r.$$

In this equation, the first term on the right is differentiated out, which results in

$$\sum_{s=1}^{N} m_s \ddot{u}_s \, \delta u_s = \sum_{r=1}^{n} \frac{d}{dt} \frac{\partial T}{\partial \dot{q}_r} \delta q_r - \sum_{r=1}^{n} \frac{\partial T}{\partial q_r} \delta q_r. \tag{13.4.5}$$

This expression is now substituted in the fundamental equation

$$\sum_{s=1}^{N} m_s \ddot{u}_s \, \delta u_s - \sum_{s=1}^{N} F_s \, \delta u_s = 0, \tag{13.4.6}$$

where we note that, by (11.2.11),

$$\sum_{s=1}^{N} F_s \, \delta u_s = \sum_{s=1}^{N} F_s \sum_{r=1}^{n} \frac{\partial u_s}{\partial q_r} \delta q_r = \sum_{r=1}^{n} Q_r \, \delta q_r, \tag{13.4.7}$$

and the last equality results from (12.3.8). Thus, we find

$$\sum_{r=1}^{n} \left(\frac{d}{dt} \frac{\partial T}{\partial \dot{q}_r} - \frac{\partial T}{\partial q_r} - Q_r \right) \delta q_r = 0, \tag{13.4.8}$$

identical with (13.1.1). Hence, Lagrange's equations follow by (13.3.2).

Inasmuch as the fundamental equation was central in this derivation, it does not differ essentially from the previous one. However, it proceeded relatively smoothly because of the convenient relations offered by the central principle.

13.5. Derivation from Hamilton's Principle

Hamilton's principle in its most general form was derived in (10.5.2) as

$$\left[\sum_{r=1}^{N/3} m_r \dot{x}^r \cdot \delta x^r\right]_{t_0}^{t_1} = \int_{t_0}^{t_1} (\delta T + \delta W) \, dt, \tag{13.5.1}$$

where δx^r is the rth component of the difference vector between the actual configuration and a simultaneous neighboring one which does not violate the constraints, i.e., the δx^r satisfy

$$\sum_{r=1}^{N/3} A_{rs} \cdot \delta x^r = 0 \qquad (s = 1, 2, \ldots, L). \tag{13.5.2}$$

Written in terms of the u_s $(s = 1, 2, \ldots, N)$, these equations become

$$\left[\sum_{s=1}^{N} m_s \dot{u}_s \, \delta u_s\right]_{t_0}^{t_1} = \int_{t_0}^{t_1} (\delta T + \delta W)\, dt \tag{13.5.3}$$

and

$$\sum_{s=1}^{N} A_{rs} \, \delta u_s = 0 \qquad (r = 1, 2, \ldots, L). \tag{13.5.4}$$

We also saw that Hamilton's principle takes on the form

$$\int_{t_0}^{t_1} (\delta T + \delta W)\, dt = 0 \tag{13.5.5}$$

when the endpoints are fixed, i.e., when the $\delta u_s(t_0) = \delta u_s(t_1) = 0$ $(s = 1, 2, \ldots, N)$. In (13.5.5), the kinetic energy is $T = \frac{1}{2}\sum_{s=1}^{N} m_s \dot{u}_s^2$, and the virtual work is $\delta W = \sum_{s=1}^{N} F_s \, \delta u_s$.

If we proceed to generalized coordinates we find

$$T = T(q_1, q_2, \ldots, q_n, \dot{q}_1, \dot{q}_2, \ldots, \dot{q}_n, t) \tag{13.5.7}$$

in accordance with (12.1.3), and

$$\delta W = \sum_{s=1}^{n} Q_s \, \delta q_s \tag{13.5.8}$$

by definition. The δq_s are the virtual displacement components which satisfy

$$\sum_{s=1}^{n} B_{rs} \, \delta q_s = 0 \qquad (r = 1, 2, \ldots, l) \tag{13.5.9}$$

in accordance with (11.2.16), and the constraints (all nonholonomic) are

$$\sum_{s=1}^{n} B_{rs}\dot{q}_s + B_r = 0 \qquad (r = 1, 2, \ldots, l) \tag{13.5.10}$$

in accordance with (11.2.15). Therefore, our problem is to determine those $q_s(t)$ $(s = 1, 2, \ldots, n)$ which satisfy

$$\int_{t_0}^{t_1} \left(\delta T + \sum_{s=1}^{n} Q_s \, \delta q_s\right) dt = 0 \tag{13.5.11}$$

when the δq_s satisfy (13.5.9), and such that the constraint equations (13.5.10) are satisfied by the q_s and the $\dot{q}_s$ for every value of t in $[t_0, t_1]$.

It can be shown that the rule of the Lagrange multipliers given in Section 13.2 is also applicable to integrals and requires that

$$\int_{t_0}^{t_1} \left(\delta T + \sum_{s=1}^{n} Q_s \, \delta q_s - \sum_{s=1}^{n} \sum_{r=1}^{l} \lambda_r B_{rs} \, \delta q_s \right) dt = 0, \tag{13.5.12}$$

where the δq_s are now completely arbitrary. Forming the variation of T as defined in (9.6.3), one finds

$$\begin{aligned} \delta T &= \sum_{s=1}^{n} \frac{\partial T}{\partial q_s} \, \delta q_s + \sum_{s=1}^{n} \frac{\partial T}{\partial \dot{q}_s} \, \delta \dot{q}_s \\ &= \sum_{s=1}^{n} \frac{\partial T}{\partial q_s} \, \delta q_s + \sum_{s=1}^{n} \frac{\partial T}{\partial \dot{q}_s} \, \frac{d}{dt} (\delta q_s), \end{aligned} \tag{13.5.13}$$

where use was made of (11.4.7).

Now, in $\int \delta T \, dt$, there arises the term

$$\begin{aligned} \int_{t_0}^{t_1} \sum_{s=1}^{n} \frac{\partial T}{\partial \dot{q}_s} \frac{d}{dt} (\delta q_s) \, dt &= \left[\sum_{s=1}^{n} \frac{\partial T}{\partial \dot{q}_s} \, \delta q_s \right]_{t_0}^{t_1} - \int_{t_0}^{t_1} \sum_{s=1}^{n} \frac{d}{dt} \left(\frac{\partial T}{\partial \dot{q}_s} \right) \delta q_s \, dt \\ &= - \int_{t_0}^{t_1} \sum_{s=1}^{n} \frac{d}{dt} \left(\frac{\partial T}{\partial \dot{q}_s} \right) \delta q_s \, dt. \end{aligned} \tag{13.5.14}$$

The first equality results from an integration by parts, and the second from the fact that the $\delta q_s(t_0) = \delta q_s(t_1) = 0$ $(s = 1, 2, \ldots, n)$; these were assumed when Hamilton's principle was written in the form (13.5.5) rather than (13.5.3).

Making the appropriate substitutions in (13.5.12), that equation becomes

$$\int_{t_0}^{t_1} \sum_{t=1}^{n} \left[- \frac{d}{dt} \left(\frac{\partial T}{\partial \dot{q}_s} \right) + \frac{\partial T}{\partial q_s} + Q_s - \sum_{r=1}^{l} \lambda_r B_{rs} \right] \delta q_s \, dt = 0. \tag{13.5.15}$$

Equation (13.5.15) has as a consequence Lagrange's equations

$$\frac{d}{dt} \frac{\partial T}{\partial \dot{q}_s} - \frac{\partial T}{\partial q_s} - Q_s + \sum_{r=1}^{l} \lambda_r B_{rs} = 0 \qquad (s = 1, 2, \ldots, n), \tag{13.5.16}$$

identical with (13.3.3). The argument on which (13.3.16) is based is similar to the proof of Theorem (13.3.2). If one, several, or all of the quantities in the square brackets of (13.5.15) are different from zero, we may choose for each of these a δq_s having the same sign as that quantity. In that case, the integral in (13.5.15) would be necessarily positive, contrary to Hamilton's principle, which requires that it be zero.

In Section 10.5 we observed that Hamilton's principle for conservative systems,

$$\int_{t_0}^{t_1} \delta(T - V)\, dt = 0, \tag{13.5.17}$$

is not a problem in the calculus of variations when the system is nonholonomic, i.e., it does not lead to the same result as the problem

$$\delta \int_{t_0}^{t_1} (T - V)\, dt = 0. \tag{13.5.18}$$

This fact is easy to demonstrate when generalized coordinates are used, i.e., when all constraints (13.5.10) are nonholonomic. In that case, the first problem, (13.5.17), leads to (13.5.15) except that the $Q_s = -\partial V/\partial q_s$, i.e., one has, instead of (13.5.15),

$$\int_{t_0}^{t_1} \sum_{s=1}^{n} \left[-\frac{d}{dt}\left(\frac{\partial T}{\partial \dot{q}_s}\right) + \frac{\partial T}{\partial q_s} - \frac{\partial V}{\partial q_s} - \sum_{r=1}^{l} \lambda_r B_{rs} \right] \delta q_s\, dt = 0. \tag{13.5.19}$$

Now, the second problem, (13.5.18), is by definition a problem in the calculus of variations. Its verbalization is that, among all $q_s(t)$ $(s = 1, 2, \ldots, n)$ which together with their time derivatives satisfy the kinematical constraint equations (13.5.10), we seek that set which gives a stationary value to the definite time integral of $T - V$. This is a classical variational problem with differential equations as side conditions. Then, the well-known rule of the Lagrange multipliers in the calculus of variations states that the solution to this variational problem is furnished by that set $\{q_s(t);\ (s = 1, 2, \ldots, n)\}$ for which

$$\delta \int_{t_0}^{t_1} \left[T - V - \sum_{s=1}^{n} \sum_{r=1}^{l} \lambda_r (B_{rs} \dot{q}_s + B_r) \right] dt = 0 \tag{13.5.20}$$

with no side conditions imposed. The standard techniques of the calculus of variations transform (13.5.20) into

$$\begin{aligned} \int_{t_0}^{t_1} \sum_{s=1}^{n} \Bigg\{ & \left[-\frac{d}{dt}\frac{\partial T}{\partial \dot{q}_s} + \frac{\partial T}{\partial q_s} - \frac{\partial V}{\partial q_s} \right] \delta q_s \\ & - \sum_{r=1}^{l} \lambda_r \left[\left(\frac{\partial B_r}{\partial q_s}\, \dot{q}_s + \frac{\partial B_r}{\partial q_s} \right) \delta q_s + B_{rs}\, \delta \dot{q}_s \right] \Bigg\}\, dt = 0, \end{aligned} \tag{13.5.21}$$

and this is seen to differ from (13.5.19); hence, (13.5.21) must be false in general. Equation (13.5.21) could still be somewhat simplified through integration by parts.

When the problem is holonomic, and generalized coordinates have been used, there are no constraint equations to be satisfied; thus, one may put B_{rs} and B_r equal to zero for all r and s in the above equations. For that case, (13.5.19) and (13.5.21) become identical, which shows that Hamilton's principle for the conservative, holonomic problem may be regarded as a variational principle.

13.6. Dynamic Coupling and Decoupling

In this section we shall show first that, in general, every one of Lagrange's equations contains all acceleration terms. When more than one acceleration component occurs in one, several, or all equations of motion, the system is said to be *dynamically coupled.* Hence, Lagrange's equations of motion will be shown to be, in general, dynamically coupled. Next, we shall show that it is always possible, in principle at least, to transform this set of equations into one which is dynamically uncoupled.

The kinetic energy was found in (12.1.3) as

$$T = \frac{1}{2} \sum_{\alpha,\beta=1}^{n} a_{\alpha\beta}\dot{q}_\alpha \dot{q}_\beta + \sum_{\alpha=1}^{n} b_\alpha \dot{q}_\alpha + c \tag{13.6.1}$$

and, in general, the $a_{\alpha\beta}$, b_α, and c are all functions of all q_s ($s = 1, 2, \ldots, n$) and of t.

For later substitution into Lagrange's equations we form the expressions

$$\begin{aligned} \frac{d}{dt}\frac{\partial T}{\partial \dot{q}_s} &= \frac{1}{2}\sum_{\beta}(a_{s\beta} + a_{\beta s})\ddot{q}_\beta + \frac{1}{2}\sum_{\beta,\gamma}\left(\frac{\partial a_{s\beta}}{\partial q_\gamma} + \frac{\partial a_{\beta s}}{\partial q_\gamma}\right)\dot{q}_\beta\dot{q}_\gamma \\ &\quad + \frac{1}{2}\sum_{\beta}\left(\frac{\partial a_{s\beta}}{\partial t} + \frac{\partial a_{\beta s}}{\partial t}\right)\dot{q}_\beta + \sum_{\gamma}\frac{\partial b_s}{\partial q_\gamma}\dot{q}_\gamma + \frac{\partial b_s}{\partial t} \end{aligned} \tag{13.6.2}$$

and

$$\frac{\partial T}{\partial q_s} = \frac{1}{2}\sum_{\alpha,\beta}\frac{\partial a_{\alpha\beta}}{\partial q_s}\dot{q}_\alpha\dot{q}_\beta + \sum_{\alpha}\frac{\partial b_\alpha}{\partial q_s}\dot{q}_\alpha + \frac{\partial c}{\partial q_s}. \tag{13.6.3}$$

Consider the term with $\dot{q}_\beta\dot{q}_\gamma$ in (13.6.2). It is equal to

$$\begin{aligned} &\frac{1}{2}\sum_{\beta,\gamma}\frac{\partial a_{s\beta}}{\partial q_\gamma}\dot{q}_\beta\dot{q}_\gamma + \frac{1}{2}\sum_{\beta,\gamma}\frac{\partial a_{\beta s}}{\partial q_\gamma}\dot{q}_\beta\dot{q}_\gamma \\ &\quad = \frac{1}{2}\sum_{\beta,\gamma}\frac{\partial a_{s\beta}}{\partial q_\gamma}\dot{q}_\beta\dot{q}_\gamma + \frac{1}{2}\sum_{\beta,\gamma}\frac{\partial a_{\gamma s}}{\partial q_\beta}\dot{q}_\beta\dot{q}_\gamma \\ &\quad = \frac{1}{2}\sum_{\beta,\gamma}\frac{\partial a_{s\beta}}{\partial q_\gamma}\dot{q}_\beta\dot{q}_\gamma + \frac{1}{2}\sum_{\beta,\gamma}\frac{\partial a_{s\gamma}}{\partial q_\beta}\dot{q}_\beta\dot{q}_\gamma. \end{aligned} \tag{13.6.4}$$

The first equality results from exchanging the indices β, γ in the second sum; this does not change its value. The second equality results from the symmetry property (12.1.5), which the $a_{\alpha\beta}$ satisfy.

It follows that

$$\begin{aligned}\frac{d}{dt}\frac{\partial T}{\partial \dot{q}_s} - \frac{\partial T}{\partial q_s} &= \frac{1}{2}\sum_{\beta}(a_{s\beta} + a_{\beta s})\ddot{q}_\beta \\ &+ \frac{1}{2}\sum_{\beta,\gamma}\left(\frac{\partial a_{s\beta}}{\partial q_\gamma} + \frac{\partial a_{s\gamma}}{\partial q_\beta} - \frac{\partial a_{\gamma\beta}}{\partial q_s}\right)\dot{q}_\beta\dot{q}_\gamma \\ &+ \frac{1}{2}\sum_{\beta}\left(\frac{\partial a_{s\beta}}{\partial t} + \frac{\partial a_{\beta s}}{\partial t}\right)\dot{q}_\beta \\ &+ 2\sum_{\gamma}\frac{1}{2}\left(\frac{\partial b_s}{\partial q_\gamma} - \frac{\partial b_\gamma}{\partial q_s}\right)\dot{q}_\gamma + \frac{\partial b_s}{\partial t} - \frac{\partial c}{\partial q_s}. \qquad (13.6.5)\end{aligned}$$

The coefficient of $\dot{q}_\beta\dot{q}_\gamma$ in this equation occurs frequently in the field of differential geometry and other branches of applied mathematics, and a special symbol has been invented for it; it is called the *Christoffel symbol of the first kind* and is written as

$$[\beta\gamma, s] = \frac{1}{2}\left(\frac{\partial a_{s\beta}}{\partial q_\gamma} + \frac{\partial a_{s\gamma}}{\partial q_\beta} - \frac{\partial a_{\gamma\beta}}{\partial q_s}\right). \qquad (13.6.6)$$

We shall also introduce the symbol

$$[\gamma, s] = \frac{1}{2}\left(\frac{\partial b_s}{\partial q_\gamma} - \frac{\partial b_\gamma}{\partial q_s}\right). \qquad (13.6.7)$$

The quantity $[\gamma, s]$ has a special meaning. The b_s are, by definition, functions of the form

$$b_s = b_s(q_1, q_2, \ldots, q_n, t) \qquad (s = 1, 2, \ldots, n).$$

These last equations may be thought to constitute transformations which map a point $(q_1, q_2, \ldots, q_n)$ into a point $(b_1, b_2, \ldots, b_n)$ for fixed t. The Jacobian of this transformation is

$$J = \frac{\partial(b_1, b_2, \ldots, b_n)}{\partial(q_1, q_2, \ldots, q_n)}.$$

This Jacobian may be separated into a symmetrical part, which is

$$S_{\gamma s} = \frac{1}{2}\left(\frac{\partial b_s}{\partial q_\gamma} + \frac{\partial b_\gamma}{\partial q_s}\right),$$

and a skew-symmetric, or alternating, part:

$$A_{\gamma s} = \frac{1}{2}\left(\frac{\partial b_s}{\partial q_\gamma} - \frac{\partial b_\gamma}{\partial q_s}\right) = [\gamma, s].$$

It is seen that only the alternating part of J enters into Lagrange's equations.

Finally, if we make use of the symmetry relation $a_{\alpha\beta} = a_{\alpha\beta}$ in the remaining terms of (13.6.5), we find

$$\frac{d}{dt}\frac{\partial T}{\partial q_s} - \frac{\partial T}{\partial q_s} = \sum_\beta a_{s\beta}\ddot{q}_\beta + \sum_{\beta,\gamma} [\beta\gamma, s]\dot{q}_\beta\dot{q}_\gamma + \sum \frac{\partial a_{s\beta}}{\partial t}\dot{q}_\beta + 2\sum_\gamma [\gamma, s]\dot{q}_\gamma + \frac{\partial b_s}{\partial t} - \frac{\partial c}{\partial q_s}. \tag{13.6.8}$$

Let us now substitute this expression into Lagrange's equations. We shall assume for the present that the q_s are generalized coordinates (no holonomic constraints are to be satisfied) and that there are no non-holonomic constraints. The latter assumption is not necessary, and the results in this section are not affected by its removal; however, it shortens the equations which follow. Under this assumption (13.6.8) is substituted into (13.3.2) with the result that

$$\sum_\beta a_{s\beta}\ddot{q}_\beta + \sum_{\beta,\gamma} [\beta\gamma, s]\dot{q}_\beta\dot{q}_\gamma + \sum_\beta \frac{\partial a_{s\beta}}{\partial t}\dot{q}_\beta + 2\sum_\gamma [\gamma, s]\dot{q}_\gamma + \frac{\partial b_s}{\partial t} - \frac{\partial c}{\partial q_s} - Q_s = 0 \qquad (s = 1, 2, \ldots, n). \tag{13.6.9}$$

This system of equations shows that, in general, every Lagrangean equation of motion contains all acceleration components, as claimed.

We wish to demonstrate next that this set of equations can always be transformed into a dynamically uncoupled set. To do this, it is evidently necessary to show that the matrix $(a_{s\beta})$ is nonsingular, or that the determinant $|a_{s\beta}| \neq 0$. This is easily seen. The kinetic energy is given in (13.6.1); that equation must hold for all bounded values of the velocity components $\dot{q}_s$. Therefore, one can always choose velocity components so large that, approximately,

$$T \simeq \frac{1}{2}\sum_{\alpha,\beta} a_{\alpha\beta}\dot{q}_\alpha\dot{q}_\beta + \cdots, \tag{13.6.10}$$

where the neglected terms are negligibly small compared to those retained. Now, it is always true that

$$T = \frac{1}{2}\sum_{s=1}^{N} m_s\dot{u}_s^2 \geq 0,$$

and this vanishes only with the velocities. Thus, T in (13.6.10) is necessarily positive-definite because $|a_{s\beta}| > 0$ is necessary for T in (13.6.10) to be positive-definite.

We now define quantities $A^{\varrho s}$ such that

$$\sum_{s=1}^{n} A^{\varrho s} a_{s\beta} = \delta_{\varrho\beta} = \begin{cases} 1 & \text{for} \quad \varrho = \beta, \\ 0 & \text{for} \quad \varrho \neq \beta, \end{cases} \tag{13.6.11}$$

where $\delta_{\varrho\beta}$ is Kronecker's delta, defined in (6.3.10). Therefore, $A^{\varrho s}$ is the cofactor of $a_{\varrho s}$ divided by $|a_{s\beta}|$.

Let us now multiply each of (13.6.9) by $A^{\varrho s}$ and then add these equations. In the sum which results we shall use the notation

$$\sum_{s=1}^{n} A^{\varrho s}[\beta\gamma, s] = \begin{Bmatrix} \varrho \\ \beta\gamma \end{Bmatrix}, \tag{13.6.12}$$

$$\sum_{s=1}^{n} A^{\varrho s}[\gamma, s] = \begin{Bmatrix} \varrho \\ \gamma \end{Bmatrix}, \tag{13.6.13}$$

$$\sum_{s=1}^{n} A^{\varrho s} Q_s = Q^{\varrho}. \tag{13.6.14}$$

The first is called the Christoffel symbol of the second kind, the second has been formed in analogy to it, and the third is the generalized force in dynamically uncoupled coordinates. On making use of the above notation, the Lagrange equations become

$$\ddot{q}_\varrho + \sum_{\beta,\gamma} \begin{Bmatrix} \varrho \\ \beta\gamma \end{Bmatrix} \dot{q}_\beta \dot{q}_\gamma + \sum_{\gamma} \left[2\begin{Bmatrix} \varrho \\ \gamma \end{Bmatrix} + \sum_{s=1}^{n} A^{\varrho s} \frac{\partial a_{s\gamma}}{\partial t} \right] \dot{q}_\gamma + \sum_{s=1}^{n} A^{\varrho s} \frac{\partial b_s}{\partial t}$$
$$- \sum_{s=1}^{n} A^{\varrho s} \frac{\partial c}{\partial q_s} - Q^{\varrho} = 0 \qquad (\varrho = 1, 2, \ldots, n). \tag{13.6.15}$$

This is a dynamically *un*coupled set of Lagrangean equations. Hence, it is always possible, as claimed, to transform a dynamically coupled set of Lagrangean equations into a set not so coupled.

It is also clear that the presence of nonholonomic constraints does not in any way affect the above statement. If the system were subject to l nonholonomic constraints, one would define

$$\sum_{s=1}^{n} A^{\varrho s} \lambda_r B_{rs} = \lambda_r B_r{}^{\varrho}, \tag{13.6.16}$$

and one would add the term $\sum_{r=1}^{l} \lambda_r B_r{}^{\varrho}$ to the left-hand side of (13.6.15).

In the special case when the kinetic energy is not an explicit function of time t, one has $b_s = c \equiv 0$ ($s = 1, 2, \ldots, n$); in view of (13.6.7) one has then also $[\gamma, s] = 0$, and this implies $\{{}^{\varrho}_{\gamma}\} = 0$; moreover, all partial derivatives with respect to t vanish identically as well. Therefore, when the kinetic energy is not an explicit function of time, the dynamically uncoupled set of Lagrangean equations reduces to

$$\ddot{q}^{\varrho} + \sum_{\beta,\gamma} \begin{Bmatrix} \varrho \\ \beta\gamma \end{Bmatrix} \dot{q}_{\beta}\dot{q}_{\gamma} - Q^{\varrho} = 0 \qquad (\varrho = 1, 2, \ldots, n) \tag{13.6.17}$$

when the system is holonomic. When it is not, the term $\sum_{r=1}^{l} \lambda_r B_r^{\varrho}$ is added to the left-hand side.

While these demonstrations serve the purpose of showing that Lagrange's equations can, in general, be dynamically decoupled, it turns out that, in applications, neither the coupled form (13.6.9) nor the dynamically decoupled form (13.6.15) is practical. It is far easier to retain the form (13.3.2) in terms of the kinetic energy. That energy is then computed for the problem in hand, and its value is substituted in these equations.

Example 13.6.1. As an example of dynamic decoupling, suppose the kinetic energy is

$$T = \tfrac{1}{2}(\alpha_{11}\dot{q}_1^2 + 2\alpha_{12}\dot{q}_1\dot{q}_2 + \alpha_{22}\dot{q}_2^2),$$

where the α_{ij} are constants. Then, as

$$\frac{d}{dt}\frac{\partial T}{\partial \dot{q}_1} = \alpha_{11}\ddot{q}_1 + \alpha_{12}\ddot{q}_2,$$

$$\frac{d}{dt}\frac{\partial T}{\partial \dot{q}_2} = \alpha_{12}\ddot{q}_1 + \alpha_{22}\ddot{q}_2,$$

the $q_{1,2}$ are dynamically coupled. Let us now introduce coordinates $q'_{1,2}$, which are related to the $q_{1,2}$ by

$$q_1 = a_1 q_1' + a_2 q_2',$$

$$q_2 = b_1 q_1' + b_2 q_2',$$

where the constants a_i and b_i are to be determined so that the coordinates are dynamically decoupled. Substitution into the kinetic energy gives

$$T = \tfrac{1}{2}\{(\alpha_{11}a_1^2 + 2\alpha_{12}a_1b_1 + \alpha_{22}b_1^2)\dot{q}_1'^2 + [2\alpha_{11}a_1a_2 + 2\alpha_{12}(a_1b_2 + a_2b_1) + 2\alpha_{22}b_1b_2]\dot{q}_1'\dot{q}_2' + (\alpha_{11}a_2^2 + 2\alpha_{12}a_2b_2 + \alpha_{22}b_2^2)\dot{q}_2'^2\}.$$

To be dynamically decoupled, the a_i and b_i must be so chosen that the coefficient of $\dot{q}_1'\dot{q}_2'$ vanishes. One such choice is

$$a_1 = 1, \qquad b_1 = 0, \qquad b_2 = -\frac{\alpha_{11}}{\alpha_{12}}a_2.$$

Then the kinetic energy becomes

$$T = \frac{1}{2}\left\{\dot{q}_1''^2 + (\alpha_{11}\alpha_{22} - \alpha_{12}^2)\frac{a_2^2}{\alpha_{12}^2}\dot{q}_2''^2\right\},$$

where $q_i'' = (\alpha_{11})^{1/2}q_i'$. It is evident that q_1'' and q_2'' are dynamically decoupled coordinates.

13.7. Special Forms of Lagrange's Equations

When a strictly Newtonian system is subject to equality constraints only (holonomic and/or nonholonomic), Lagrange's equations are

$$\frac{d}{dt}\frac{\partial T}{\partial \dot{q}_s} - \frac{\partial T}{\partial q_s} - Q_s + \sum_{r=1}^{l}\lambda_r B_{rs} = 0 \qquad (s = 1, 2, \ldots, \bar{n}), \tag{13.7.1}$$

where the q_s $(s = 1, 2, \ldots, \bar{n})$ are any set of coordinates which define the configuration of the system uniquely. If n is the number of generalized coordinates, one has necessarily $\bar{n} \geq n$, and $\bar{n} - n = l'$ of the constraints are holonomic.

(a) Existence of a Potential

When *some* of the given forces are derivable from a potential

$$V = V(q_1, q_2, \ldots, q_{\bar{n}}),$$

Lagrange's equations become, in view of (12.4.2),

$$\frac{d}{dt}\frac{\partial T}{\partial \dot{q}_s} - \frac{\partial T}{\partial q_s} + \frac{\partial V}{\partial q_s} - Q_s + \sum_{r=1}^{l}\lambda_r B_{rs} = 0 \qquad (s = 1, 2, \ldots, \bar{n}), \tag{13.7.2}$$

where the Q_s are now the components of the *non*potential given forces.

When *all* given forces are potential, the equations become

$$\frac{d}{dt}\frac{\partial T}{\partial \dot{q}_s} - \frac{\partial T}{\partial q_s} + \frac{\partial V}{\partial q_s} + \sum_{r=1}^{l}\lambda_r B_{rs} = 0 \qquad (s = 1, 2, \ldots, \bar{n}). \tag{13.7.3}$$

(b) Holonomic Systems

When the system is *holonomic, and when the q_s are generalized coordinates*, but only then, the above equations are valid if one

(i) replaces $\bar{n}$ by n;

(ii) sets $\lambda_r \equiv 0$ $(r = 1, 2, \ldots, l)$.

(c) Rayleigh's Dissipation Function

Let us consider a function

$$D = \frac{1}{2} \sum_{\alpha=1}^{\bar{n}} \sum_{\beta=1}^{\bar{n}} \bar{d}_{\alpha\beta} \dot{q}_\alpha \dot{q}_\beta , \tag{13.7.4}$$

where the $\bar{d}_{\alpha\beta}$ are constants. Suppose every $\bar{d}_{\alpha\beta}$ could be expressed as

$$\bar{d}_{\alpha\beta} = d_{\beta\alpha} + d'_{\alpha\beta} , \tag{13.7.5}$$

where

$$\left. \begin{aligned} d_{\alpha\beta} &= d_{\beta\alpha} , \\ d'_{\alpha\beta} &= -d'_{\beta\alpha} \end{aligned} \right\} \qquad (\alpha, \beta = 1, 2, \ldots, \bar{n}). \tag{13.7.6}$$

Then, the partial derivative of D with respect to a velocity component $\dot{q}_s$ is

$$\frac{\partial D}{\partial \dot{q}_s} = \frac{1}{2} \left\{ \left[\sum_{\beta=1}^{\bar{n}} d_{s\beta} \dot{q}_\beta + \sum_{\alpha=1}^{\bar{n}} d_{\alpha s} \dot{q}_\alpha \right] + \left[\sum_{\beta=1}^{\bar{n}} d'_{s\beta} \dot{q}_\beta + \sum_{\alpha=1}^{\bar{n}} d'_{\alpha s} \dot{q}_\alpha \right] \right\}$$

or, in view of the second equation of (13.7.6),

$$\frac{\partial D}{\partial \dot{q}_s} = \sum_{\alpha=1}^{\bar{n}} d_{s\alpha} \dot{q}_\alpha . \tag{13.7.7}$$

But this is the same result as one would obtain from a function

$$D = \frac{1}{2} \sum_{\alpha=1}^{\bar{n}} \sum_{\beta=1}^{\bar{n}} d_{\alpha\beta} \dot{q}_\alpha \dot{q}_\beta , \tag{13.7.8}$$

where the $d_{\alpha\beta}$ satisfy the first equation of (13.7.6).

We now define:

When the components Q_s^D of a given force are derivable from a function D as defined in (13.7.8) *or*

$$Q_s^D = \frac{\partial D}{\partial \dot{q}_s} \qquad (s = 1, 2, \ldots, \bar{n}),$$

D is called Rayleigh's dissipation function.

Evidently, these force components are derivable from a dissipation function much in the same way as the components of a potential force are derivable from a potential function, i.e., the negative of the potential energy. Moreover, the form of the dissipation function is strongly reminiscent of that of the kinetic energy of scleronomic systems. The Q_s derivable from D are always linear in the velocity; hence, they have the character of damping forces if the $d_{\alpha\beta}$ are negative. It is the energy-dissipating property of damping forces which has given the name "dissipation function" to D; its introduction is due to Lord Rayleigh. It should be noted that forces for which the $d_{\alpha\beta}$ are positive are also derivable from D, provided that the coefficients satisfy the first equation of (13.7.6). In these cases the forces release energy to the system rather than absorb energy from it.

When some of the given forces are derivable from a dissipation function Lagrange's equations may be written as

$$\frac{d}{dt}\frac{\partial T}{\partial \dot{q}_s} - \frac{\partial T}{\partial q_s} - \frac{\partial D}{\partial \dot{q}_s} - Q_s + \sum_{r=1}^{l} \lambda_r B_{rs} = 0 \qquad (s = 1, 2, \ldots, \bar{n}), \tag{13.7.9}$$

where the Q_s are all those components of given forces which are not derivable from a dissipation function.

Let us consider a force component corresponding to the *first* equation of (13.7.6), or

$$Q_s^D = \frac{\partial D}{\partial \dot{q}_s} = \sum_{\alpha=1}^{\bar{n}} d_{s\alpha}\dot{q}_\alpha. \tag{13.7.10}$$

The virtual work done by this force component in a virtual displacement $\delta q = (\delta q_1, \delta q_2, \ldots, \delta q_{\bar{n}})$ is

$$\delta W_s = Q_s^D\,\delta q_s = \sum_{\alpha=1}^{\bar{n}} d_{s\alpha}\dot{q}_\alpha\,\delta q_s \qquad (s = 1, 2, \ldots, \bar{n}). \tag{13.7.11}$$

If the system is *catastatic* the class of virtual and possible displacements coincides, i.e., the work done in a *possible* displacement is then

$$dW_s = \sum_{\alpha=1}^{\bar{n}} d_{\alpha s}\dot{q}_\alpha\,dq_s. \tag{13.7.12}$$

It follows that the rate of work done by the force component Q_s is

$$\dot{W}_s = \sum_{\alpha=1}^{\bar{n}} d_{\alpha s}\dot{q}_\alpha\dot{q}_s. \tag{13.7.13}$$

The rate of work done by the force $Q^D = (Q_1^D, Q_2^D, \ldots, Q_{\bar{n}}^D)$ is, then,

$$\dot{W} = \sum_{s=1}^{\bar{n}} \dot{W}_s = \sum_{\alpha=1}^{\bar{n}}\sum_{s=1}^{\bar{n}} d_{\alpha s}\dot{q}_\alpha\dot{q}_s = 2D. \tag{13.7.14}$$

Therefore, the rate of work done by this dissipation force is twice the value of the dissipation functiŏns from which it is derivable.

Let us next suppose that there exist given forces for which the *second* equation of (13.7.6) is satisfied. The rate of work done by them is

$$
\begin{aligned}
2D' &= \sum_{\alpha,s=1}^{\bar{n}} d'_{\alpha s}\dot{q}_\alpha\dot{q}_s \\
&= \left\{\frac{1}{2}\sum_{\alpha,s} d'_{\alpha s}\dot{q}_\alpha\dot{q}_s + \frac{1}{2}\sum_{\alpha,s} d'_{\alpha s}\dot{q}_\alpha\dot{q}_s\right\} \\
&= \left\{\frac{1}{2}\sum_{\alpha,s} d'_{\alpha s}\dot{q}_\alpha\dot{q}_s + \frac{1}{2}\sum_{s,\alpha} d'_{s\alpha}\dot{q}_s\dot{q}_\alpha\right\} \\
&= \left\{\frac{1}{2}\sum_{\alpha,s} d'_{\alpha s}\dot{q}_\alpha\dot{q}_s - \frac{1}{2}\sum_{s,\alpha} d'_{\alpha s}\dot{q}_s\dot{q}_\alpha\right\} \equiv 0,
\end{aligned}
\tag{13.7.15}
$$

where use was made of the second equation of (13.7.6). Therefore, if given forces that are linear in the velocities act on a catastatic system and the coefficients satisfy the second equation of (13.7.6), these forces cannot be derived from a dissipation function, and the work done by them in a possible displacement is zero. Such forces are called *gyroscopic forces.*

As these forces do no work in possible displacements, their presence does not affect energy conservation. Hence, if energy is conserved in the absence of gyroscopic forces, it is also conserved in their presence even though these forces resemble damping forces.

(d) The Dissipation Function of Lur'e

Rayleigh's dissipation function is a quadratic form in the generalized velocity components. In consequence, the dissipation forces derivable from it are always linear in the velocities. However, it is frequently useful to consider nonlinear dissipation forces such as, for instance, "velocity-squared" and Coulomb damping. Such dissipation forces are not derivable from Rayleigh's dissipation function.

Lur'e (p. 232) has generalized Rayleigh's dissipation function to one which admits nonlinear damping forces whose components in the u_i direction are of the form

$$
F_i = k_i(u_1, u_2, \ldots, u_N)f_i(\dot{u}_i) \qquad (i = 1, 2, \ldots, N), \tag{13.7.16}
$$

where the k_i are positive functions of the u_i $(i = 1, 2, \ldots, N)$, and

$$
\dot{u}_i f_i(\dot{u}_i) \geq 0.
$$

Here the u_i are the Cartesian components of the N-dimensional configuration space. Utilizing (12.3.8) to form generalized damping force components, one finds for the force component in the q_s direction

$$Q_s^L = \sum_{i=1}^{N} k_i(q_1, q_2, \ldots, q_n, t) f_i(\dot{u}_i) \frac{\partial u_i}{\partial q_s},$$

which may be written, because of (12.2.3), as

$$Q_s^L = \sum_{i=1}^{N} k_i(q_1, q_2, \ldots, q_n, t) f_i(\dot{u}_i) \frac{\partial \dot{u}_i}{\partial \dot{q}_s}. \tag{13.7.17}$$

Lur'e's dissipation function is

$$D_L = \sum_{i=1}^{N} k_i(q, t) \int_0^{\dot{u}_i} f_i(v)\, dv. \tag{13.7.18}$$

It is easily verified from (13.7.18) that

$$\frac{\partial D_L}{\partial \dot{q}_s} = Q_s^L \qquad (s = 1, 2, \ldots, n), \tag{13.7.19}$$

with Q_s^L, as defined in (13.7.17).

To show that Rayleigh's dissipation function is a special case of Lur'e's, let

$$f_i(\dot{u}_i) = \dot{u}_i, \qquad k_i = \text{const.}$$

Then, the dissipation function becomes in this case

$$D_L = \frac{1}{2} \sum_{i=1}^{N} k_i \dot{u}_i^2.$$

Proceeding to generalized coordinates in which the relations between the u_i and q_s are time-*in*dependent, one finds

$$D_L = \frac{1}{2} \sum_{\alpha=1}^{n} \sum_{\beta=1}^{n} \left[\sum_{i=1}^{N} k_i \frac{\partial u_i}{\partial q_\alpha} \frac{\partial u_i}{\partial q_\beta} \right] \dot{q}_\alpha \dot{q}_\beta.$$

Now, if the relations between the u_i and the q_s are linear, or

$$u_i = \sum_{s=1}^{n} a_{is} q_s \qquad (i = 1, 2, \ldots, N),$$

where the a_{is} are constants, one has

$$\sum_{i=1}^{N} k_i \frac{\partial u_i}{\partial q_\alpha} \frac{\partial u_i}{\partial q_\beta} = \sum_{i=1}^{N} a_{i\alpha} a_{i\beta} k_i.$$

If that quantity is denoted by $d_{\alpha\beta}$ and substituted in D_L, Rayleigh's dissipation function (13.7.8) results.

More generally, let the damping force be proportional to a power of the magnitude of the velocity, or

$$f_i(\dot{u}_i) = |\dot{u}_i|^m. \tag{13.7.20}$$

Then, the substitution of this function in (13.7.18) results in

$$D_L = \frac{1}{m+1} \sum_{i=1}^{N} k_i(q) \, |\dot{u}_i|^{m+1}. \tag{13.7.21}$$

Under the restriction that the relations between the Cartesian and generalized coordinates do not contain the time t explicitly, the above formula becomes

$$D_L = \frac{1}{m+1} \sum_{i=1}^{N} k_i(q_i) \left| \sum_{s=1}^{n} \frac{\partial u_i}{\partial q_s} \dot{q}_s \right|^{m+1} \operatorname{sgn}(\dot{u}_i). \tag{13.7.22}$$

We now suppose that the system is catastatic. Then the rate of change of work done by the damping forces is [see (13.7.11) to (13.7.14)]

$$\dot{W} = \sum_{s=1}^{n} Q_s^L \dot{q}_s = \sum_{s=1}^{n} \frac{\partial D_L}{\partial \dot{q}_s} \dot{q}_s, \tag{13.7.23}$$

where use was made of (13.7.19). But D_L in (13.7.22) is a homogeneous function of degree $m + 1$ in the $\dot{q}_s$. Let $\varphi(\dot{q}_1, \dot{q}_2, \ldots, \dot{q}_n)$ be such a function. This means that, for any constant K,

$$\varphi(K\dot{q}_1, K\dot{q}_2, \ldots, K\dot{q}_n) = K^{m+1}\varphi(\dot{q}_1, \dot{q}_2, \ldots, \dot{q}_n). \tag{13.7.24}$$

Differentiating both sides with respect to K and subsequently putting $K = 1$ gives the well-known result

$$\sum_{s=1}^{n} \frac{\partial \varphi}{\partial \dot{q}_s} \dot{q}_s = (m+1)\varphi. \tag{13.7.25}$$

This is known as Euler's theorem for homogeneous functions. Applying it to (13.7.23) results in

$$\dot{W} = (m+1)D_L, \tag{13.7.26}$$

a result demonstrated by Lur'e. It reduces to (13.7.14) when $m = 1$. The case of "velocity-squared" damping results from setting $m = 2$, and Coulomb damping from $m = 0$.

Example 13.7.1. (Lur'e, p. 233). Calculate Lur'e's dissipation function and the damping force when the system has only one generalized coordinate, and m is an even integer.

We have $n = 1$ and $m = 2\alpha$, where α is an integer. Then,

$$D_L = \frac{k}{2\alpha + 1} \, | \dot{q} \, |^{2\alpha+1} \operatorname{sgn}(\dot{q}),$$

and

$$Q^L = k\dot{q}^{2\alpha} \frac{\partial}{\partial \dot{q}} \, | \, \dot{q} \, | = k\dot{q}^{2\alpha} \operatorname{sgn}(\dot{q}).$$

13.8. The Principle of Least Action Reconsidered

In Section 10.7 we showed that Lagrange's principle of least action can be found from Hamilton's principle provided the energy level is constant throughout the motion and is the same for all admissible trajectories passing through the same end configurations. Moreover, in Section 13.5 we derived Lagrange's equations from Hamilton's principle. Can we also derive Lagrange's equations for conservative systems from the principle of least action? If the answer is affirmative, we shall have found a new and indirect way of establishing that principle.

The principle of least action requires that we find those $q_s(t)$ $(s = 1, 2, \ldots, n)$ for which the action integral

$$\int_{t_0}^{t_1} 2T \, dt \tag{13.8.1}$$

is stationary under a noncontemporaneous variation δ_t, and under the side condition that

$$T + V - h = 0, \tag{13.8.2}$$

and where

$$\begin{aligned} q(t_0) &= q^0 = (q_1^{\,0}, q_2^{\,0}, \ldots, q_n^{\,0}), \\ q(t_1) &= q^1 = (q_1^{\,1}, q_2^{\,1}, \ldots, q_n^{\,1}) \end{aligned} \tag{13.8.3}$$

are prescribed terminal configurations. Then, the rule of the Lagrange multipliers states that the $q_s(t)$ are those for which

$$\delta_t J = \delta_t \int_{t_0}^{t_1} F(q, \dot{q}) \, dt = 0, \tag{13.8.4}$$

where

$$F = 2T(q, \dot{q}) + \lambda(t)[T(q, \dot{q}) + V(q) - h], \tag{13.8.5}$$

and where δ_t is the operator of the noncontemporaneous variation described in Section 10.6; the quantity $\lambda(t)$ is an undetermined Lagrange multiplier which is, in general, a function of the independent variable, and t_0 and t_1 are not fixed.

As in Section 10.6, we introduce an auxiliary time τ, $0 \le \tau \le 1$, such that

$$t = t(\tau), \qquad \frac{dt}{d\tau} > 0 \qquad (t_0 \le t \le t_1). \tag{13.8.6}$$

Under these conditions the inverse function $\tau(t)$ exists on $t_0 \le t \le t_1$, and

$$\tau(t_0) = 0, \qquad \tau(t_1) = 1.$$

Denoting differentiation with respect to τ by a prime, we may write in place of (13.8.4)

$$\delta_t J = \delta_t \int_0^1 F\left(q, \frac{q'}{t'}\right) t' \, d\tau = 0 \tag{13.8.7}$$

or, exchanging the order of variation and integration and applying to $F \cdot t'$ the rules of varying a product, we find

$$\delta_t J = \int_0^1 \delta_t(F) t' \, d\tau + \int_0^1 F \, \delta_t(t') \, d\tau = 0. \tag{13.8.8}$$

Since

$$F = F(q, \dot{q}) = F\left(q, \frac{q'}{t'}\right),$$

we find

$$\begin{aligned} \delta_t F &= \sum_{r=1}^n \frac{\partial F}{\partial q_r} \delta_t q_r + \sum_{r=1}^n \frac{\partial F}{\partial \dot{q}_r} \frac{\partial \dot{q}_r}{\partial q_r'} \delta_t q_r' + \sum_{r=1}^n \frac{\partial F}{\partial \dot{q}_r} \frac{\partial \dot{q}_r}{\partial t'} \delta_t t' \\ &= \sum_{r=1}^n \frac{\partial F}{\partial q_r} \delta_t q_r + \sum_{r=1}^n \frac{\partial F}{\partial \dot{q}_r} \frac{1}{t'} \delta_t q_r' - \sum_{r=1}^n \frac{\partial F}{\partial \dot{q}_r} \frac{q_r'}{(t')^2} \delta_t t'. \end{aligned} \tag{13.8.9}$$

It follows that the first integral on the right-hand side of (13.8.8) is

$$\begin{aligned} \int_0^1 \delta_t(F) t' \, dt = \int_{t_0}^{t_1} \sum_{r=1}^n \frac{\partial F}{\partial q_r} \delta_t q_r \, dt + \int_0^1 \sum_{r=1}^n \frac{\partial F}{\partial \dot{q}_r} \frac{d}{d\tau} (\delta_t q_r) \, d\tau \\ - \int_0^1 \sum_{r=1}^n \frac{\partial F}{\partial \dot{q}_r} \dot{q}_r \, \delta_t t' \, d\tau. \end{aligned}$$

When the second integral on the right-hand side of the above equation is integrated by parts, as was done in (13.5.14) for instance, (13.8.8) becomes

$$\delta_t J = \sum_{r=1}^{n}\left[\frac{\partial F}{\partial \dot{q}_r}\,\delta_t q_r\right]_{q_r^0}^{q_r^1} + \int_{t_0}^{t_1}\sum_{r=1}^{n}\left(\frac{\partial F}{\partial q_r} - \frac{d}{dt}\,\frac{\partial F}{\partial \dot{q}_r}\right)\delta_t q_r\,dt$$
$$- \int_0^1\left[\sum_{r=1}^{n}\frac{\partial F}{\partial \dot{q}_r}\,\dot{q}_r - F\right]\delta_t t'\,d\tau = 0. \qquad (13.8.10)$$

The first term on the right-hand side of (13.8.10) is zero because the endpoints q^0 and q^1 are fixed so that $\delta_t q_r$ vanishes in the upper and lower limits. Also, because the variations $\delta_t q_r$ and $\delta_t t'$ are arbitrary it follows from the argument used in connection with (13.5.16) that one must have for all t, $t_0 \leq t \leq t_1$,

$$\frac{d}{dt}\left(\frac{\partial F}{\partial \dot{q}_r}\right) - \frac{\partial F}{\partial q_r} = 0 \qquad (r = 1, 2, \ldots, n), \qquad (13.8.11)$$

and for all τ, $0 \leq \tau \leq 1$,

$$F - \sum_{r=1}^{n}\frac{\partial F}{\partial \dot{q}_r}\,\dot{q}_r = 0. \qquad (13.8.12)$$

Substituting the definition of F in (13.8.11), that equation becomes

$$\frac{d}{dt}\left[(2+\lambda)\frac{\partial T}{\partial \dot{q}_r}\right] = (2+\lambda)\frac{\partial T}{\partial q_r} + \lambda\frac{\partial V}{\partial q_r} \qquad (r = 1, 2, \ldots, n),$$

which may be rewritten as

$$(2+\lambda)\left[\frac{d}{dt}\left(\frac{\partial T}{\partial \dot{q}_r}\right) - \frac{\partial(T-V)}{\partial q_r}\right] = 2(1+\lambda)\frac{\partial V}{\partial q_r} - \frac{d\lambda}{dt}\,\frac{\partial T}{\partial \dot{q}_r}$$
$$(r = 1, 2, \ldots, n). \qquad (13.8.13)$$

The evaluation of λ is made from (13.8.12). When F as given in (13.8.5) is substituted in it, it becomes

$$2T + \lambda(T + V - h) - (2+\lambda)\sum_{r=1}^{n}\frac{\partial T}{\partial \dot{q}_r}\,\dot{q}_r = 0. \qquad (13.8.14)$$

Since energy is conserved and the system is holonomic, it is necessarily also scleronomic, and in scleronomic systems

$$T = \frac{1}{2}\sum_{\alpha=1}^{n}\sum_{\beta=1}^{n} a_{\alpha\beta}\dot{q}_\alpha\dot{q}_\beta.$$

Then, an easy calculation shows that†

$$\sum_{r=1}^{n} \frac{\partial T}{\partial \dot{q}_r} \dot{q}_r = 2T. \tag{13.8.15}$$

The substitution of (13.8.2) and (13.8.15) in (13.8.14) gives

$$-2(1 + \lambda)T = 0. \tag{13.8.16}$$

Since $T \neq 0$, we must have $\lambda = -1$, and the substitution of $\lambda = -1$ in (13.8.13) yields the Lagrangean equations of motion of a conservative system. This is what we wished to show.

13.9. Problems

13.1. The three weights of mass m_1, m_2, and m_3, respectively, are the only massive elements of the system of weights, pulleys, and inextensible strings shown.

(a) Write down the equation(s) of constraint satisfied by the coordinates x_1, x_2, and x_3 shown.

(b) Calculate the x_i in terms of the q_j and show that the q_j satisfy the equation(s) of constraint identically. Hence, the q_j are generalized coordinates.

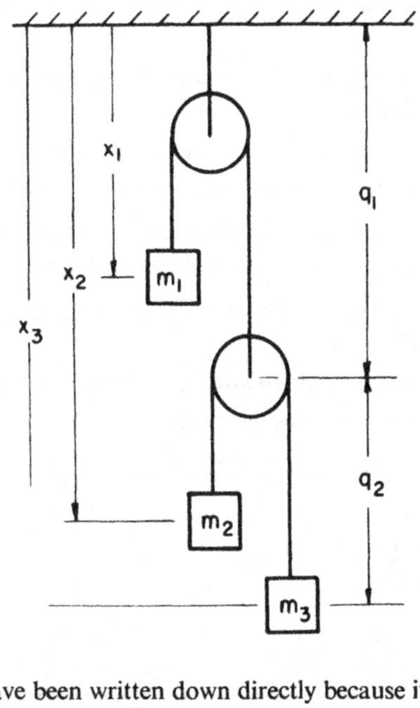

† This equation could have been written down directly because it is merely an application of Euler's theorem derived in (13.7.25).

(c) Construct Lagrange's equations of motion.

(d) Dynamically uncouple Lagrange's equations when $m_1 = 6$, $m_2 = 1$, $m_3 = 5$.

To solve Problems 13.2 to 13.6 refer to Problem 7.1.

13.2. One extremity of a heavy, uniform straight rod of length $2l$ and mass M can slide without friction along a vertical line. The other extremity is connected to one end of a massless, inextensible string of length $2l$ whose other end is tied to a fixed point O on the vertical line. Let θ be the angle subtended by the vertical line and the string, and let φ be the angle which the yz plane makes with the plane formed by the rod and string, as shown.

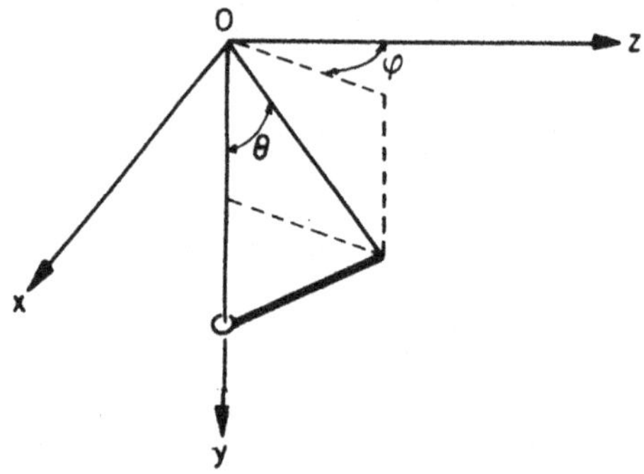

(a) How many and which are the constraints on the Cartesian coordinates of the endpoints of the rod? How many degrees of freedom does the rod have?

(b) Construct Lagrange's equations in terms of the variables θ and φ and their time derivatives.

13.3. Answer the same questions as in Problem 13.2 when the string is replaced by a heavy, uniform, straight rod of length $2l$ and mass M'.

13.4. Three heavy, uniform rods of lengths l_1, l_2, l_3 and of masses M_1, M_2, M_3, respectively, are linked together and can move in a vertical plane as shown.

(a) Which are the constraints on the Cartesian coordinates of the end points of the rods?

(b) Is it true that the angles θ and φ, as shown, are generalized coordinates?

(c) Construct Lagrange's equations.

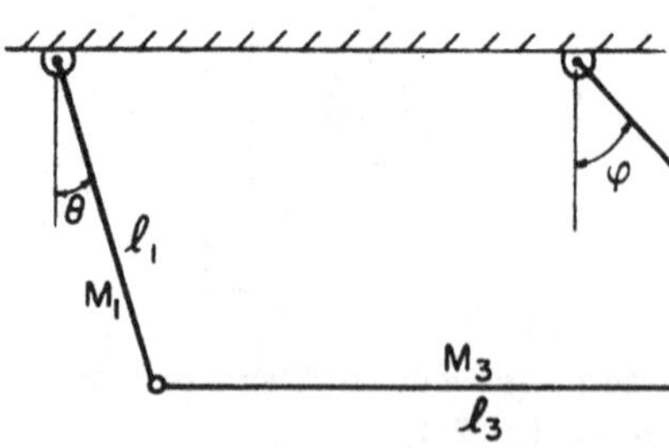

13.5. Two heavy, uniform rods of lengths l_1, l_2 and of masses M_1 and M_2, respectively, form a plane, compound double pendulum, as shown. Let θ be the angle between l_1 and the vertical, and let φ be the angle between l_1 and l_2.

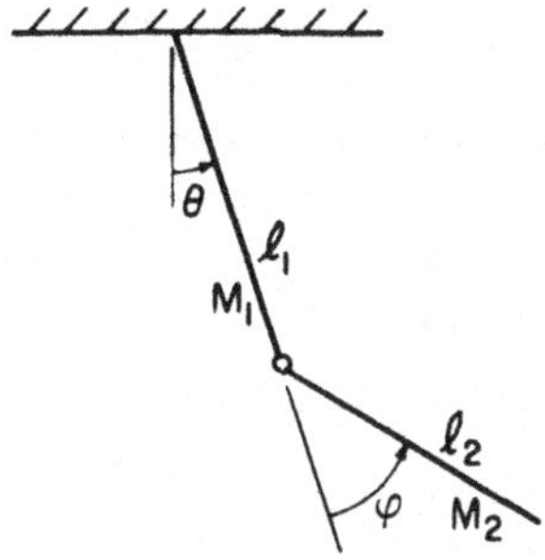

(a) What constraints do the Cartesian coordinates of the endpoints satisfy?

(b) Show that θ and φ are generalized coordinates.

(c) Construct Lagrange's equations in terms of the Cartesian coordinates of the endpoints.

(d) Construct Lagrange's equations in terms of θ and φ.

13.6. Four identical, uniform, heavy rods are hinged together to form a rhombus $OBAC$, as shown. This rhombus is constrained to move in the vertical plane. One of the corners of the rhombus is hinged at the fixed point O, and all hinges are frictionless. In addition to the force of gravity, there

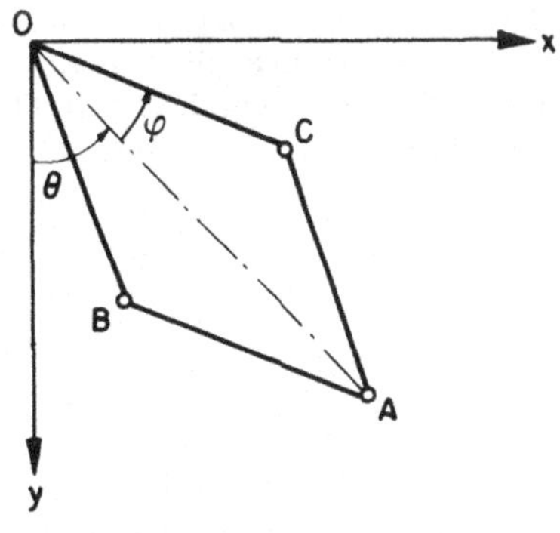

is an attractive force between the points O and A which is linearly proportional to the distance $\overline{OA}$. Let this force be written in the form $Mg/kl\,\overline{OA}$, where M is the mass of one of the rods, l is its length, and k is a constant. Construct the equations of motion in terms of the angles θ and φ.

13.7. Two particles of mass m_1 and m_2, respectively, are connected by springs and dampers to each other and to fixed points, as shown. They can only move to the right or left. Choose a suitable system of generalized coordinates and construct the dissipation function for the following cases:

(a) The dampers are linear; there is no friction between the particles and the ground.

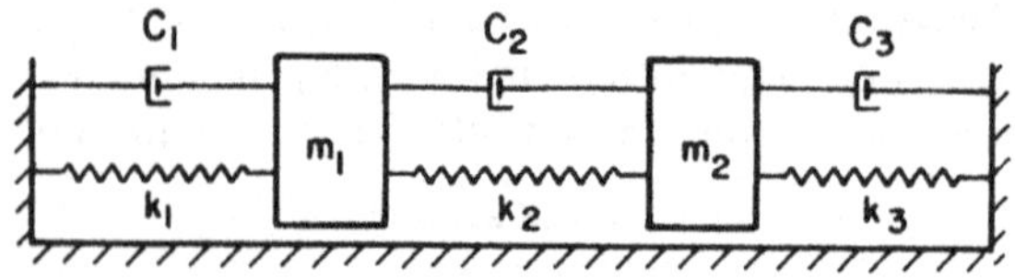

(b) The dampers are of the "velocity-squared" type; there is no friction between the particles and the ground.

(c) The above two cases when there is Coulomb friction between the particles and the ground.

Write Lagrange's equations for all cases.

13.8. Generalize Problem 13.7 for the case of n particles.

13.9. A smooth thin ring is mounted in the vertical plane on a smooth horizontal table so that it can rotate freely about its vertical diameter. A straight uniform rod of length l and mass m passes through the ring. The rod is set into motion in any way, but so that it remains with all its points on the table. What are Lagrange's equations of motion of the rod so long as it does not slip out of the ring?

13.10. A uniform rod of length l and mass m lies on a smooth horizontal table. Each mass element of the rod is attracted to a fixed line in the table with a force which is directly proportional to its distance from the line and to the mass of the element. What are Lagrange's equations of motion of the rod?

13.11. A heavy homogeneous hoop of negligible thickness, of mass m and of radius a, is free to move in a vertical plane. A ring having the same mass as the hoop slides without friction along the hoop. Determine the motion of this system under arbitrary initial conditions.

14

Embedding Constraints

14.1. Introductory Remarks

We saw that, when a system is subject to equality constraints given by

$$\sum_{s=1}^{\bar{n}} B_{rs}\dot{q}_s + B_r = 0 \qquad (r = 1, 2, \ldots, l) \tag{14.1.1}$$

(which may be holonomic or nonholonomic), the motion which satisfies the dynamical laws and the constraints is governed by the differential equations of motion

$$\frac{d}{dt}\frac{\partial T}{\partial \dot{q}_s} - \frac{\partial T}{\partial q_s} - Q_s + \sum_{r=1}^{l} \lambda_r B_{rs} = 0 \qquad (s = 1, 2, \ldots, \bar{n}). \tag{14.1.2}$$

When the constraints are introduced into the formulation of the problem by means of the Lagrangean multipliers λ_r $(r = 1, 2, \ldots, l)$, the equations of constraint are said to be *adjoined* to the problem. The λ_r are auxiliary variables, introduced into the problem in order to ensure that the actual motion does not violate the constraints.

When the formulation of the constrained problem is achieved without the use of auxiliary variables, the constraints are said to be *embedded* in (rather than adjoined to) the problem. In this chapter we examine means of embedding equality constraints.

14.2. A Fallacy

It may be thought that the constraint equations (14.1.1) can be introduced into the kinetic energy T, thereby transforming it into a function

$T^\dagger$ of $\bar{n} - l$ velocity components,* and that the kinetic energy simplified in this manner may be utilized in Lagrange's equations of motion. We shall now show by an example that this procedure may lead to wrong results.

Example 14.2.1. Consider the force-free constrained motion, i.e. free of given forces, of a single particle. Let the coordinates be

$$q_1 = x, \qquad q_2 = y, \qquad q_3 = z. \tag{14.2.1}$$

Since the motion is force-free, the given force is

$$Q = (X, Y, Z) = 0.$$

The kinetic energy is

$$T = \tfrac{1}{2}m(\dot{x}^2 + \dot{y}^2 + \dot{z}^2). \tag{14.2.2}$$

Let the system be subject to the constraint

$$\dot{y} - z\dot{x} = 0. \tag{14.2.3}$$

This equation implies

$$\delta y - z\,\delta x = 0. \tag{14.2.4}$$

We shall first formulate the problem by adjoining the constraint. From (14.2.2) we have

$$\frac{d}{dt}\frac{\partial T}{\partial \dot{x}} = m\ddot{x}, \qquad \frac{d}{dt}\frac{\partial T}{\partial \dot{y}} = m\ddot{y}, \qquad \frac{d}{dt}\frac{\partial T}{\partial \dot{z}} = m\ddot{z}, \qquad \frac{\partial T}{\partial x} = \frac{\partial T}{\partial y} = \frac{\partial T}{\partial z} = 0.$$

Substitution in (14.1.2) gives

$$\begin{aligned} m\ddot{x} - \lambda z &= 0, \\ m\ddot{y} + \lambda &= 0, \\ m\ddot{z} &= 0. \end{aligned} \tag{14.2.5}$$

These three equations together with (14.2.3) constitute the problem formulation by adjoining the constraint.

If one eliminates λ between the first two relations of (14.2.5), one finds

$$\begin{aligned} \ddot{x} + z\ddot{y} &= 0, \\ \ddot{z} &= 0, \\ \dot{y} - z\dot{x} &= 0. \end{aligned} \tag{14.2.6}$$

These are three equations in three unknowns. Differentiation of the last equation of (14.2.6) and subsequent multiplication by z gives

$$z\ddot{y} = z^2\ddot{x} + z\dot{x}\dot{z}, \tag{14.2.7}$$

* When the constraints are embedded directly into the kinetic energy T, the resulting expression will be written as $T^\dagger$. (See also Example 10.5.2.)

and the substitution of this expression in (14.2.6) reduces that system to two equations in two unknowns, i.e.,

$$\begin{aligned}(1 + z^2)\ddot{x} + z\dot{x}\dot{z} &= 0,\\ \ddot{z} &= 0.\end{aligned} \tag{14.2.8}$$

It is tempting to believe that this result could have been obtained more easily and directly by substituting (14.2.3) in (14.2.2), giving

$$T^\dagger = \tfrac{1}{2}m[(1 + z^2)\dot{x}^2 + \dot{z}^2]. \tag{a}$$

Then

$$\frac{\partial T^\dagger}{\partial \dot{x}} = m\dot{x}(1 + z^2), \qquad \frac{d}{\partial t}\frac{\partial T^\dagger}{\partial \dot{x}} = m\ddot{x}(1 + z^2) + 2mz\dot{x}\dot{z},$$

$$\frac{d}{dt}\frac{\partial T^\dagger}{\partial \dot{z}} = m\ddot{z}, \qquad \frac{\partial T^\dagger}{\partial x} = 0, \qquad \frac{\partial T^\dagger}{\partial z} = mz\dot{x}^2.$$

However, the substitution of these expressions into Lagrange's equations results in

$$\begin{aligned}(1 + z^2)\ddot{x} + 2z\dot{x}\dot{z} &= 0,\\ \ddot{z} - z\dot{x}^2 &= 0.\end{aligned} \tag{b}$$

These equations are different from (14.2.8) and, hence *in*correct because equations (14.2.8) were obtained by a previously established, correct method.

It should not be thought that this problem cannot be formulated by embedding. We shall now show how this may be done. If one substitutes T, as given in (14.2.2), in the fundamental equation (13.1.1) one finds

$$m\ddot{x}\,\delta x + m\ddot{y}\,\delta y + m\ddot{z}\,\delta z = 0. \tag{14.2.9}$$

One now substitutes (14.2.4) in this equation, resulting in

$$(\ddot{x} + z\ddot{y})\,\delta x + \ddot{z}\,\delta z = 0. \tag{14.2.10}$$

Finally, the substitution of (14.2.7) gives

$$[(1 + z^2)\ddot{x} + z\dot{x}\dot{z}]\,\delta x + \ddot{z}\,\delta z = 0. \tag{14.2.11}$$

But, for (14.2.11) to hold, one must have

$$\begin{aligned}(1 + z^2)\ddot{x} + z\dot{x}\dot{z} &= 0,\\ \ddot{z} &= 0,\end{aligned} \tag{14.2.12}$$

identical with (14.2.8).

It was shown above that wrong results may be obtained when the velocity constraint is embedded in the kinetic energy, and the resulting expression is then substituted in Lagrange's equations. Nevertheless, it is also quite possible that the result, so found, may be correct. We demonstrate this by:

Example 14.2.2. Consider the same problem as before, except that here the constraint is

$$\dot{y} - k\dot{x} = 0 \tag{14.2.13}$$

where k is a constant. This constraint implies

$$\delta y - k\,\delta x = 0. \tag{14.2.14}$$

We shall formulate the problem first by adjoining the constraints by means of a Lagrangean multiplier. Then, one finds, analogous to (14.2.5)

$$\begin{aligned} m\ddot{x} - \lambda k &= 0, \\ m\ddot{y} + \lambda &= 0, \\ m\ddot{z} &= 0. \end{aligned} \tag{14.2.15}$$

Eliminating λ between the first two, one has

$$\begin{aligned} \ddot{x} + k\ddot{y} &= 0, \\ \ddot{z} &= 0. \end{aligned} \tag{14.2.16}$$

From the contraint equation (14.2.13), it follows that

$$k\ddot{y} = k^2\ddot{x}, \tag{14.2.17}$$

and the substitution of that equation in (14.2.16) gives the problem formulation as

$$\begin{aligned} (1 + k^2)\ddot{x} &= 0, \\ \ddot{z} &= 0. \end{aligned} \tag{14.2.18}$$

This result could have been found more simply and more directly by substituting (14.2.13) in the kinetic energy (14.2.2), giving

$$T^\dagger = \tfrac{1}{2}m[(1 + k^2)\dot{x}^2 + \dot{z}^2]. \tag{14.2.19}$$

The substitution of $T^\dagger$ in Lagrange's equation gives

$$\begin{aligned} (1 + k^2)\ddot{x} &= 0, \\ \ddot{z} &= 0, \end{aligned} \tag{14.2.20}$$

identical with (14.2.18).

Why did the identical procedure lead to a correct result in Example 14.2.2, but not in Example 14.2.1?

The basic difference between these is that in the first the constraint is nonholonomic, while in the second, it is holonomic.

Whether the fundamental equation or Hamilton's principle is used to derive Lagrange's equations, both these dynamical principles are based on work done in *virtual displacements*; hence, any constraint on these displacements must be embedded in the problem in order to obtain a correct problem formulation. In Example 14.2.1, that constraint is (14.2.4) and it was embedded when (14.2.10) was deduced from (14.2.9). However, (14.2.4) does *not* imply (14.2.3), and it was the latter that was embedded in the problem when (a) was deduced from (14.2.2). Moreover, at no subsequent time was the constraint on the virtual displacement utilized in deriving (b) from (a). Hence, the basic reason for the failure of this method in Example 14.2.1 stems from the fact that the incorrect constraint was embedded in the problem.

When the constraint is holonomic, (14.2.13) implies (14.2.14) and, conversely, (14.2.14) implies (14.2.13). Moreover, both imply

$$y - kx + c = 0, \tag{14.2.21}$$

where c is a constant. Therefore, the embedding of (14.2.13) or (14.2.14) or (14.2.21) are all equivalent operations. It is for this reason that the procedure used in deducing (14.2.10) leads to correct results.

14.3. Embedding of Nonholonomic Constraints

Here, we show how nonholonomic equality constraints may be embedded. While the theory given below may also be used for embedding holonomic constraints, this is not the preferred procedure. It is far easier to embed holonomic constraints in their integrated form; this has the effect that the remaining coordinates are a set of generalized coordinates. Thus, we shall assume that all constraints are nonholonomic. They comprise $l < n$ independent equations of the form

$$\sum_{s=1}^{n} B_{rs}\dot{q}_s + B_r = 0 \qquad (r = 1, 2, \ldots, l), \tag{14.3.1}$$

and the q_s $(s = 1, 2, \ldots, n)$, are generalized coordinates. These equations imply that the virtual displacements are constrained by

$$\sum_{s=1}^{n} B_{rs}\,\delta q_s = 0 \qquad (r = 1, 2, \ldots, l). \tag{14.3.2}$$

Evidently these equations may be written in the matrix form

$$\begin{bmatrix} B_{11} & B_{12} & \cdots & B_{1l} \\ B_{21} & B_{22} & \cdots & B_{2l} \\ \vdots & & & \\ B_{l1} & B_{l2} & \cdots & B_{ll} \end{bmatrix} \begin{bmatrix} \delta q_1 \\ \delta q_2 \\ \vdots \\ \delta q_l \end{bmatrix} = - \begin{bmatrix} B_{1,l+1} & B_{1,l+2} & \cdots & B_{1,n} \\ B_{2,l+1} & B_{2,l+2} & \cdots & B_{2,n} \\ \vdots & & & \\ B_{l,l+1} & B_{l,l+2} & \cdots & B_{l,n} \end{bmatrix} \begin{bmatrix} \delta q_{l+1} \\ \delta q_{l+2} \\ \vdots \\ \delta q_n \end{bmatrix}. \tag{14.3.3}$$

Moreover, the square matrix on the left-hand side is nonsingular; if this were not so, the equations of constraint would not be independent of each other.

It follows that the equations of constraint may be solved for l virtual displacements δq_s as linear combinations of the remaining $n - l$ ones, or

$$\delta q_i = \sum_{j=l+1}^{n} a_{ij}\,\delta q_j \qquad (i = 1, 2, \ldots, l), \tag{14.3.4}$$

where the a_{ij} are, in general, functions of the q_s ($s = 1, 2, \ldots, n$), and of t. They are the elements of the matrix product

$$-\begin{bmatrix} B_{11} & B_{12} & \cdots & B_{1l} \\ B_{21} & B_{22} & \cdots & B_{2l} \\ \vdots & & & \vdots \\ B_{l1} & B_{l2} & \cdots & B_{ll} \end{bmatrix}^{-1} \cdot \begin{bmatrix} B_{1,l+1} & B_{1,l+2} & \cdots & B_{1,n} \\ B_{2,l+1} & B_{2,l+2} & \cdots & B_{2,n} \\ \vdots & & & \vdots \\ B_{l,l+1} & B_{l,l+2} & \cdots & B_{l,n} \end{bmatrix}$$

Let us now write the fundamental equation (13.1.1) in the form

$$\sum_{i=1}^{l}\left\{\frac{d}{dt}\frac{\partial T}{\partial \dot q_i} - \frac{\partial T}{\partial q_i} - Q_i\right\}\delta q_i + \sum_{j=l+1}^{n}\left\{\frac{d}{dt}\frac{\partial T}{\partial \dot q_j} - \frac{\partial T}{\partial q_j} - Q_j\right\}\delta q_j = 0. \tag{14.3.5}$$

For simplicity, we put

$$\frac{d}{dt}\frac{\partial T}{\partial \dot q_i} - \frac{\partial T}{\partial q_i} - Q_i = R_i \qquad (i = 1, 2, \ldots, n).$$

Then, (14.3.5) is written as

$$\sum_{i=1}^{l} R_i\,\delta q_i + \sum_{j=l+1}^{n} R_j\,\delta q_j = 0. \tag{14.3.6}$$

If we substitute (14.3.4) in this equation and exchange the order of summation in the first term of the left-hand side we find

$$\sum_{j=l+1}^{n}\left(\sum_{i=1}^{l} R_i a_{ij} + R_j\right)\delta q_j = 0. \tag{14.3.7}$$

In this expression, the δq_j are free; hence, we deduce from it

$$\sum_{i=1}^{l} R_i a_{ij} + R_j = 0 \qquad (j = l+1, l+2, \ldots, n) \tag{14.3.8}$$

or, restoring the meaning of the R_i,

$$\sum_{i=1}^{l} \left\{a_{ij}\left(\frac{d}{dt}\frac{\partial T}{\partial \dot{q}_i} - \frac{\partial T}{\partial q_i} - Q_i\right)\right\} + \frac{d}{dt}\frac{\partial T}{\partial \dot{q}_j} - \frac{\partial T}{\partial q_j} - Q_j = 0$$
$$(j = l + 1, l + 2, \ldots, n). \qquad (14.3.9)$$

These equations, together with

$$\sum_{s=1}^{n} B_{rs}\dot{q}_s + B_r = 0 \qquad (r = 1, 2, \ldots, l),$$

constitute the formulation of the problem by embedding. They form a system of n equations in the n unknowns q_s $(s = 1, 2, \ldots, n)$, and they involve no Lagrangean multipliers.

Example 14.3.1. Using Cartesian coordinates

$$q_1 = x, \qquad q_2 = y, \qquad q_3 = z, \tag{a}$$

formulate by embedding the problem of the motion of a particle under the given force

$$Q = (X, Y, Z) \tag{b}$$

when the motion is subject to the constraints

$$\begin{aligned} \dot{x} + x\dot{y} + y\dot{z} &= 0, \\ y\dot{x} + z\dot{y} + \dot{z} &= 0. \end{aligned} \tag{c}$$

From (c) we deduce

$$\begin{aligned} \delta x + x\,\delta y + y\,\delta z &= 0, \\ y\,\delta x + z\,\delta y + \delta z &= 0, \end{aligned} \tag{d}$$

or, in matrix form,

$$\begin{bmatrix} 1 & x \\ y & z \end{bmatrix} \begin{bmatrix} \delta x \\ \delta y \end{bmatrix} = \begin{bmatrix} -y\,\delta z \\ -\delta z \end{bmatrix}.$$

One finds readily

$$\begin{bmatrix} \delta x \\ \delta y \end{bmatrix} = \frac{1}{z - xy} \begin{bmatrix} z & -x \\ -y & 1 \end{bmatrix} \begin{bmatrix} -y\,\delta z \\ -\delta z \end{bmatrix},$$

or,

$$\begin{aligned} \delta x &= \frac{-zy + x}{z - xy}\,\delta z, \\ \delta y &= \frac{y^2 - 1}{z - xy}\,\delta z. \end{aligned} \tag{e}$$

The kinetic energy is

$$T = \tfrac{1}{2}m(\dot{x}^2 + \dot{y}^2 + \dot{z}^2)$$

and, therefore, the fundamental equation is

$$(m\ddot{x} - X)\,\delta x + (m\ddot{y} - Y)\,\delta y + (m\ddot{z} - Z)\,\delta z = 0. \tag{f}$$

The embedding consists of substituting (e) in (f). Since δz is then arbitrary, one has

$$(m\ddot{x} - X)(x - yz) + (m\ddot{y} - Y)(y^2 - 1) + (m\ddot{z} - Z)(z - xy) = 0. \tag{g}$$

This equation together with (c) constitutes a system of three equations in three unknowns. It is the formulation of this problem by embedding.

14.4. Problems

14.1. Show that the formulation found by adjoining the constraints by means of Lagrangean multipliers leads to a result which is reducible to the system (c) and (g) of Example 14.3.1.

14.2. Show that adjoining and embedding of constraints leads to the same results in any of the problems at the end of Chapter 13.

15

Formulating Problems by Lagrange's Equations

15.1. General Remarks

Lagrange's equations formulate the general, constrained dynamics problem without the need for fixing the coordinate system at the outset, and in terms of general constraints [provided only that they belong to the broad class of constraints defined in (11.2.14) and (11.2.15)]. This is, undoubtedly, the greatest contribution of Lagrangean mechanics to dynamics. However, an important feature, and one cherished greatly by the practitioner concerned with solving problems, is that Lagrange's equations provide a very simple, and a nearly foolproof device for the mathematical formulation of a physical problem. This formulation must, of course, precede any attempts at the solution, and it can be difficult. Consider for instance, the problem of the simple plane double pendulum illustrated in Fig. 15.1.1. If one chooses

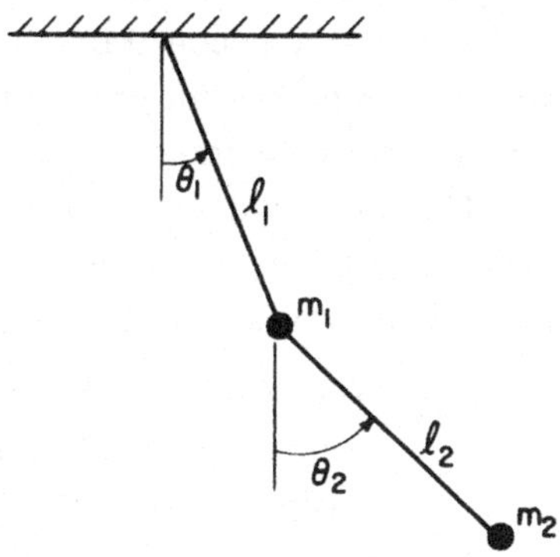

Fig. 15.1.1. Plane, simple double pendulum.

the coordinates θ_1 and θ_2 to describe the configuration, the correct equations of motion are

$$(m_1 + m_2)l_1^2\ddot{\theta}_1 + m_2l_1l_2\cos(\theta_1 - \theta_2)\ddot{\theta}_2 + m_2l_1l_2\dot{\theta}_2^2\sin(\theta_1 - \theta_2) + (m_1 + m_2)gl_1\sin\theta_1 = 0,$$

and

$$m_2l_2^2\ddot{\theta}_2 + m_2l_1l_2\cos(\theta_1 - \theta_2)\ddot{\theta}_1 - m_2l_1l_2\dot{\theta}_1^2\sin(\theta_1 - \theta_2) + m_2gl_2\sin\theta_2 = 0.$$

These are difficult to write down by inspection and appeal to Newton's second law, but they are readily found by means of Lagrange's equations (see Example 15.5.2).

In using Lagrange's equations to formulate a given problem, the first step is to choose a set of coordinates which describes uniquely the configuration of the system with respect to an inertial frame. Next, the kinetic energy and the constraints are written in terms of these coordinates, and the force components of the given forces along these coordinates are computed. Then, the substitution of these quantities in Lagrange's equations results in the mathematical formulation of the problem under study.

As there exists an infinity of sets of generalized coordinates (and of course, also an infinity of sets which are not *generalized* coordinates) it is clear that one can formulate the problem in infinitely many ways. It turns out frequently that some coordinate systems result in a simple and transparent formulation of the problem while others do not; thus, the choice of coordinates can have a profound effect on one's ability to solve the problem.

The best choice of coordinates, or even merely a good one, depends significantly on the problem under study. The formulation of the general problem contains terms arising from the kinetic energy, components of given forces, and constraint terms. It is very common that one choice of coordinates makes one group of terms particularly simple while complicating others, and a different choice may have the opposite effect.

Certain recommendations are given below for choosing suitable coordinate systems; however, they are not hard and fast rules, and they may have to be revised in specific cases.

A. When *generalized* coordinates are used the number of equations of motion and the number of constraints is reduced to the least possible; this is always desirable.

B. When a holonomically constrained problem is considered, and a system of generalized coordinates exists which is "natural to the constraints," it should be used. For instance, in the case of the spherical pendulum, spherical coordinates should be used; for motion in a cylindrical surface, cylindrical coordinates should be used, etc.

C. When a set of generalized coordinates renders the equations of motion linear, it should be chosen.

D. When there is no indication as to choice of coordinates, and when the Cartesian coordinates are also generalized coordinates, they should be used.

These recommendations are best illustrated by examples; therefore we shall now consider a number of specific problems.

15.2. The Unconstrained Particle

Three coordinates are necessary and sufficient to specify the position of a single unconstrained particle in 3-space. Hence the Cartesian coordinates are generalized coordinates, or

$$q_1 = x, \qquad q_2 = y, \qquad q_3 = z. \tag{15.2.1}$$

The kinetic energy is

$$T = \tfrac{1}{2}m(\dot{x}^2 + \dot{y}^2 + \dot{z}^2). \tag{15.2.2}$$

If the Cartesian force components are

$$Q_1 = X, \qquad Q_2 = Y, \qquad Q_3 = Z, \tag{15.2.3}$$

the substitution of (15.2.2) and (15.2.3) in Lagrange's equations results in

$$m\ddot{x} - X = 0, \qquad m\ddot{y} - Y = 0, \qquad m\ddot{z} - Z = 0. \tag{15.2.4}$$

The position of the particle in 3-space can, of course, also be described uniquely by the cylindrical coordinates

$$q_1 = \varrho, \qquad q_2 = \varphi, \qquad q_3 = z. \tag{15.2.5}$$

They are connected to the Cartesian coordinates by the transformations

$$x = \varrho \cos \varphi, \qquad y = \varrho \sin \varphi, \qquad z = z. \tag{15.2.6}$$

It follows from (15.2.6) that

$$\begin{aligned} \dot{x} &= \dot{\varrho}\cos\varphi - \varrho\dot{\varphi}\sin\varphi, \\ \dot{y} &= \dot{\varrho}\sin\varphi + \varrho\dot{\varphi}\cos\varphi, \\ \dot{z} &= \dot{z}, \end{aligned} \tag{15.2.7}$$

and the substitution of these in the kinetic energy results in

$$T = \tfrac{1}{2}m(\dot{\varrho}^2 + \varrho^2\dot{\varphi}^2 + \dot{z}^2). \tag{15.2.8}$$

If we write the generalized forces as

$$Q_1 = R, \qquad Q_2 = \Phi, \qquad Q_3 = Z$$

and note that, as defined in (12.3.8),

$$Q_s = \sum_{r=1}^{N} F_r \frac{\partial u_r}{\partial q_s} \qquad (s = 1, 2, \ldots, n), \tag{15.2.9}$$

we find

$$\begin{aligned} R &= X\cos\varphi + Y\sin\varphi, \\ \Phi &= -X\varrho\sin\varphi + Y\varrho\cos\varphi, \\ Z &= Z. \end{aligned} \tag{15.2.10}$$

Since

$$\begin{aligned} \frac{d}{dt}\frac{\partial T}{\partial\dot{\varrho}} &= m\ddot{\varrho}, \\ \frac{d}{dt}\frac{\partial T}{\partial\dot{\varphi}} &= 2m\varrho\dot{\varrho}\dot{\varphi} + m\varrho^2\ddot{\varphi}, \\ \frac{d}{dt}\frac{\partial T}{\partial\dot{z}} &= m\ddot{z}, \\ \frac{\partial T}{\partial\varrho} &= m\varrho\dot{\varphi}^2, \\ \frac{\partial T}{\partial\varphi} &= \frac{\partial T}{\partial z} = 0, \end{aligned}$$

the Lagrange equations in cylindrical coordinates become

$$\begin{aligned} m\ddot{\varrho} - m\varrho\dot{\varphi}^2 - R &= 0, \\ m\varrho^2\ddot{\varphi} + 2m\varrho\dot{\varrho}\dot{\varphi} - \Phi &= 0, \\ m\ddot{z} - Z &= 0, \end{aligned} \tag{15.2.11}$$

where R and Φ are given in (15.2.10).

From a comparison between (15.2.4) and (15.2.11) one may be tempted to conclude that Cartesian coordinates always result in a much simpler formulation of the problem of the unconstrained particle than do cylindrical coordinates. In particular, the equations (15.2.11) are always nonlinear while the terms in (15.2.4) arising from the kinetic energy are always linear. Nevertheless, cases may arise where the formulation in cylindrical coordinates is preferable. Suppose, for instance, that the Cartesian force components are

$$X = \frac{Ax - By}{(x^2 + y^2)^{1/2}}, \qquad Y = \frac{Bx + Ay}{(x^2 + y^2)^{1/2}}, \qquad Z = C, \tag{15.2.12}$$

where A, B, and C are real constants. In that case, direct substitution of (15.2.12) in (15.2.10) gives

$$R = A, \qquad \Theta = B\varrho, \qquad Z = C. \tag{15.2.13}$$

We observe that the generalized forces are *nonlinear* functions of the Cartesian coordinates, but they are either constants or linear in the cylindrical coordinates. We have here a case where Cartesian coordinates simplify the inertia terms (those arising from the kinetic energy), but they complicate the forces, whereas cylindrical coordinates have the opposite effect; the equations of motion are nonlinear in either coordinates.

Let us consider the unconstrained single particle in the spherical coordinates

$$q_1 = r, \qquad q_2 = \theta, \qquad q_3 = \varphi, \tag{15.2.14}$$

which are connected to the Cartesian coordinates by the transformations

$$\begin{aligned} x &= r \sin\theta \cos\varphi, \\ y &= r \sin\theta \sin\varphi, \\ z &= r \cos\theta. \end{aligned} \tag{15.2.15}$$

The substitution of these in (15.2.2) gives the kinetic energy as

$$T = \tfrac{1}{2}m(\dot{r}^2 + r^2 \sin^2\theta\dot{\varphi}^2 + r^2\dot{\theta}^2). \tag{15.2.16}$$

If the generalized force components are denoted by

$$Q_1 = R, \qquad Q_2 = \Theta, \qquad Q_3 = \Phi, \tag{15.2.17}$$

one finds from utilizing (15.2.9) that

$$\begin{aligned} R &= X \sin\theta \cos\varphi + Y \sin\theta \sin\varphi + Z \cos\theta, \\ \Theta &= Xr \cos\theta \cos\varphi + Yr \cos\theta \sin\varphi - Zr \sin\theta, \\ \Phi &= -Xr \sin\theta \sin\varphi + Yr \sin\theta \cos\varphi, \end{aligned} \tag{15.2.18}$$

where X, Y, and Z are the Cartesian components of the given force. The Lagrangean equations are found to be

$$\begin{aligned} \frac{d}{dt}(m\dot{r}) - mr \sin^2\theta\dot{\varphi}^2 - mr\dot{\theta}^2 - R &= 0, \\ \frac{d}{dt}(mr^2\dot{\theta}) - mr^2 \sin\theta\cos\theta\dot{\varphi}^2 - \Theta &= 0, \\ \frac{d}{dt}(mr^2 \sin^2\theta\dot{\varphi}) - \Phi &= 0. \end{aligned} \tag{15.2.19}$$

These equations would take on a more involved appearance still if the differentiations on the left-hand side were carried out. Even as they stand, they appear to be much more complicated than (15.2.4).

Here however, there exists a classical example—the central force problem—for which spherical coordinates give the simplest known formulation to the problem. (See Section 19.2.)

The central force problem is characterized by the fact that the Cartesian components of the given forces are of the form

$$\begin{aligned} X &= \frac{xF(\sqrt{x^2 + y^2 + z^2})}{\sqrt{x^2 + y^2 + z^2}}, \\ Y &= \frac{yF(\sqrt{x^2 + y^2 + z^2})}{\sqrt{x^2 + y^2 + z^2}}, \\ Z &= \frac{zF(\sqrt{x^2 + y^2 + z^2})}{\sqrt{x^2 + y^2 + z^2}}, \end{aligned} \tag{15.2.20}$$

where F is any bounded function of the argument $(x^2 + y^2 + z^2)^{1/2}$. If one substitutes the transformations (15.2.15) in (15.2.20), and the resulting equations in (15.2.18), the generalized force components become

$$R = F(r), \qquad \Theta = \Phi \equiv 0.$$

The substitution of these in (15.2.19) results in equations which are easily integrated. In the central force problem, spherical coordinates are the best set of generalized coordinates.

15.3. The Holonomically Constrained Particle

The holonomically constrained particle moves on a surface if it is subject to only one constraint, or in the intersection of two surfaces, i.e., a curve, if it is subject to two constraints. These surfaces or curves are rigid if both constraints are scleronomic; they move or warp if one or both constraints are rheonomic.

Consider a particle moving in a smooth surface defined by

$$z = f(x, y) \tag{15.3.1}$$

where f is a C^2 function of x and y. Then, one finds

$$\dot{z} = \frac{\partial f}{\partial x}\dot{x} + \frac{\partial f}{\partial y}\dot{y}, \tag{15.3.2}$$

and this constraint may be embedded directly in the kinetic energy

$$T = \tfrac{1}{2}m(\dot{x}^2 + \dot{y}^2 + \dot{z}^2)$$

because it is holonomic. The substitution of (15.3.2) in the kinetic energy gives

$$T^{\dagger} = \tfrac{1}{2}m\dot{x}^2(1 + f_x^2) + m\dot{x}\dot{y}f_xf_y + \tfrac{1}{2}m\dot{y}^2(1 + f_y^2) \tag{15.3.3}$$

where

$$f_x = \frac{\partial f}{\partial x}, \qquad f_y = \frac{\partial f}{\partial y}.$$

When this last relation is substituted in Lagrange's equations, one finds

$$\begin{aligned} &\ddot{x}(1 + f_x^2) + \ddot{y}f_xf_y + \dot{x}^2f_xf_{xx} + 2\dot{x}\dot{y}f_xf_{xy} + \dot{y}^2f_xf_{yy} - X/m = 0, \\ &\ddot{x}f_xf_y + \ddot{y}(1 + f_y^2) + \dot{x}^2f_yf_{xx} + 2\dot{x}\dot{y}f_yf_{xy} + \dot{y}^2f_yf_{yy} - Y/m = 0, \end{aligned} \tag{15.3.4}$$

where

$$f_{xx} = \frac{\partial^2 f}{\partial x^2}, \qquad f_{xy} = \frac{\partial^2 f}{\partial x\,\partial y}, \qquad f_{yy} = \frac{\partial^2 f}{\partial y^2},$$

and X and Y are the Cartesian components of the given forces. These are the equations of motion; they are seen to be dynamically coupled (see Section 13.6) as both accelerations occur in each equation of motion.

Let us suppose the constraint to be rheonomic. In other terms, let the surface be defined by

$$z = g(x, y, t), \tag{15.3.5}$$

where g is a C^2 function in all arguments. Then, we may substitute

$$\dot{z} = g_x\dot{x} + g_y\dot{y} + g_t \tag{15.3.6}$$

in the kinetic energy with the result that

$$\begin{aligned} T^\dagger = {} & \tfrac{1}{2}m(1 + g_x{}^2)\dot{x}^2 + mg_xg_y\dot{x}\dot{y} + \tfrac{1}{2}m(1 + g_y{}^2)\dot{y}^2 \\ & + m(g_xg_t\dot{x} + g_yg_t\dot{y}) + \tfrac{1}{2}mg_t{}^2. \end{aligned} \tag{15.3.7}$$

It is seen that the kinetic energy is of the general form (12.1.3), i.e., it contains terms which are linear in the velocities, and terms independent of the velocities. The problem is formulated when $d(\partial T/\partial\dot{x})/dt$, $\partial T/\partial x$, and similar terms in y are computed and substituted into Lagrange's equations.

In actual problems, the functions defining the surfaces are specified at the outset and are substituted in (15.3.7) before Lagrange's equations are constructed.

When the problem is holonomically constrained, it is very common that the surface of constraint dictates the best choice of generalized coordinates. We shall demonstrate this by an example.

Example 15.3.1. Formulate the problem of a particle constrained to move on the surface of a right circular cylinder.

Cartesian coordinates are quite unsuitable to formulate this problem. We use them nevertheless to show how awkward the formulation becomes compared to that in cylindrical coordinates. Thus we take

$$q_1 = x, \qquad q_2 = y, \qquad q_3 = z, \tag{15.3.8}$$

subject to the constraint

$$x^2 + y^2 - \varrho^2 = 0, \tag{15.3.9}$$

where $\varrho = \text{const}$ is the radius of the cylinder. We begin by adjoining the constraint. The infinitesimal constraint is

$$x\,dx + y\,dy = 0, \tag{15.3.10}$$

and the kinetic energy is

$$T = \tfrac{1}{2}m(\dot{x}^2 + \dot{y}^2 + \dot{z}^2). \tag{15.3.11}$$

Therefore, substitution in Lagrange's equations gives

$$\begin{aligned} m\ddot{x} - X + \lambda x &= 0, \\ m\ddot{y} - Y + \lambda y &= 0, \\ m\ddot{z} - Z &= 0, \end{aligned} \tag{15.3.12}$$

where X, Y, and Z are the Cartesian components of the given force, and λ is a Lagrange multiplier. Equations (15.3.9) and (15.3.12) are four equations in the four unknowns x, y, z, and λ.

As the problem is holonomic, one may directly embed the constraint (15.3.9) in the kinetic energy. One has from (15.3.9)

$$x\dot{x} + y\dot{y} = 0$$

or

$$\dot{y} = -\frac{x}{y}\dot{x}, \tag{15.3.13}$$

and also

$$y = \pm(\varrho^2 - x^2)^{1/2}. \tag{15.3.14}$$

Combining them gives

$$\dot{y}^2 = \frac{x^2\dot{x}^2}{\varrho^2 - x^2}. \tag{15.3.15}$$

Substitution of this equation in the kinetic energy gives

$$T^\dagger = \frac{1}{2}m\left[\left(1 + \frac{x^2}{\varrho^2 - x^2}\right)\dot{x}^2 + \dot{z}^2\right], \tag{15.3.16}$$

in which

$$q_1 = x, \qquad q_2 = z$$

are generalized coordinates. The substitution of (15.3.16) in Lagrange's equations results in

$$\begin{aligned} m\left[\left(1 + \frac{x^2}{\varrho^2 - x^2}\right)\ddot{x} + \frac{2\varrho^2 x\dot{x}^2}{(\varrho^2 - x^2)^2}\right] - X &= 0, \\ m\ddot{z} - Z &= 0. \end{aligned} \tag{15.3.17}$$

If the given force components X and Z depend on y and/or $\dot{y}$, these variables may be eliminated by utilizing (15.3.13) and (15.3.14).

For this problem, it is much more natural to use cylindrical coordinates

$$q_1 = \varphi, \qquad q_2 = z.$$

Here, ϱ = const and is, therefore, not a generalized coordinate. The formulation in these coordinates is found directly from (15.2.11) by setting the derivatives of ϱ identically equal to zero in that equation. This results in

$$\begin{aligned} m\varrho^2\ddot{\varphi} - \Phi &= 0, \\ m\ddot{z} - Z &= 0. \end{aligned} \tag{15.3.18}$$

These are the equations of motion in cylindrical coordinates. The latter formulation is not only much simpler than the one in Cartesian coordinates, but (15.3.18) are linear equations whenever Φ and Z are linear functions of φ and z and their time-derivatives, while the first of (15.3.17) is never linear.

Example 15.3.2. To consider a rheonomic example, let us suppose that the radius of the cylinder of Example 15.3.1 increases with time in a prescribed manner:

$$\varrho = \varrho(t), \qquad \dot{\varrho} > 0. \tag{15.3.19}$$

From (15.2.8), the kinetic energy is

$$T^{\dagger} = \tfrac{1}{2}m(\dot{\varrho}^2 + \varrho^2\dot{\varphi}^2 + \dot{z}^2), \tag{15.3.20}$$

where the generalized coordinates are

$$q_1 = \varphi, \qquad q_2 = z.$$

The quantity ϱ is *not* a generalized coordinate; it is a prescribed function of time.

It follows from (15.2.11) that the Lagrangean equations of motion are now

$$\begin{aligned} m\varrho^2\ddot{\varphi} + 2m\varrho\dot{\varrho}\dot{\varphi} - \Phi &= 0, \\ m\ddot{z} - Z &= 0. \end{aligned} \tag{15.3.21}$$

This formulates the problem.

15.4. The Nonholonomically Constrained Particle

In nonholonomic problems, the number of generalized coordinates always exceeds the number of degrees of freedom. The reason for this is that nonholonomic constraints do not affect the accessibility of configurations (see Section 4.6). Hence, the particle in 3-space, subject to nonholonomic constraints only, may occupy any position whatever, and three numbers are required to specify that position.

Here, we consider nonholonomic *equality* constraints only. These may always be adjoined by the method of Lagrange multipliers (see Section 13.2) or they may be embedded by the method of Section 14.3 but they may never be embedded by the method utilized for holonomic constraints in Section 15.3; it was already indicated in Section 14.2 that that procedure leads to wrong results. Most nonholonomic constraints arising in physical problems are constraints on orientation or direction of velocities. But, by definition, a particle has no orientation; therefore, nonholonomic problems usually involve systems of particles (for instance, the line connecting two particles does have orientation) or rigid bodies. Nevertheless, meaningful nonholonomic problems in the motion of a single particle do occur. Here we discuss one catastatic and one acatastatic problem.

Example 15.4.1. Let the horizontal velocity components $\dot{x}$ and $\dot{y}$ of a particle be controlled by an altitude-dependent device in such a way that the velocity ratio $\dot{y}/\dot{x}$ is directly proportional to the altitude z. This mechanism may be regarded as a steering mechanism: When $z = 0$ the projection of the trajectory on the xy plane is parallel to the x axis, but it becomes more nearly parallel to the y axis, the more the altitude increases. This is illustrated in Fig. 15.4.1.

From $\dot{y}/\dot{x} = z$ we find the Pfaffian form

$$dy - z\,dx = 0, \tag{15.4.1}$$

and this is not integrable, as was already shown in Example 4.2.3. The generalized coordinates are

$$q_1 = x, \qquad q_2 = y, \qquad q_3 = z.$$

and the constraint on the virtual displacement is

$$\delta y - z\,\delta x = 0. \tag{15.4.2}$$

From (13.1.3)

$$B_{11} = -z, \qquad B_{12} = 1, \qquad (r = 1) \tag{15.4.3}$$

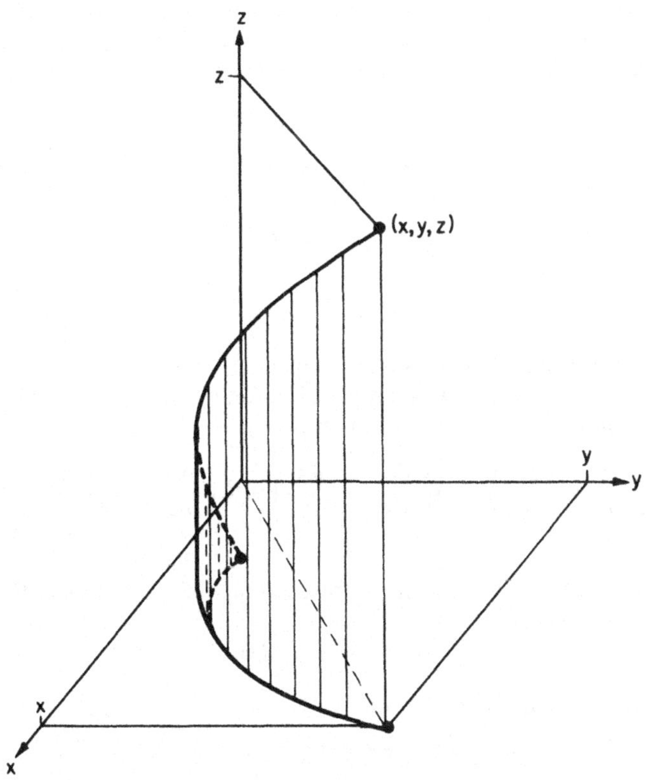

Fig. 15.4.1. Catastatic, nonholonomic constraint of Example 15.4.1.

in $\sum_s B_{rs}\,\delta q_s = 0$. The kinetic energy is

$$T = \tfrac{1}{2}m(\dot{x}^2 + \dot{y}^2 + \dot{z}^2). \tag{15.4.4}$$

Then, if the Cartesian force components are X, Y, and Z, substitution in Lagrange's equations (13.3.3) gives

$$\begin{aligned} m\ddot{x} - X - \lambda z &= 0, \\ m\ddot{y} - Y + \lambda &= 0, \\ m\ddot{z} - Z &= 0. \end{aligned} \tag{15.4.5}$$

These equations together with

$$\dot{y} - z\dot{x} = 0 \tag{15.4.6}$$

determine x, y, z, and λ.

In this problem it is simple to eliminate the Lagrange multiplier between the first two equations of (15.4.5). One finds

$$m\ddot{x} - X + z(m\ddot{y} - Y) = 0, \tag{15.4.7}$$

and this equation, (15.4.6), and the last relation of (15.4.5) formulate the problem.

Example 15.4.2. As an example of an acatastatic, nonholonomic problem we treat a generalization of Example 10.5.2. Let a particle moving in the plane under Cartesian force components X and Y be subject to the constraint

$$\dot{x} - t\dot{y} = a, \tag{15.4.8}$$

where $a \neq 0$ is a constant. (Putting $a = 0$ gives the constraint in Example 10.5.2.) The Pfaffian form of (15.4.8) is

$$dx - t\,dy - a\,dt = 0, \tag{15.4.9}$$

and it implies

$$\delta x - t\,\delta y = 0 \tag{15.4.10}$$

for the constraints on the virtual displacements.

It is evident that (15.4.9) is not integrable, for, on substituting it in the integrability condition (4.5.7) with $z = t$, one finds

$$A\left(\frac{\partial B}{\partial t} - \frac{\partial C}{\partial y}\right) + B\left(\frac{\partial C}{\partial x} - \frac{\partial A}{\partial t}\right) + C\left(\frac{\partial A}{\partial y} - \frac{\partial B}{\partial x}\right) = -1,$$

and not zero.

We use the embedding technique of Section 14.3 to formulate the problem. The kinetic energy is

$$T = \tfrac{1}{2}m(\dot{x}^2 + \dot{y}^2). \tag{15.4.11}$$

Proceeding as in Example 14.3.1, one finds

$$mt\ddot{x} - tX + m\ddot{y} - Y = 0. \tag{15.4.12}$$

This equation together with (15.4.8) determines x and y.

15.5. Systems of Particles and Rigid Bodies

The meaning of the phrase "systems of particles" was defined in Section 2.2. However, the difference between systems of many particles on the one hand and of a single particle on the other is physical, rather than mathematical, in nature. This is seen when one writes the kinetic energy of $N/3$ particles:

$$T = \frac{1}{2} \sum_{s=1}^{N} m_s \dot{u}_s^2. \tag{15.5.1}$$

Then, the transformations

$$w_s = m_s^{1/2} u_s \qquad (s = 1, 2, \ldots, N) \tag{15.5.2}$$

change (15.5.1) into

$$T = \frac{1}{2} \sum_{s=1}^{N} \dot{w}_s^2. \tag{15.5.3}$$

For $N = 3$, this is the kinetic energy of a single particle of unit mass moving in 3-space, and for $N > 3$ it may be regarded as the kinetic energy of a single particle (of unit mass) moving in N-space. When the transformations (15.5.2) and their time derivatives are substituted in the force components, there results equations of motion of the form

$$\ddot{w}_s = f_s\left(\frac{w_1}{m_1^{1/2}}, \frac{w_2}{m_2^{1/2}}, \ldots, \frac{w_N}{m_N^{1/2}}; \frac{\dot{w}_1}{m_1^{1/2}}, \frac{\dot{w}_2}{m_2^{1/2}}, \ldots, \frac{\dot{w}_N}{m_N^{1/2}}; t\right) \qquad (s = 1, 2, \ldots, N), \tag{15.5.4}$$

and these may be regarded as the equations of motion of a single unit mass in N-space, subject to a force whose component in the w_s direction is f_s. This view of the dynamics problem has the effect of endowing the C trajectories with a certain measure of pseudo-reality because, under the scale changes (15.5.2), the C trajectories of Section 3.1 become in fact "trajectories" in the sense that they are the paths traced out by the unit mass. Thus, the transformations (15.5.2) are often helpful in arriving at an interpretation of the solution of a problem. But they are rarely helpful in formulating it.

Perhaps, the best way of treating the formulation of problems involving many particles is by example; therefore, we now give a number of these.

Example 15.5.1. Formulate the problem of two particles of masses m_1 and m_2, respectively, moving in the plane in such a way that the line connecting their positions passes for all time through a fixed point in the plane. [We shall return again to this problem in the discussion of methods of integration (see Sections 16.3 and 16.6).]

We use this example to illustrate the advantages which accrue from choosing a good set of generalized coordinates. We begin by using Cartesian coordinates, a choice which is not very good.

Let the fixed point in the plane be the origin of the xy system. Let a particle of mass m_i have position (x_i, y_i), and let the Cartesian components of the force on m_i be X_i, Y_i $(i = 1, 2)$; all this is shown in Fig. 15.5.1.

The kinetic energy is

$$T = \tfrac{1}{2}[m_1(\dot{x}_1^2 + \dot{y}_1^2) + m_2(\dot{x}_2^2 + \dot{y}_2^2)], \tag{15.5.5}$$

and the constraint which ensures that both particles are on a line passing through the origin is

$$y_1/x_1 = y_2/x_2$$

or

$$x_1 y_2 - x_2 y_1 = 0. \tag{15.5.6}$$

The Pfaffian forms are

$$\begin{aligned} y_2\, dx_1 - y_1\, dx_2 - x_2\, dy_1 + x_1\, dy_2 &= 0, \\ y_2\, \delta x_1 - y_1\, \delta x_2 - x_2\, \delta y_1 + x_1\, \delta y_2 &= 0, \end{aligned} \tag{15.5.7}$$

and the possible velocities satisfy

$$y_2\dot{x}_1 - y_1\dot{x}_2 - x_2\dot{y}_1 + x_1\dot{y}_2 = 0. \tag{15.5.8}$$

The system has three degrees of freedom because the four Cartesian coordinates must satisfy one constraint, equation (15.5.6). The procedure for formulating this problem in generalized coordinates x_2, y_1, y_2 is to solve (15.5.6) for x_1, i.e.,

$$x_1 = x_2 y_1 / y_2, \tag{15.5.9}$$

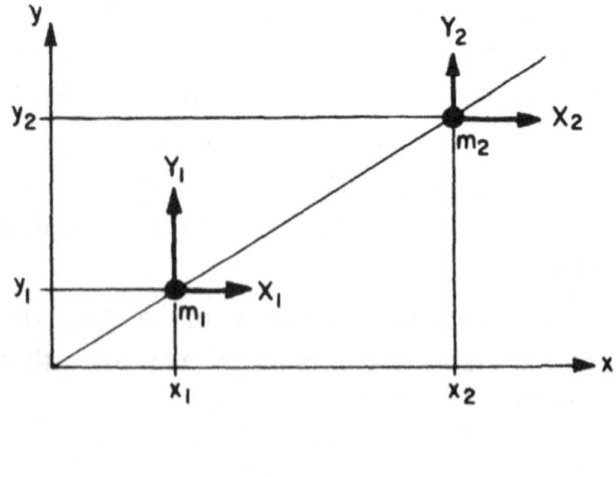

Fig. 15.5.1. Constrained motion of two particles of Example 15.5.1.

and to substitute this in (15.5.8), solved for $\dot{x}_1$, i.e.,

$$\dot{x}_1 = (y_1\dot{x}_2 + x_2\dot{y}_1 - x_2y_1\dot{y}_2/y_2)/y_2. \tag{15.5.10}$$

When the square of (15.5.10) is substituted in the kinetic energy, one finds

$$T^\dagger = \frac{1}{2}\left[\left(m_1\frac{y_1^2}{y_2^2} + m_2\right)\dot{x}_2^2 + m_1\left(\frac{x_2^2}{y_2^2} + 1\right)\dot{y}_1^2 + \left(m_1\frac{x_2^2y_1^2}{y_2^4} + m_2\right)\dot{y}_2^2 + 2m_1\left(\frac{x_2y_1}{y_2^2}\dot{x}_2\dot{y}_1 - \frac{x_2y_1^2}{y_2^3}\dot{x}_2\dot{y}_2 - \frac{x_2^2y_1}{y_2^3}\dot{y}_1\dot{y}_2\right)\right].$$

Of course, (15.5.9) and (15.5.10) must also be substituted in the force components if these depend on x_1 and/or $\dot{x}_1$. Finally, $T^\dagger$ and the force components are substituted in Lagrange's equations with $q_1 = x_2$, $q_2 = y_1$, $q_3 = y_2$. The kinetic energy contains not only the squares of the velocity components but all their cross products, and it depends on all three coordinates as well. Therefore, the use of Cartesian coordinates has resulted in a very cumbersome formulation. This is not due to the inherent difficulty of the problem but, rather, to the poor choice of coordinates.

As both masses are always on a straight line through the origin, the use of polar coordinates r_i, θ_i $(i = 1, 2)$ is indicated. They are connected with the Cartesian coordinates by

$$x_i = r_i\cos\theta_i, \qquad y_i = r_i\sin\theta_i \qquad (i = 1, 2), \tag{15.5.11}$$

and differentiating these and substituting in the kinetic energy gives

$$T^\dagger = \tfrac{1}{2}[m_1(\dot{r}_1^2 + r_1^2\dot{\theta}_1^2) + m_2(\dot{r}_2^2 + r_2^2\dot{\theta}_2^2)]. \tag{15.5.12}$$

The constraint equation is seen from the geometry of Fig. 15.5.1 to be

$$\theta_1 - \theta_2 = 0$$

or

$$\theta_1 = \theta_2 = \theta. \tag{15.5.13}$$

This can also be derived formally by substituting (15.5.11) in the constraint equation (15.5.6); this gives

$$\sin\theta_1\cos\theta_2 - \cos\theta_1\sin\theta_2 = 0,$$

which may be written as

$$\sin(\theta_1 - \theta_2) = 0;$$

this equation implies (15.5.13). The substitution of (15.5.13) in the kinetic energy results in

$$T^\dagger = \tfrac{1}{2}[(m_1\dot{r}_1^2 + m_2\dot{r}_2^2) + (m_1r_1^2 + m_2r_2^2)\dot{\theta}^2]. \tag{15.5.14}$$

This expression could have been written down by inspection because the first term is the kinetic energy due to translation of the masses along the rotating line, and the second is that due to the rotation.

From (15.5.14), we have

$$\frac{d}{dt}\frac{\partial T^\dagger}{\partial \dot r_i} = m_i\ddot r_i, \qquad \frac{\partial T^\dagger}{\partial r_i} = m_i r_i\dot\theta^2 \qquad (i = 1, 2),$$

$$\frac{d}{dt}\frac{\partial T^\dagger}{\partial \dot\theta} = (m_1 r_1^2 + m_2 r_2^2)\ddot\theta + 2(m_1 r_1\dot r_1 + m_2 r_2 \dot r_2)\dot\theta, \qquad \frac{\partial T^\dagger}{\partial \theta} = 0.$$

If we write the generalized forces as R_1, R_2, Θ, the Lagrange equations are

$$\begin{aligned} &m_i\ddot r_i - m_i r_i\dot\theta^2 - R_i = 0 \qquad (i = 1, 2),\\ &(m_1 r_1^2 + m_2 r_2^2)\ddot\theta + 2(m_1 r_1\dot r_1 + m_2 r_2\dot r_2)\dot\theta - \Theta = 0, \end{aligned} \tag{15.5.15}$$

and these formulate the problem.

Example 15.5.2. As a second example of a multiparticle problem, let us formulate the problem of the simple plane double pendulum, acted on by the gravitational force only.

As the force of gravity is a potential force and the problem is holonomic, the applicable form of Lagrange's equation is

$$\frac{d}{dt}\frac{\partial T}{d\dot q_s} - \frac{\partial T}{\partial q_s} + \frac{\partial V}{\partial q_s} = 0 \qquad (s = 1, 2, \ldots, n),$$

provided generalized coordinates are used. To give the problem greater generality we shall derive the kinetic and potential energies for an N-tuple plane, simple pendulum. We use as generalized coordinates the angles θ_i $(i = 1, 2, \ldots, N)$ which the pendulums make with the y axis (see Fig. 15.5.2).

In general, the position of the bob of the ith pendulum is

$$\begin{aligned} x_i &= \sum_{\alpha=1}^{i} l_\alpha \sin\theta_\alpha,\\ y_i &= \sum_{\alpha=1}^{i} l_\alpha \cos\theta_\alpha, \end{aligned} \tag{15.5.16}$$

as is evident from Fig. 15.5.2. Then, by differentiation one has

$$\dot x_i = \sum_{\alpha=1}^{i} l_\alpha\dot\theta_\alpha \cos\theta_\alpha,$$

$$\dot y_i = -\sum_{\alpha=1}^{i} l_\alpha\dot\theta_\alpha \sin\theta_\alpha,$$

and the squares of these velocity components may be written as

$$\begin{aligned} \dot x_i^2 &= \sum_{\alpha=1}^{i}\sum_{\beta=1}^{i} l_\alpha l_\beta\dot\theta_\alpha\dot\theta_\beta \cos\theta_\alpha\cos\theta_\beta,\\ \dot y_i^2 &= \sum_{\alpha=1}^{i}\sum_{\beta=1}^{i} l_\alpha l_\beta\dot\theta_\alpha\dot\theta_\beta \sin\theta_\alpha\sin\theta_\beta. \end{aligned} \tag{15.5.17}$$

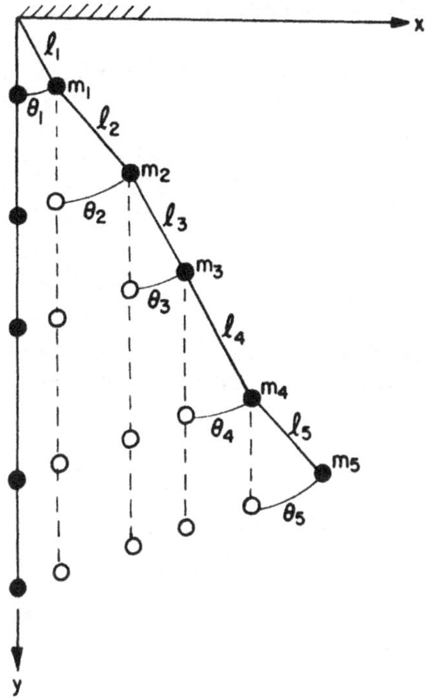

Fig. 15.5.2. *N*-tuple pendulum of Example 15.5.2.

Now, the kinetic energy is

$$T = \frac{1}{2} \sum_{i=1}^{N} m_i(\dot{x}_i^2 + \dot{y}_i^2),$$

and the substitution of (15.5.17) in this expression gives the equation

$$T = \frac{1}{2} \sum_{i=1}^{N} \sum_{\alpha=1}^{i} \sum_{\beta=1}^{i} m_i l_\alpha l_\beta \dot{\theta}_\alpha \dot{\theta}_\beta \cos(\theta_\alpha - \theta_\beta). \tag{15.5.18}$$

This formula may be put into a somewhat different form. First, let us separate out the terms for which $\alpha = \beta$, so that

$$T = \frac{1}{2} \sum_{i=1}^{N} m_i \left\{ \sum_{\alpha=1}^{i} l_\alpha^2 \dot{\theta}_\alpha^2 + \sum_{\alpha=1}^{i} \sum_{\substack{\beta=1 \\ \beta\neq\alpha}}^{i} l_\alpha l_\beta \dot{\theta}_\alpha \dot{\theta}_\beta \cos(\theta_\alpha - \theta_\beta) \right\}.$$

Now, the double sum in the braces remains unchanged by exchanging the subscripts α and β because the cosine is an even function. In consequence, every term occurs twice, i.e.,

$$\sum_{\alpha=1}^{i} \sum_{\substack{\beta=1 \\ \beta\neq\alpha}}^{i} l_\alpha l_\beta \dot{\theta}_\alpha \dot{\theta}_\beta \cos(\theta_\alpha - \theta_\beta) = 2 \sum_{\alpha=1}^{i-1} \sum_{\beta=\alpha+1}^{i} l_\alpha l_\beta \dot{\theta}_\alpha \dot{\theta}_\beta \cos(\theta_\alpha - \theta_\beta).$$

Hence, the kinetic energy becomes

$$T = \frac{1}{2}\left\{\sum_{i=1}^{N}\sum_{\alpha=1}^{i} m_i l_\alpha^2 \dot{\theta}_\alpha^2 + 2\sum_{i=1}^{N}\sum_{\alpha=1}^{i-1}\sum_{\beta=\alpha+1}^{i} m_i l_\alpha l_\beta \dot{\theta}_\alpha \dot{\theta}_\beta \cos(\theta_\alpha - \theta_\beta)\right\}. \tag{15.5.19}$$

To compute the potential energy, we regard the equilibrium position as the datum. Suppose, the first pendulum is given a deflection θ_1, and all other pendulums remain vertical. Then the system acquires the potential energy

$$\sum_{i=1}^{N} m_i g l_1 (1 - \cos\theta_1)$$

because all pendulum bobs are raised by $l_1(1 - \cos\theta_1)$. Therefore, when each pendulum is given a deflection θ_α, the total potential energy will be

$$V = \sum_{\alpha=1}^{N}\sum_{i=\alpha}^{N} m_i g l_\alpha (1 - \cos\theta_\alpha). \tag{15.5.20}$$

As we wish to consider the double pendulum we write (15.5.19) and (15.5.20) for $N = 2$. This results in

$$T = \tfrac{1}{2}\{(m_1 + m_2) l_1^2 \dot{\theta}_1^2 + m_2[l_2^2 \dot{\theta}_2^2 + 2 l_1 l_2 \dot{\theta}_1 \dot{\theta}_2 \cos(\theta_1 - \theta_2)]\}, \tag{15.5.21}$$

and

$$V = g[(m_1 + m_2) l_1 (1 - \cos\theta_1) + m_2 l_2 (1 - \cos\theta_2)]. \tag{15.5.22}$$

Then, one finds by direct differentiation

$$\frac{d}{dt}\frac{\partial T}{\partial \dot{\theta}_1} = (m_1 + m_2) l_1^2 \ddot{\theta}_1 + m_2 l_1 l_2 \cos(\theta_1 - \theta_2)\ddot{\theta}_2 - m_2 l_1 l_2 \dot{\theta}_2 (\dot{\theta}_1 - \dot{\theta}_2)\sin(\theta_1 - \theta_2),$$

$$\frac{d}{dt}\frac{\partial T}{\partial \dot{\theta}_2} = m_2 l_2^2 \ddot{\theta}_2 + m_2 l_1 l_2 \cos(\theta_1 - \theta_2)\ddot{\theta}_1 - m_2 l_1 l_2 \dot{\theta}_1 (\dot{\theta}_1 - \dot{\theta}_2)\sin(\theta_1 - \theta_2),$$

$$\frac{\partial T}{\partial \theta_1} = -\, m_2 l_1 l_2 \dot{\theta}_1 \dot{\theta}_2 \sin(\theta_1 - \theta_2),$$

$$\frac{\partial T}{\partial \theta_2} = m_2 l_1 l_2 \dot{\theta}_1 \dot{\theta}_2 \sin(\theta_1 - \theta_2),$$

$$\frac{\partial V}{\partial \theta_1} = g(m_1 + m_2) l_1 \sin\theta_1,$$

$$\frac{\partial V}{\partial \theta_2} = g m_2 l_2 \sin\theta_2.$$

The substitution of these quantities in Lagrange's equations gives

$$\begin{aligned}(m_1 + m_2) l_1^2 \ddot{\theta}_1 + m_2 l_1 l_2 \cos(\theta_1 - \theta_2)\ddot{\theta}_2 + m_2 l_1 l_2 \dot{\theta}_2^2 \sin(\theta_1 - \theta_2)&\\ + (m_1 + m_2) g l_1 \sin\theta_1 = 0,&\end{aligned} \tag{15.5.23}$$

and

$$m_2 l_2^2 \ddot{\theta}_2 + m_2 l_1 l_2 \cos(\theta_1 - \theta_2)\ddot{\theta}_1 - m_2 l_1 l_2 \dot{\theta}_1^2 \sin(\theta_1 - \theta_2) + m_2 g l_2 \sin\theta_2 = 0. \tag{15.5.24}$$

These are the equations of motion of the simple plane double pendulum under the action of gravity.

As a nonholonomic problem in rigid body dynamics we consider the classical example:

Example 15.5.3. (see also Example 7.5.1). Formulate the equations of motion of a homogeneous disk of radius r that rolls without sliding on the horizontal plane.

Consider Fig. 15.5.3; let the coordinates of the contact point be x, z, and let the angle between the contact tangent and the positive z axis be φ. Let θ be the inclination of the disk to the horizontal, and let ψ be the angle of rolling measured, for instance, as the angle between the contact radius and the radius to some point P fixed on the rim; let all angles be positive, as shown. Then the five coordinates x, z, φ, θ, and ψ specify the configuration of the disk uniquely, and they are not holonomically constrained; thus, they are generalized coordinates.

The condition of pure rolling ensures that the instantaneous rim velocity equals the instantaneous velocity with which the contact point, belonging to the xz plane, translates, i.e.,

$$r\dot{\psi} = -v.$$

From the geometry of Fig. 15.5.3, we have

$$\dot{x} = v \sin \varphi, \qquad \dot{z} = v \cos \varphi.$$

Therefore, the equations of rolling constraint are

$$\begin{aligned} \dot{x} + r\dot{\psi} \sin \varphi &= 0, \\ \dot{z} + r\dot{\psi} \cos \varphi &= 0, \end{aligned} \tag{15.5.25}$$

or, in Pfaffian form,

$$\begin{aligned} dx + r \sin \varphi \, d\psi &= 0, \\ dz + r \cos \varphi \, d\psi &= 0. \end{aligned} \tag{15.5.26}$$

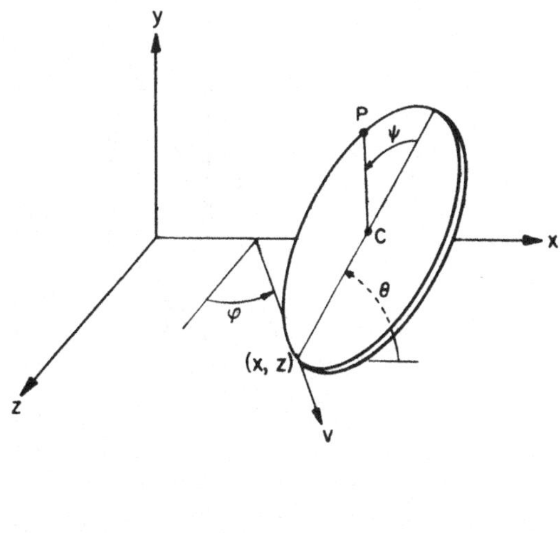

Fig. 15.5.3. Rolling disk of Example 15.5.3.

The knife edge constraint (see Example 9.4.2) is contained in (15.5.26); it results from dividing the first equation into the second.

We compute the kinetic energy in two parts: that due to translation of the mass center, and that due to rotation. From Fig. 15.5.3, the coordinates x_0, y_0, z_0 of the mass center are

$$\begin{aligned} x_0 &= x + r\cos\theta\cos\varphi, \\ z_0 &= z - r\cos\theta\sin\varphi, \\ y_0 &= r\sin\theta. \end{aligned}$$

On differentiating, squaring, and adding them, we find

$$\begin{aligned} \dot{x}_0^2 + \dot{y}_0^2 + \dot{z}_0^2 = \dot{x}^2 + \dot{z}^2 + r^2\dot{\theta}^2 + r^2\dot{\varphi}^2\cos^2\theta + 2r(&-\dot{x}\dot{\theta}\sin\theta\cos\varphi \\ &- \dot{x}\dot{\varphi}\cos\theta\sin\varphi + \dot{z}\dot{\theta}\sin\theta\sin\varphi - \dot{z}\dot{\varphi}\cos\theta\cos\varphi). \end{aligned}$$

When that quantity is multiplied by one-half of the disk mass it furnishes the kinetic energy without rotations.

To calculate the kinetic energy due to rotation, consider Fig. 15.5.4. The lines with single arrow heads are axes which are so chosen that the $x'y'$ plane is that of the disk, the x' axis remains parallel to the xz plane, and the z' axis completes the right-handed coordinate system. One sees from Fig. 15.5.4 that

$$\begin{aligned} \omega_{x'} &= \dot{\theta}, \\ \omega_{y'} &= \dot{\varphi}\sin\theta, \\ \omega_{z'} &= \dot{\psi} + \dot{\varphi}\cos\theta. \end{aligned} \tag{15.5.27}$$

It is seen that θ, φ, and ψ are, in fact, Euler angles because (15.5.27) are precisely the equations (6.8.14) when one puts $\psi = 0$ in them.

In this example we have departed from our usual practice of orienting the (right-handed) coordinate system so that the positive z axis points vertically up. Also, we have defined that direction of $\dot{\psi}$ as positive which gives rise to negative $\dot{x}$ and $\dot{z}$ components. Both these steps were taken because we wished to utilize Euler angles to describe the rotations. Of course, the fact that $\dot{\psi}$ produces any translational

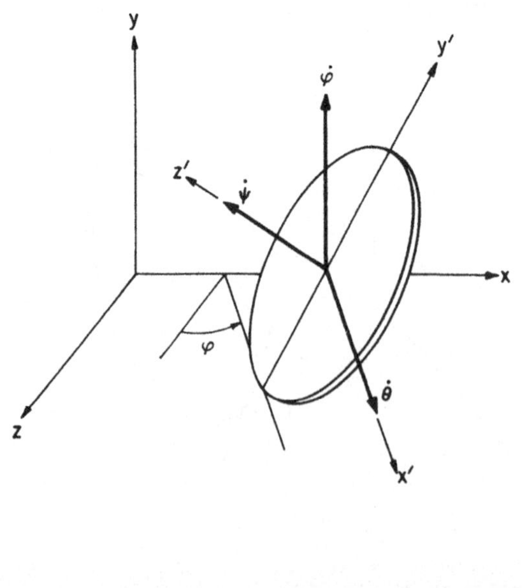

Fig. 15.5.4. Angular velocity vector of rolling disk of Example 15.5.3.

velocities at all is a consequence of the rolling constraint. Euler angles were not conceived in particular to treat problems with rolling constraints; therefore, the feature that some velocity in the positive direction produces others that are negative is not common to Euler angles; it is peculiar to this problem.

It is evident that the x', y' and z' axes are principal axes, and the moments of inertia are

$$I_{x'x'} = I_{y'y'} = I, \qquad I_{z'z'} = J.$$

Therefore, the kinetic energy of the rolling disk is

$$\begin{aligned} T = \tfrac{1}{2}m[\dot{x}^2 + \dot{z}^2 + r^2\dot{\theta}^2 + r^2\dot{\varphi}^2\cos^2\theta + 2r(-\,\dot{x}\dot{\theta}\sin\theta\cos\varphi - \dot{x}\dot{\varphi}\cos\theta\sin\varphi \\ + \dot{z}\dot{\theta}\sin\theta\sin\varphi - \dot{z}\dot{\varphi}\cos\theta\cos\varphi)] + \tfrac{1}{2}I(\dot{\theta}^2 + \dot{\varphi}^2\sin^2\theta) + \tfrac{1}{2}J(\dot{\psi} + \dot{\varphi}\cos\theta)^2. \end{aligned} \tag{15.5.28}$$

The only given force is the force of gravity, and it is derivable from the potential energy

$$V = mgr\sin\theta\,, \tag{15.5.29}$$

where we have used the horizontal position of the disk as the datum i.e., the x, z plane.

From (15.5.26), the constraints on the virtual displacements are

$$\begin{aligned} \delta x + r\sin\varphi\,\delta\psi &= 0, \\ \delta z + r\cos\varphi\,\delta\psi &= 0. \end{aligned}$$

If we define

$$q_1 = x, \qquad q_2 = z, \qquad q_3 = \theta, \qquad q_4 = \varphi, \qquad q_5 = \psi,$$

the B_{rs} in the constraint equations are

$$\begin{aligned} B_{11} &= 1, \qquad B_{15} = r\sin\varphi, \qquad B_{12} = B_{13} = B_{14} = 0, \\ B_{22} &= 1, \qquad B_{25} = r\cos\varphi, \qquad B_{21} = B_{23} = B_{24} = 0. \end{aligned}$$

Substitution in Lagrange's equations (13.7.3) gives

$$\begin{aligned} &\frac{d}{dt}[m\dot{x} + mr(-\,\dot{\theta}\sin\theta\cos\varphi - \dot{\varphi}\cos\theta\sin\varphi] + \lambda_1 = 0, \\ &\frac{d}{dt}[m\dot{z} + mr(\dot{\theta}\sin\theta\sin\varphi - \dot{\varphi}\cos\theta\cos\varphi)] + \lambda_2 = 0, \\ &\frac{d}{dt}[mr^2\dot{\theta} + mr(-\,\dot{x}\sin\theta\cos\varphi + \dot{z}\sin\theta\sin\varphi) + I\dot{\theta}] + mr^2\dot{\varphi}^2\cos\theta\sin\theta \\ &\qquad + mr(\dot{x}\dot{\theta}\cos\theta\cos\varphi - \dot{x}\dot{\varphi}\sin\theta\sin\varphi - \dot{z}\dot{\theta}\cos\theta\sin\varphi - \dot{z}\dot{\varphi}\sin\theta\cos\varphi) \\ &\qquad - I\dot{\varphi}^2\sin\theta\cos\theta + J(\dot{\psi} + \dot{\varphi}\cos\theta)\dot{\varphi}\sin\theta + mgr\cos\theta = 0, \\ &\frac{d}{dt}[mr^2\dot{\varphi}\cos^2\theta + mr(-\,\dot{x}\cos\theta\sin\varphi - \dot{z}\cos\theta\cos\varphi) \\ &\qquad + I\dot{\varphi}\sin^2\theta + J(\dot{\psi} + \dot{\varphi}\cos\theta)\cos\theta] + mr(-\,\dot{x}\dot{\theta}\sin\theta\sin\varphi \\ &\qquad + \dot{x}\dot{\varphi}\cos\theta\cos\varphi - \dot{z}\dot{\theta}\sin\theta\cos\varphi - \dot{z}\dot{\varphi}\cos\theta\sin\varphi) = 0, \\ &\frac{d}{dt}[J(\dot{\psi} + \dot{\varphi}\cos\theta)] + \lambda_1 r\sin\varphi + \lambda_2 r\cos\varphi = 0. \end{aligned} \tag{15.5.30}$$

These equations of motion and the kinematical constraints (15.5.25) are seven equations in the five generalized coordinates and the two Lagrange multipliers.

As a second example of a nonholonomic system we discuss a mechanism which was first analyzed by Novoselov† and is quoted here from Lur'e (p. 406).

Example 15.5.4. Discuss the mechanism shown in Fig. 15.5.5. Its purpose is to transmit the rotation of a driving shaft S_1 to a driven shaft S_2, and to have the speed of the driven shaft remain sensibly constant even when that of the driving shaft is not.

The vertical driving shaft has a rigidly attached horizontal disk of axial moment of inertia I_1. An intermediate horizontal shaft S_3 has a thin disk of radius a attached to it. Its axial moment of inertia is I_3, and its mass (including that of the shaft) is m_3. The shaft S_3 can translate along its axis of rotation; to do so it must overcome the force from a linear spring of rate k_3. The disk of S_3 rides on that of S_1 in the manner shown; the distance from the center of S_1 to the rim of the disk on S_3 is ϱ, as shown. The disk on S_3 drives a drum of radius R, which is rigidly attached to the driven shaft S_2; the axial moment of inertia of the drum is I_2. Also mounted on S_2 is a centrifugal governor, each of whose two weights has mass m_1. The angle between a link of the governor and the axis of rotation is θ. The sliding sleeve of the governor has mass m_2. When the sleeve translates, it does so against a linear spring of rate k_2. The position of the sleeve relative to some fixed point is x. This sleeve is connected by an inextensible cable, running over two pulleys, to the intermediate shaft in such a way that the translation of the shaft is the same as that of the sleeve.

The mechanism keeps the angular speed of the driven shaft constant. For, suppose the driving shaft speeds up; this speeds up the driven shaft. Hence, the governor opens, x and thus ϱ are reduced, and this reduces the speed of the driven shaft.

We note that x and ϱ are connected by

$$x - \varrho = c = \text{const} \tag{15.5.31}$$

because a change in x causes an equal change in ϱ, but they need not be zero simultaneoulsy.

The tangential velocity of disk 1 at the contact point with disk 3 is $\varrho\dot{\varphi}_1$. Under pure rolling without slipping the tangential rim velocity of disk 3, which is $a\dot{\varphi}_3$, is equal to both $\varrho\dot{\varphi}_1$ and to $-R\dot{\varphi}_2$. Hence, the constraint is

$$\varrho\dot{\varphi}_1 = -R\dot{\varphi}_2$$

or, with (15.5.31), in Pfaffian form

$$(x - c)\, d\varphi_1 + R\, d\varphi_2 = 0. \tag{15.5.32}$$

† V. S. Novoselov, "An Example of a Nonholonomic, Nonlinear System Not of the Chetaev Type," *Vestnik Leningradskogo Universiteta*, No. 19 (1957).

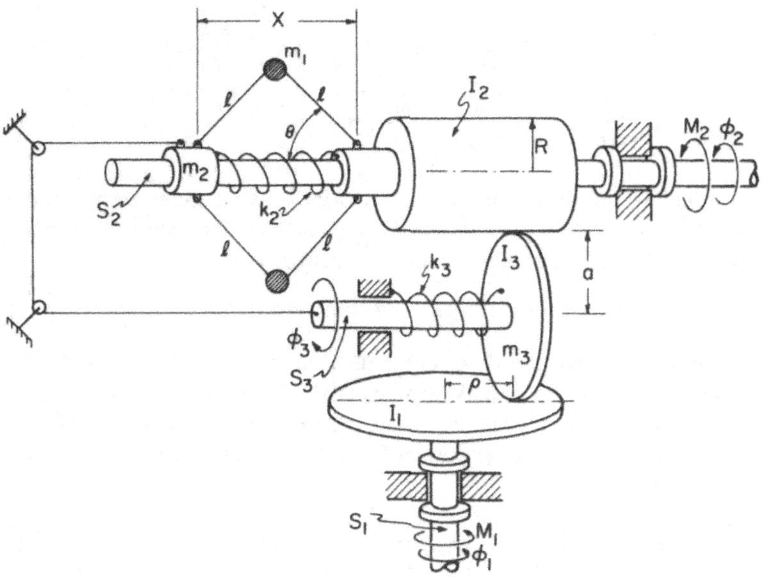

Fig. 15.5.5. Constant speed drive of Example 15.5.4.

This is a nonholonomic constraint of the form already seen in Example 4.2.3. Finally, x is related to θ by

$$x = 2l\cos\theta, \tag{15.5.33}$$

from which one finds

$$\dot\theta^2 = \frac{\dot x^2}{4l^2 - x^2}. \tag{15.5.34}$$

The kinetic energy due to $\dot\varphi_2$ is

$$T_1 = \tfrac{1}{2}[I_2 + \tfrac{1}{2}m_1(4l^2 - x^2)]\dot\varphi_2{}^2.$$

The kinetic energy due to $\dot x$ is

$$T_2 = \left(\frac{m_1 l^2}{4l^2 - x^2} + \frac{m_2}{2} + \frac{m_3}{2}\right)\dot x^2.$$

The kinetic energy due to $\dot\varphi_1$ is

$$T_3 = \frac{1}{2}\left[I_1 + I_3\frac{(x-c)^2}{a^2}\right]\dot\varphi_1{}^2.$$

Therefore, the total kinetic energy is

$$T = \frac{1}{2}\left[I_1 + I_3\frac{(x-c)^2}{a^2}\right]\dot{\varphi}_1^2 + \frac{1}{2}\left[I_2 + \frac{m_1}{2}(4l^2 - x^2)\right]\dot{\varphi}_2^2 + \left(\frac{m_1 l^2}{4l^2 - x^2} + \frac{m_2}{2} + \frac{m_3}{2}\right)\dot{x}^2. \tag{15.5.35}$$

To calculate the potential energy, we suppose that there exists some steady state in which $\varphi_1 = \varphi_{10}$, $\varphi_2 = \varphi_{20}$, $x = x_0$, and we calculate the change of potential energy when this steady state is disturbed. It is

$$V = \tfrac{1}{2}(k_2 + k_3)(x - x_0)^2. \tag{15.5.36}$$

Let us suppose that the disk of radius a on the intermediate shaft is so thin that its moment of inertia I_3 may be neglected compared to I_1. This supposition does no violence to the mechanism, and it simplifies the equations. Then, using the embedding technique of Section 14.3 the reader may verify that the equations of motion become

$$\left[\frac{I_1 R^2}{(x-c)^2} + I_2 + \frac{m_1}{2}(4l^2 - x^2)\right]\dot{\omega} - \left[\frac{I_1 R^2}{(x-c)^3} + m_1 x\right]\dot{x}\omega = -\frac{M_1 R}{x - c} + M_2, \tag{15.5.37}$$

and

$$2\left(\frac{m_1 l^2}{4l^2 - x^2} + \frac{m_2}{2} + \frac{m_3}{2}\right)\ddot{x} + \frac{2m_1 l^2}{(4l^2 - x^2)^2}x\dot{x}^2 + \frac{1}{2}m_1 x\omega^2 + (k_2 + k_3)x = (k_2 + k_3)x_0, \tag{15.5.38}$$

where we have put

$$\dot{\varphi}_2 = \omega.$$

This formulates the problem.

15.6. Problems

15.1. An unconstrained particle of mass m moves in 3-space under a force

$$F = X_0\hat{\imath} + Y_0\hat{\jmath} + Z_0\hat{k},$$

where X_0, Y_0, Z_0 are constants. Write the Lagrangean equations of motion in the generalized coordinates ξ, η, ζ, which are connected to the Cartesian coordinates by

$$x = l\cosh\xi\cos\eta, \qquad y = l\sinh\xi\sin\eta, \qquad z = \zeta,$$

and state why ξ, η, and ζ are called elliptic, cylindrical coordinates. Calculate an arc length ds in terms of ξ, η, and ζ. Denote the generalized force components by Ξ, H, and Z, respectively.

15.2. Answer the same questions as in Problem 15.1 for the biaxial, cylindrical coordinates

$$x = \frac{l \sinh \xi}{\cosh \xi + \cos \eta}, \qquad y = \frac{l \sin \eta}{\cosh \xi + \cos \eta}, \qquad z = \zeta.$$

15.3. Answer the same questions as in Problem 15.1 for the parabolic, cylindrical coordinates

$$x = l(\xi^2 - \eta^2), \qquad y = 2l\xi\eta, \qquad z = \zeta.$$

15.4. A heavy particle moving in 3-space is connected by a massless linear spring to a smooth vertical rod on which the spring can slide. Formulate the equations of motion in Cartesian coordinates, and in the generalized coordinates which are natural to the force system.

15.5. A heavy particle moves on a smooth surface of revolution. Formulate Lagrange's equations in Cartesian and cylindrical coordinates without using multipliers. Show that the second result follows formally from the first by utilizing the transformations from Cartesian to cylindrical coordinates.

15.6. In Example 4.2.2, a gutter was defined by a parabola whose apex descends in a prescribed manner. Let this gutter be smooth and let a heavy particle slide in it. Derive Lagrange's equations in suitable generalized coordinates.

15.7. Let the Cartesian coordinates of a 4-space be w, x, y, z. A particle of unit mass moves on the surface of a four-dimensinal sphere of radius R under a potential force, and the potential energy is constant on the cylindrical surface $w^2 + x^2 = \text{const}$. If θ, φ, and ψ are connected to w, x, y, and z by

$$w = R\cos\theta\cos\varphi, \qquad x = R\cos\theta\sin\varphi,$$
$$y = R\sin\theta\cos\psi, \qquad z = R\sin\theta\sin\psi,$$

show that θ, φ, and ψ are suitable generalized coordinates, and construct Lagrange's equations in θ, φ, and ψ.

15.8. A heavy eccentric disk can rotate about a fixed, smooth, horizontal axis at O. Let its mass moment of inertia about the axis of rotation be I, and let its mass center G be a distance s from the axis of rotation. A massless connecting rod of length l is smoothly hinged to the disk at a point P a

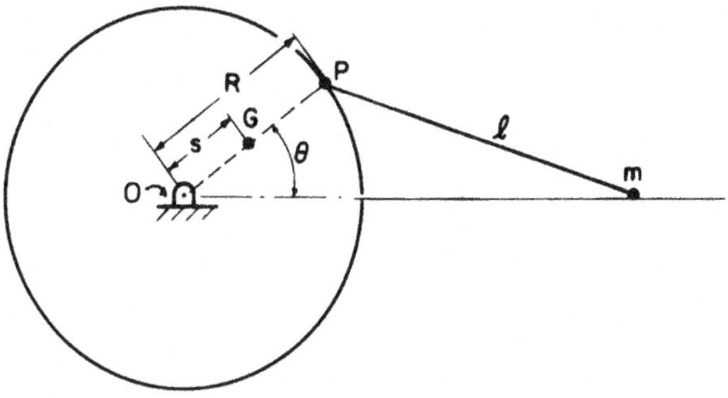

distance R from the axis of rotation, and connected to a particle of mass m, which is constrained to move on a smooth horizontal surface as shown. O, G, and P lie on a straight line. If gravity is the only force acting on the system, define suitable coordinates and construct Lagrange's equations of motion for this system.

15.9. A heavy bead of mass m slides on a smooth rod that rotates with constant angular velocity Ω about a fixed point lying on the rod centerline, as shown. What are Lagrange's equations of motion of the bead in suitably chosen generalized coordinates?

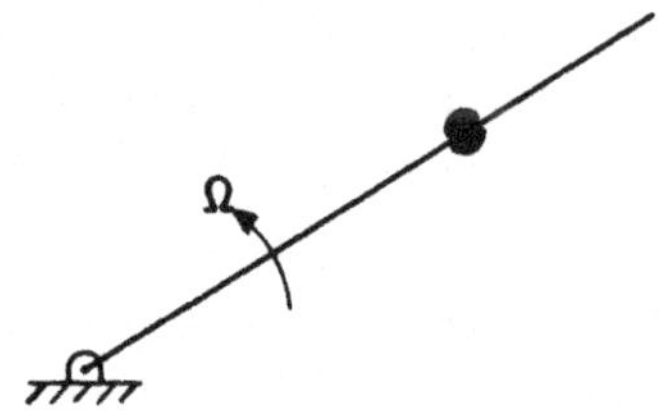

15.10. Two particles of masses m_1 and m_2, respectively, move on the surface of the four-dimensional sphere described in Problem 15.7. The only force acting on each particle is the Newtonian gravitational attraction from the other. Construct Lagrange's equations of motion without the use of multipliers.

15.11. Generalize Problem 15.10 for the case of n particles.

15.12. Construct Lagrange's equations of motion of a particle of mass m in 2-space without the use of multipliers using r, φ, and A, where A is the double of the area swept out by the radius vector; the three coordinates are connected by the nonholonomic constraint

$$\dot{A} - r^2\dot{\varphi} = 0.$$

16

The Integration

16.1. The Meaning of an Integral

We saw that the strictly Newtonian problem has the Lagrangean formulation

$$\frac{d}{dt}\frac{\partial T}{\partial \dot{q}_s} - \frac{\partial T}{\partial q_s} - Q_s + \sum_{r=1}^{l} \lambda_r B_{rs} = 0 \qquad (s = 1, 2, \ldots, n), \tag{16.1.1}$$

where the equality constraints are

$$\sum_{s=1}^{n} B_{rs}\dot{q}_s + B_r = 0 \qquad (r = 1, 2, \ldots, l). \tag{16.1.2}$$

These are n second-order and l first-order ordinary differential equations in the generalized coordinates $q_1, q_2, \ldots, q_n$. The first set is easily reduced to $2n$ first-order equations. Let us write

$$v_1 = q_1,\ v_2 = q_2,\ \ldots,\ v_n = q_n,$$
$$v_{n+1} = \dot{q}_1 = \dot{v}_1,\ v_{n+2} = \dot{v}_2,\ \ldots,\ v_{2n} = \dot{v}_n.$$

Then, the kinetic energy becomes

$$T = T(v_1, v_2, \ldots, v_{2n}, t)$$

and Lagrange's equations become

$$\frac{d}{dt}\frac{\partial T}{\partial v_{n+s}} - \frac{\partial T}{\partial v_s} - Q_s + \sum_{r=1}^{l} \lambda_r B_{rs} = 0 \qquad (s = 1, 2, \ldots, n). \tag{16.1.3}$$

These are n first-order equations in the v_p $(p = 1, 2, \ldots, 2n)$ because the partial derivatives $\partial T/\partial v_{n+s}$ are linear functions of the v_p. Therefore, (16.1.3) together with

$$\frac{dv_s}{dt} = v_{n+s} \qquad (s = 1, 2, \ldots, n) \tag{16.1.4}$$

are $2n$ first-order equations in the v_p. Using these same transformations on the constraint equations, one finds

$$\sum_{s=1}^{n} B_{rs}v_{n+s} + B_r = 0 \qquad (r = 1, 2, \ldots, l), \tag{16.1.5}$$

where the B_{rs} and the B_r are, in general, functions of the $v_1, v_2, \ldots, v_n$, and t. Thus, the constraint equations are finite rather than differential equations in the new variables.

Example 16.1.1. Let us formulate the steered particle of Example 15.4.1 in terms of first-order differential equations.

The Lagrangean equations are the second-order equations (15.4.5), and the constraint equation is the first-order equation (15.4.6).

Let

$$\dot{x} = u, \qquad \dot{y} = v, \qquad \dot{z} = w. \tag{a}$$

Then, the equations of motion become

$$\begin{aligned} m\dot{u} &- X(x, y, z, u, v, w, t) - \lambda z = 0, \\ m\dot{v} &- Y(x, y, z, u, v, w, t) + \lambda \;= 0, \\ m\dot{w} &- Z(x, y, z, u, v, w, t) \qquad\;\; = 0, \end{aligned} \tag{b}$$

and the constraint equation becomes

$$v - zu = 0. \tag{c}$$

The system (a) and (b) consists of six first-order equations in the dependent variables x, y, z, u, v, w and the constraint equation (c) is an algebraic relation between some of these variables.

We now give the definition of an "integral" of the equations of motion.

If one can find a function $F_\alpha(v_1, v_2, \ldots, v_{2n}, t)$ which has the property that

$$\frac{dF_\alpha}{dt} = \sum_{p=1}^{2n} \frac{\partial F_\alpha}{\partial v_p}\dot{v}_p + \frac{\partial F_\alpha}{\partial t} = 0$$

whenever the v_p $(p = 1, 2, \ldots, 2n)$ satisfy the $2n$ first-order equations

(16.1.3) *and* (16.1.4) *as well as the constraint equations* (16.1.5), *the relation*

$$F_\alpha(v_1, v_2, \ldots, v_{2n}, t) = C_\alpha = \text{const}$$

is called an integral of the motion, *and* C_α *is called a* constant of integration *or* a constant of the motion.

The system consists of $2n$ first-order differential equations; thus there exist $2n$ integrals

$$F_\alpha(v_1, v_2, \ldots, v_{2n}, t) = C_\alpha \qquad (\alpha = 1, 2, \ldots, 2n). \tag{16.1.6}$$

When all these functions are known, the system is said to be "completely integrated."

As the problems of Lagrangean mechanics are formulated in terms of second-order rather than first-order equations in generalized coordinates, it is useful to define integrals in terms of generalized coordinates. Then, the first half of (16.1.6) becomes

$$F_\beta(q_1, q_2, \ldots, q_n; \dot{q}_1, \dot{q}_2, \ldots, \dot{q}_n; t) = C_\beta \qquad (\beta = 1, 2, \ldots, n). \tag{16.1.7}$$

These are n first-order equations; hence, finding n so-called "first integrals" reduces the order of the system to n. Then, the missing integrals are the integrals of the first-order equations (16.1.7). They are of the form

$$G_\gamma(q_1, q_2, \ldots, q_n; t; c_1, c_2, \ldots, c_n) = C_\gamma \qquad (\gamma = 1, 2, \ldots, n). \tag{16.1.8}$$

The process of "solving" a dynamics problem completely is one of finding all integrals (16.1.6). It is clear that (16.1.8) is equivalent to (16.1.6) because both sets of equations contain $2n$ constants of integration. To give a concrete example of an integral we note:

Example 16.1.2. In problems in which energy is conserved, a first integral is known because

$$T(q_1, q_2, \ldots, q_n; \dot{q}_1, \dot{q}_2, \ldots, \dot{q}_n; t) + V(q_1, q_2, \ldots, q_n) = h = \text{const}$$

is of the form (16.1.7).

In the remaining sections of this chapter we describe some methods for finding integrals.

16.2. Jacobi's Integral

Jacobi's integral is a generalized form of the energy integral. Necessary conditions for the existence of an energy integral are that the system be catastatic and that all given forces be derivable from a potential energy

$$V = V(q_1, q_2, \ldots, q_n). \tag{16.2.1}$$

In that case, we may utilize the Lagrangean function

$$L = T - V, \tag{16.2.2}$$

which was already introduced in (8.5.7).

In terms of the Lagrangean function the fundamental equation becomes

$$\sum_{s=1}^{n} \left[\frac{d}{dt} \left(\frac{\partial L}{\partial \dot{q}_s} \right) - \frac{\partial L}{\partial q_s} \right] \delta q_s = 0 \tag{16.2.3}$$

and, when the q_s are generalized coordinates, Lagrange's equations become for holonomic systems

$$\frac{d}{dt} \frac{\partial L}{\partial \dot{q}_s} = \frac{\partial L}{\partial q_s} \qquad (s = 1, 2, \ldots, n), \tag{16.2.4}$$

and for nonholonomic systems

$$\frac{d}{dt} \frac{\partial L}{\partial \dot{q}_s} = \frac{\partial L}{\partial q_s} - \sum_{r=1}^{l} \lambda_r B_{rs} \qquad (s = 1, 2, \ldots, n). \tag{16.2.5}$$

As indicated in (16.2.1), the potential function is usually a function of the generalized coordinates, but not of the generalized velocities, nor of time.

In Example 10.5.3, we did consider a potential function that depended explicitly on time. Now, when the potential V is time-dependent, but not velocity-dependent and when all given forces are defined by

$$Q_s = -\partial V/\partial q_s,$$

then it can be easily verified that

$$\frac{d}{dt} \frac{\partial L}{\partial \dot{q}_s} - \frac{\partial L}{\partial q_s} = \frac{d}{dt} \frac{\partial T}{\partial \dot{q}_s} - \frac{\partial T}{\partial q_s} + \frac{\partial V}{\partial q_s}. \tag{16.2.6}$$

Therefore, (16.2.4) and (16.2.5) are then still applicable.

It turns out that these equations have greater generality, still. As was shown in Section 12.5, it is possible to have velocity and/or time-dependent potentials V, and if all given forces can be computed from the equation

$$Q_s = \frac{d}{dt}\frac{\partial V}{\partial \dot{q}_s} - \frac{\partial V}{\partial q_s} \tag{16.2.7}$$

then Lagrange's equations may still be written in the forms (16.2.4) and (16.2.5). However, as shown in (12.5.6), the functional dependence of the potential energy on the velocity components must be *linear* because otherwise the terms $d(\partial L/\partial \dot{q}_s)/dt$, and thus the Q_s, would involve acceleration components. This was shown (see Example 2.5.1) to be inadmissible in Newtonian particle mechanics.

Let us assume that the Lagrangean function does not contain time explicitly and that the system is catastatic. Because of the latter assumption we may write the fundamental equation in the form

$$\sum_{s=1}^{n}\left[\frac{d}{dt}\left(\frac{\partial L}{\partial \dot{q}_s}\right) - \frac{\partial L}{\partial q_s}\right]\dot{q}_s = 0. \tag{16.2.8}$$

Now, the time derivative

$$\frac{d}{dt}\left[\sum_{s=1}^{n}\dot{q}_s\frac{\partial L}{\partial \dot{q}_s}\right] = \sum_{s=1}^{n}\dot{q}_s\frac{d}{dt}\left(\frac{\partial L}{\partial \dot{q}_s}\right) + \sum_{s=1}^{n}\ddot{q}_s\frac{\partial L}{\partial \dot{q}_s}.$$

The substitution of this expression into (16.2.8) gives

$$\frac{d}{dt}\left[\sum_{s=1}^{n}\dot{q}_s\frac{\partial L}{\partial \dot{q}_s}\right] - \sum_{s=1}^{n}\frac{\partial L}{\partial \dot{q}_s}\ddot{q}_s - \sum_{s=1}^{n}\frac{\partial L}{\partial q_s}\dot{q}_s = 0. \tag{16.2.9}$$

Then, as L depends in general on the q_s and $\dot{q}_s$, the last two terms in (16.2.9) are the time derivative of L. Therefore, (16.2.9) is an exact differential and can be integrated to give

$$\sum_{s=1}^{n}\dot{q}_s\frac{\partial L}{\partial \dot{q}_s} - L = h = \text{const.} \tag{16.2.10}$$

This is *Jacobi's integral.* It is easily brought into a familiar form as we shall now show. It follows from (12.1.3) and (12.1.4) that the kinetic energy has the general form

$$T = T_2 + T_1 + T_0, \tag{16.2.11}$$

where

$$\begin{aligned} T_2 &= \frac{1}{2}\sum_{\alpha=1}^{n}\sum_{\beta=1}^{n} a_{\alpha\beta}\dot{q}_\alpha\dot{q}_\beta, \\ T_1 &= \sum_{\alpha=1}^{n} b_\alpha\dot{q}_\alpha, \\ T_0 &= c, \end{aligned} \tag{16.2.12}$$

and where $a_{\alpha\beta}$, b_α, and c are functions of the q_s and possibly of t, but not of the velocity components. Therefore, T_2 is a homogeneous *quadratic* form, and T_1 is a homogeneous *linear* form in the velocity components.

It follows from the definition of homogeneous functions [see Section 13.7(d)] that

$$\sum_{s=1}^{n} \frac{\partial T_2}{\partial \dot{q}_s} \dot{q}_s = 2T_2. \tag{16.2.13}$$

and

$$\sum_{s=1}^{n} \frac{\partial T_1}{\partial \dot{q}_s} \dot{q}_s = T_1. \tag{16.2.14}$$

Now, Jacobi's integral (16.2.10) in expanded form is

$$\sum_{s=1}^{n} \dot{q}_s\left(\frac{\partial T_2}{\partial \dot{q}_s} + \frac{\partial T_1}{\partial \dot{q}_s} + \frac{\partial T_0}{\partial \dot{q}_s}\right) - T_2 - T_1 - T_0 + V = h. \tag{16.2.15}$$

If one substitutes (16.2.13) and (16.2.14) in (16.2.15) and notes that $\partial T_0/\partial \dot{q}_s = 0$, there results

$$T_2 + V - T_0 = h. \tag{16.2.16}$$

We have implicitly supposed that V is a function of the q_s, but not of the $\dot{q}_s$, and, by definition, T_0 is never a function of the $\dot{q}_s$. Therefore, V and T_0 are functions of the same arguments, and the effect of these two terms in the equations of motion is indistinguishable.

We have assumed that the Lagrangean function, and hence T, does not depend explicitly on time. It might be thought that, in consequence, $T_1 = T_0 \equiv 0$. But, a glance at Section 7.1 shows that T_1 and T_0 will be different from zero whenever the relation between the Cartesian and the generalized coordinates depends explicitly on time, i.e., when

$$u_r = u_r(q_1, q_2, \ldots, q_n, t) \qquad (r = 1, 2, \ldots, N)$$

but these equations do not necessarily imply that T depends explicitly on time.

Example 16.2.1. (Pars, p. 82). Let the inertial Cartesian coordinates of a particle of mass m be x, y, z, and let the x', y', z' system of coordinates rotate with constant angular velocity ω about the z' axis. Then, the coordinate transformation is

$$x = x' \cos \omega t - y' \sin \omega t,$$
$$y = x' \sin \omega t + y' \cos \omega t,$$
$$z = z'.$$

An easy calculation shows that

$$\dot{x}^2 + \dot{y}^2 + \dot{z}^2 = (\dot{x}' - y'\omega)^2 + (\dot{y}' + x'\omega)^2 + \dot{z}'^2$$

or

$$T = \tfrac{1}{2}m(\dot{x}'^2 + \dot{y}'^2 + \dot{z}'^2) + \omega m(x'\dot{y}' - y'\dot{x}') + \tfrac{1}{2}\omega^2 m(x'^2 + y'^2).$$

Therefore, in this case the kinetic energy does not depend explicitly on time, but T_1 and T_0 are not zero; in fact,

$$\begin{aligned} T_2 &= \tfrac{1}{2}m(\dot{x}'^2 + \dot{y}'^2 + \dot{z}'^2), \\ T_1 &= m\omega(x'\dot{y}' - y'\dot{x}'), \\ T_0 &= \tfrac{1}{2}m\omega^2(x'^2 + y'^2). \end{aligned}$$

If all given forces acting on the particle are derivable from a potential function of the form

$$V = V(x'^2 + y'^2, z'),$$

then V does not depend explicitly on time either, and the conditions for Jacobi's integral are satisfied. That integral is then

$$\tfrac{1}{2}m[\dot{x}'^2 + \dot{y}'^2 + \dot{z}'^2 - \omega^2(x'^2 + y'^2)] + V(x'^2 + y'^2, z') = h.$$

When the relation between the Cartesian and the generalized coordinates does not depend explicitly on time (as is usually the case in holonomic, scleronomic problems), one has

$$T_1 = T_0 \equiv 0, \qquad T = T_2,$$

and Jacobi's integral takes on the familiar form

$$T + V = h. \tag{16.2.17}$$

From the preceding theory it is clear that the existence of a Jacobi integral requires that all given forces be potential. Where this is not the case, one need not attempt to seek such an integral, for none exists.

16.3. The Routhian Function and the Momentum Integrals

The momentum integrals of dynamical systems are those which assert that certain momenta are conserved. They arise from so-called "ignorable coordinates," and they simplify the subsequent integration of the equations of motion by a process called "ignoration of coordinates."

The generalized momentum components are defined as

$$p_s = \frac{\partial T}{\partial \dot{q}_s} \qquad (s = 1, 2, \ldots, n); \tag{16.3.1}$$

they play a central role in Hamiltonian mechanics.

Consider a holonomic system in which all given forces are potential, and in which the Lagrangean function is *not* an explicit function of the coordinates q_β $(\beta = 1, 2, \ldots, b \leq n)$, but the remaining coordinates q_α $(\alpha = b+1, b+2, \ldots, n)$ *do* occur explicitly, as do all velocity components $\dot{q}_s$ $(s = 1, 2, \ldots, n)$. Thus

$$L = L(q_{b+1}, q_{b+2}, \ldots, q_n; \dot{q}_1, \dot{q}_2, \ldots, \dot{q}_n).$$

Then, Lagrange's equations may be written as

$$\begin{aligned} &\frac{d}{dt}\frac{\partial L}{\partial \dot{q}_\beta} = \dot{p}_\beta = \frac{\partial L}{\partial q_\beta} = 0 \qquad (\beta = 1, 2, \ldots, b), \\ &\frac{d}{dt}\frac{\partial L}{\partial \dot{q}_\alpha} = \dot{p}_\alpha = \frac{\partial L}{\partial q_\alpha} \qquad (\alpha = b+1, b+2, \ldots, n), \end{aligned} \tag{16.3.2}$$

and the first of these integrates to

$$\frac{\partial L}{\partial \dot{q}_\beta} = p_\beta = c_\beta = \text{const} \qquad (\beta = 1, 2, \ldots, b). \tag{16.3.3}$$

To describe the procedure called "ignoration of coordinates," it is helpful to use the Legendre transformation.

(a) The Legendre Transformation

The Legendre transformation is a special type of transformation from variables $x_1, x_2, \ldots, x_n$ to new variables $y_1, y_2, \ldots, y_n$; it is designed to transform a function $L(x_1, x_2, \ldots, x_n)$ to another function $\tilde{L}(y_1, y_2, \ldots, y_n)$ called the Legendre *transform* of L. We begin by illustrating the Legendre transformation by the simplest case, involving only one variable.

Let the transformation from x to y be defined by the twice differentiable function $y = \varphi(x)$ possessing the inverse $x = x(y)$. Then we define a function

$$V(x) = \int_0^x y(u)\, du, \tag{16.3.4}$$

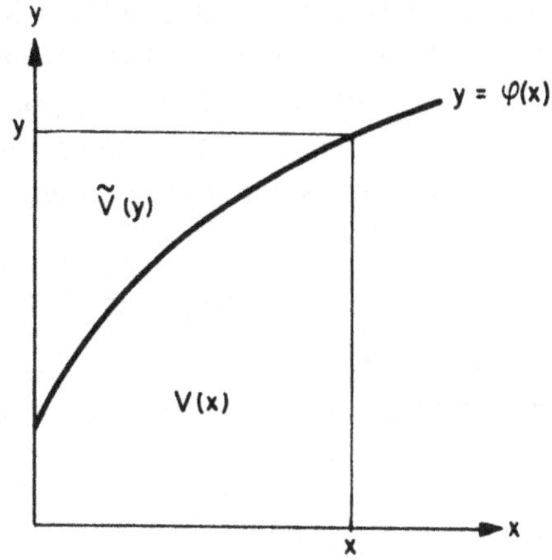

Fig. 16.3.1. $\tilde{V}(y)$ is a Legendre transform of $V(x)$.

from which we find

$$y(x) = dV/dx. \tag{16.3.5}$$

We now define the Legendre transform of $V(x)$ as

$$\tilde{V}(y) = xy - V(x) = x(y)y - V(x(y)). \tag{16.3.6}$$

A simple geometrical interpretation of the meaning of the Legendre transform emerges from Fig. 16.3.1. It is seen that $V(x)$ is the area bounded by the x axis, the y axis, the curve $y(x)$, and the line at x parallel to the y axis. Then, it follows from (16.3.6) that $\tilde{V}(y)$ is the area that remains when one subtracts $V(x)$ from the rectangle xy. This area may also be written as

$$\tilde{V}(y) = \int_0^y x(v)\, dv, \tag{16.3.7}$$

which implies

$$x(y) = d\tilde{V}/dy. \tag{16.3.8}$$

The equations (16.3.5) and (16.3.8) define the Legendre *transformation* in terms of the old function $V(x)$ and the new function $\tilde{V}(y)$. In fact, they satisfy the reciprocal relation that the new variable is the derivative of the old function, and the old variable is the derivative of the new function.

We can recover (16.3.8) without having to go through the integral representation of $\tilde{V}(y)$. Thus, forming the differential of (16.3.6) we have

$$d\tilde{V}(y) = x\, dy + y\, dx - \frac{dV}{dx}\, dx = x\, dy, \tag{16.3.9}$$

where the second equality follows from the fact that the second and third terms in $d\tilde{V}(y)$ cancel because of (16.3.5).

To introduce the general Legendre transformation, consider a function $\varphi(x_1, x_2, \ldots, \alpha_n)$ which is twice differentiable.

Then, let us define new variables

$$y_s = \frac{\partial \varphi}{\partial x_s} \qquad (s = 1, 2, \ldots, n) \tag{16.3.10}$$

on a domain where the Jacobian of this transformation does not vanish, or

$$J(y) = \begin{vmatrix} \dfrac{\partial y_1}{\partial x_1} & \cdots & \dfrac{\partial y_1}{\partial x_n} \\ \vdots & & \vdots \\ \dfrac{\partial y_n}{\partial x_1} & \cdots & \dfrac{\partial y_n}{\partial x_n} \end{vmatrix} = H(\varphi) = \begin{vmatrix} \dfrac{\partial^2 \varphi}{\partial x_1{}^2} & \cdots & \dfrac{\partial^2 \varphi}{\partial x_n \, \partial x_1} \\ \vdots & & \vdots \\ \dfrac{\partial^2 \varphi}{\partial x_1 \, \partial x_n} & \cdots & \dfrac{\partial^2 \varphi}{\partial x_n{}^2} \end{vmatrix} \neq 0.$$

The second determinant is called the Hessian of φ, and its nonvanishing guarantees that the inverses of (16.3.10) exist, or that

$$x_s = x_s(y_1, y_2, \ldots, y_n) \qquad (s = 1, 2, \ldots, n). \tag{16.3.11}$$

Then, the Legendre transform of φ is defined as

$$\tilde{\varphi}(y_1, y_2, \ldots, y_n) = \sum_{s=1}^{n} x_s y_s - \varphi, \tag{16.3.12}$$

from which we obtain

$$\frac{\partial \tilde{\varphi}}{\partial y_s} = x_s + \sum_{r=1}^{n} \frac{\partial x_r}{\partial y_s} y_r - \sum_{r=1}^{n} \frac{\partial \varphi}{\partial x_r} \frac{\partial x_r}{\partial y_s} = x_s, \tag{16.3.13}$$

where the second equality follows because the sums in (16.3.12) cancel each other in virtue of (16.3.10).

A slight generalization of the previous case occurs if we consider a function of the variables $F(x_1, x_2, \ldots, x_m; y_1, y_2, \ldots, y_n)$, and we wish to form a Legendre transform $\tilde{F}(x_1, x_2, \ldots, x_m; z_1, z_2, \ldots, z_n)$, where we have replaced the variables y_s by z_s while retaining the x_r $(r = 1, 2, \ldots, m)$. Then

$$\tilde{F}(x_1, x_2, \ldots, x_m; z_1, z_2, \ldots, z_n) = \sum_{s=1}^{n} y_s z_s - F, \tag{16.3.14}$$

and the Legendre transform is defined by

$$z_s = \frac{\partial F}{\partial y_s}, \qquad y_s = \frac{\partial \tilde{F}}{\partial z_s} \qquad (s = 1, 2, \ldots, n). \tag{16.3.15}$$

Example 16.3.1. A very interesting example of the Legendre transform is furnished by the definition of the generalized potential force

$$Q_s = -\frac{\partial V}{\partial q_s} = \frac{\partial U}{\partial q_s} \qquad (s = 1, 2, \ldots, n), \tag{a}$$

where U is the potential function, and it is the negative of the potential energy V.

From the viewpoint of the Legendre transformation, let us form the Legendre transform $\tilde{U}$ of the function U, and in the new function, the generalized coordinates q_s are replaced by the generalized forces Q_s. Thus, we form in accordance with (16.3.12)

$$\tilde{U} = \sum_{s=1}^{n} Q_s q_s - U \tag{b}$$

with

$$q_s = \frac{\partial \tilde{U}}{\partial Q_s} \qquad (s = 1, 2, \ldots, n). \tag{c}$$

The equations (a) and (c) are the Legendre transforms, and the transformation $\tilde{U}$ is called the "complementary work."

(b) The Routhian Function

Consider the Lagrangean function

$$L(q_{b+1}, q_{b+2}, \ldots, q_n; \dot{q}_1, \dot{q}_2, \ldots, \dot{q}_b, \dot{q}_{b+1}, \ldots, \dot{q}_n), \tag{16.3.16}$$

in which the coordinates q_β ($\beta = 1, 2, \ldots, b$) are ignorable. Then, it was shown in (16.3.3) that the momentum components p_β are constants. We now form a Legendre transform of L in which the velocities $\dot{q}_\beta$ are replaced by the constant momenta c_β, and this is the Routhian function; thus,

$$R = \tilde{L} = R(q_{b+1}, q_{b+2}, \ldots, q_n; c_1, c_2, \ldots, c_b; \dot{q}_{b+1}, \dot{q}_{b+2}, \ldots, \dot{q}_n) \tag{16.3.17}$$

and, in accordance with (16.3.14),

$$R = \tilde{L} = \sum_{\beta=1}^{b} \dot{q}_\beta c_\beta - L. \tag{16.3.18}$$

The Legendre transformation is defined by

$$\frac{\partial R}{\partial c_\beta} = \dot{q}_\beta, \qquad \frac{\partial L}{\partial \dot{q}_\beta} = c_\beta \qquad (\beta = 1, 2, \ldots, b). \tag{16.3.19}$$

Moreover, it follows from (16.3.18) that

$$\left.\begin{aligned}\frac{\partial R}{\partial q_\alpha} &= -\frac{\partial L}{\partial q_\alpha}\\ \frac{\partial R}{\partial \dot{q}_\alpha} &= -\frac{\partial L}{\partial \dot{q}_\alpha}\end{aligned}\right\} \quad (\alpha = b+1, b+2, \dots, n). \tag{16.3.20}$$

When these are substituted in Lagrange's equations, one finds for the non-ignorable coordinates

$$\frac{d}{dt}\frac{\partial R}{\partial \dot{q}_\alpha} - \frac{\partial R}{\partial q_\alpha} = 0 \qquad (\alpha = b+1, b+2, \dots, n), \tag{16.3.21}$$

and from the first equation of (16.3.19) for the ignorable coordinates

$$q_\beta = \int \frac{\partial R}{\partial c_\beta}\, dt \qquad (\beta = 1, 2, \dots, b). \tag{16.3.22}$$

Thus, the nonignorable coordinates are found by solving a set of Lagrange equations in the Routhian rather than the Lagrangean function, and the ignorable coordinates by quadratures. The process of proceeding from the n Lagrange equations in the Lagrangean to $n - b$ equations in the Routhian is called "ignoration of coordinates."

To illustrate conservation of momentum we return to Example 15.5.1.

Example 16.3.2. Consider the force-free motion of a system of two particles which move in a plane such that the line connecting their postions passes for all time through a fixed point.

We showed that the kinetic energy in polar coordinates is

$$T = \tfrac{1}{2}[(m_1\dot{r}_1^2 + m_2\dot{r}_2^2) + (m_1r_1^2 + m_2r_2^2)\dot{\theta}^2].$$

As no given forces act, the Lagrangean function is

$$L = T.$$

Since θ does not occur in L, it is an ignorable coordinate, or

$$p_\theta = \frac{\partial L}{\partial \dot{\theta}} = (m_1r_1^2 + m_2r_2^2)\dot{\theta} = c = \text{const.}$$

It is seen that $(m_1r_1^2 + m_2r_2^2)\dot{\theta}$ is the angular momentum of the particles about the origin.

Actually, this problem also has the energy integral

$$\tfrac{1}{2}[(m_1\dot{r}_1^2 + m_2\dot{r}_2^2) + (m_1r_1^2 + m_2r_2^2)\dot{\theta}^2] = h = \text{const.}$$

As a second example we treat a case in which the complete motion is found by the use of the Routhian function.

Example 16.3.3. Find the motion of the system for which

$$T = \frac{1}{2} \frac{\dot{q}_1^2}{a + bq_2^2} + \frac{1}{2} \dot{q}_2^2, \qquad V = c + dq_2,$$

where a, b, c, and d are real positive constants.

The Lagrangean function is

$$L = \frac{1}{2} \frac{\dot{q}_1^2}{a + bq_2^2} + \frac{1}{2} \dot{q}_2^2 - c - dq_2 \tag{a}$$

and q_1 is ignorable. Therefore,

$$R = \dot{q}_1 p_1 - L, \tag{b}$$

and

$$p_1 = c_1 = \frac{\partial L}{\partial \dot{q}_1} = \frac{\dot{q}_1}{a + bq_2^2}, \tag{c}$$

where $c_1 = \text{const}$. Substituting (a) and (c) in (b),

$$R = \tfrac{1}{2}c_1^2(a + bq_2^2) - \tfrac{1}{2}\dot{q}_2^2 + c + dq_2^2. \tag{d}$$

In conformity with (16.3.20) we find for the coordinate q_2

$$\ddot{q}_2 + (2d + bc_1^2)q_2 = 0, \tag{e}$$

which has the integral

$$q_2 = A \sin[(2d + bc_1^2)^{1/2} t + \alpha], \tag{f}$$

where A and α are constants of integration.

Then, because of (16.3.21),

$$q_1 = \int \frac{\partial R}{\partial c_1} dt = \int (a + bq_2^2)c_1 \, dt. \tag{g}$$

When q_2 from (f) is substituted in (g),

$$q_1 = c_1\left[a + \frac{1}{2} bA^2\right]t - \frac{c_1 ba^2}{4(2d + bc_1^2)^{1/2}} \sin[2(2d + bc_1^2)^{1/2} t + \alpha] + \beta, \tag{h}$$

where β is a constant of integration.

16.4. Partial and Complete Separation of Variables[†]

In general, every Lagrangean equation involves all generalized coordinates, e.g., the equations of motion are coupled. When each equation involves one variable only the system is "uncoupled," and the variables are "separated."

We assume in this section that a given system is holonomic and scleronomic, and that all given forces are derivable from a potential energy V which is a function of the coordinates only. If the energies have, or can be put into, the form

$$T = \frac{1}{2} \sum_{s=1}^{n} v_s(q_s)\dot{q}_s^2, \qquad V = \sum_{s=1}^{n} w_s(q_s), \tag{16.4.1}$$

where each v_s and w_s is an arbitrary function of q_s only, it is easy to show that Lagrange's equations are uncoupled. Substitution gives

$$v_s(q_s)\ddot{q}_s + \tfrac{1}{2}v_s'(q_s)\dot{q}_s^2 = -w_s'(q_s) \qquad (s = 1, 2, \ldots, n), \tag{16.4.2}$$

where primes denote differentiation with respect to q_s.

Not only are these equations uncoupled, but they are perfect differentials as well. They have the first integrals

$$\tfrac{1}{2}v_s(q_s)\dot{q}_s^2 + w_s(q_s) = c_s = \text{const} \qquad (s = 1, 2, \ldots, n). \tag{16.4.3}$$

"Separation of variables" is the term used to describe a technique which transforms the equations of motion into a system of uncoupled equations. It is clear that not all systems of equations can be uncoupled.

Liouville has shown (1849) that the results of (16.4.3) can be generalized in that all systems are separable whose energies have, or can be put into, the form

$$T = \frac{1}{2}\left\{\sum_{\alpha=1}^{n} \tilde{u}_\alpha(q_\alpha)\right\}\left\{\sum_{\beta=1}^{n} v_\beta(q_\beta)\dot{q}_\beta^2\right\}, \qquad V = \frac{\sum_{\alpha=1}^{n} \tilde{w}_\alpha(q_\alpha)}{\sum_{\beta=1}^{n} \tilde{u}_\beta(q_\beta)}, \tag{16.4.4}$$

where $\tilde{u}_s$, v_s, and $\tilde{w}_s$ are arbitrary functions of their arguments.

† Whittaker, pp. 67–69.

The transformations defined by

$$q_s^* = \int_0^{q_s} [v_s(q_s)]^{1/2}\, dq_s = q_s^*(q_s) \qquad (s = 1, 2, \ldots, n) \tag{16.4.5}$$

possess an inverse

$$q_s = q_s(q_s^*), \tag{16.4.6}$$

and they imply the relations

$$\dot{q}_s^{*2} = v_s(q_s)\dot{q}_s^2 \qquad (s = 1, 2, \ldots, n),$$

so that (16.4.4) may be written in the form

$$T = \frac{1}{2}\left\{\sum_{\alpha=1}^{n} u_\alpha(q_\alpha^*)\right\}\left\{\sum_{\beta=1}^{n} \dot{q}_\beta^{*2}\right\},$$

$$V = \frac{\sum_{\alpha=1}^{n} w_\alpha(q_\alpha^*)}{\sum_{\beta=1}^{n} u_\beta(q_\beta^*)},$$

where we have used the notation

$$\left.\begin{aligned} u_s(q_s^*) &= \tilde{u}_s(q_s(q_s^*)), \\ w_s(q_s^*) &= \tilde{w}_s(q_s(q_s^*)) \end{aligned}\right. \qquad (s = 1, 2, \ldots, n). \tag{16.4.7}$$

For convenience, we omit the asterisks henceforth. To shorten the equations, we write the energies as

$$\begin{aligned} T &= \frac{1}{2} u \sum_{\beta=1}^{n} \dot{q}_\beta^2, \\ V &= \frac{1}{u} \sum_{\alpha=1}^{n} w_\alpha(q_\alpha), \end{aligned} \tag{16.4.8}$$

where

$$u = \sum_{\beta=1}^{n} u_\beta(q_\beta).$$

Lagrange's equation for the coordinate q_k is found by direct substitution. It is

$$\frac{d}{dt}(u\dot{q}_k) - \frac{1}{2}\frac{\partial u}{\partial q_k}(\dot{q}_1^2 + \dot{q}_2^2 + \cdots + \dot{q}_n^2) = -\frac{\partial V}{\partial q_k}.$$

Multiplying this equation by $2u\dot{q}_k$, and noting that

$$\frac{d}{dt}(u^2\dot{q}_k^{\,2}) = 2u\dot{q}_k\frac{d}{dt}(u\dot{q}_k),$$

we obtain

$$\frac{d}{dt}(u^2\dot{q}_k^{\,2}) - u\dot{q}_k\frac{\partial u}{\partial q_k}(\dot{q}_1^{\,2} + \dot{q}_2^{\,2} + \cdots + \dot{q}_n^{\,2}) = -2u\dot{q}_k\frac{\partial V}{\partial q_k}. \tag{16.4.9}$$

Under our initial assumptions, the system possesses the energy integral

$$\tfrac{1}{2}u(\dot{q}_1^{\,2} + \dot{q}_2^{\,2} + \cdots + \dot{q}_n^{\,2}) + V = h = \text{const.}$$

Therefore, (16.4.9) may be written as

$$\begin{aligned}\frac{d}{dt}(u^2\dot{q}_k^{\,2}) &= 2(h - V)\dot{q}_k\frac{\partial u}{\partial q_k} - 2u\dot{q}_k\frac{\partial V}{\partial q_k}\\ &= 2\dot{q}_k\frac{\partial}{\partial q_k}[(h - V)u]\\ &= 2\dot{q}_k\frac{\partial}{\partial q_k}\left[hu - \sum_{\alpha=1}^{n} w_\alpha(q_\alpha)\right]\\ &= 2\dot{q}_k\frac{d}{dq_k}[hu_k(q_k) - w_k(q_k)],\end{aligned} \tag{16.4.10}$$

and such equations exist for every $k = 1, 2, \ldots, n$.

Now, the right-hand side involves q_k only, but the left-hand side contains all q_s ($s = 1, 2, \ldots, n$). Therefore, the equations are at most what might be called "partially decoupled." Nevertheless, they can be integrated directly because

$$\dot{q}_k\frac{d}{dq_k}[F(q_k)] = \frac{d}{dt}[F(q_k)].$$

It follows that the equations (16.4.10) are exact differentials and can be integrated to give

$$\tfrac{1}{2}u^2\dot{q}_k^{\,2} = hu_k(q_k) - w_k(q_k) + c_k \qquad (k = 1, 2, \ldots, n), \tag{16.4.11}$$

where c_k is a constant of integration.

If we sum (16.4.11) over all k and divide the resulting equation by u, we find

$$\frac{1}{2}u\sum_{k=1}^{n}\dot{q}_k^{\,2} = h - \frac{1}{u}\sum_{k=1}^{n} w_k(q_k) + \frac{1}{u}\sum_{k=1}^{n} c_k.$$

But in view of (16.4.8) this is precisely

$$T + V = h + \frac{1}{u}\sum_{k=1}^{n} c_k.$$

Inasmuch as the total energy is h, the constants of integration must satisfy the relation

$$\sum_{k=1}^{n} c_k = 0.$$

This is one relation between n constants. Hence, the Liouville system permits the determination of n constants of integration $c_1, c_2, \ldots, c_{n-1}, h$. The reason why the integration procedure described here has not yielded $n+1$ constants of integration is that the energy integral was utilized to integrate the Liouville system.

Example 16.4.1. As an example of a system that can be separated consider Fig. 16.4.1. Two particles, each of mass m, can slide without friction along a wire. They are attached, respectively, to identical, linear anchor springs S_1, and to a nonlinear coupling spring S_2, as shown. We suppose that an equilibrium configuration exists, and the departure from it is given by the generalized coordinates x and y.

The kinetic energy is

$$T = \tfrac{1}{2}m(\dot{x}^2 + \dot{y}^2) \tag{a}$$

and the potential energy is

$$V = \tfrac{1}{2}kx^2 + \tfrac{1}{2}ky^2 + F(x - y), \tag{b}$$

where k is a positive constant, and F is a function of $x - y$ which vanishes with its argument.

It should be noted that T has the form of the first equation of (16.4.4), but V does not have the required form. When these energies are substituted in Lagrange's equations, one finds

$$\begin{aligned} m\ddot{x} + kx + F'(x - y) &= 0, \\ m\ddot{y} + ky - F'(x - y) &= 0, \end{aligned}$$

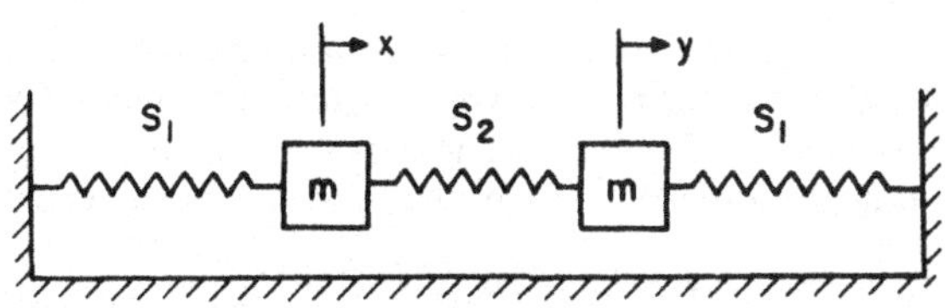

Fig. 16.4.1. The spring–mass system of Example 16.4.1.

where the prime denotes $d/d(x-y)$. It is seen that the variables are not separated. However, under the transformations

$$x = q_1 + q_2, \qquad y = q_1 - q_2, \tag{c}$$

the energies go over into

$$\begin{aligned} T &= \tfrac{1}{2}m(\dot{q}_1^2 + \dot{q}_2^2), \\ V &= \tfrac{1}{2}kq_1^2 + \tfrac{1}{2}kq_2^2 + F(2q_2). \end{aligned} \tag{d}$$

These have the required form with

$$\begin{aligned} &u_1 = u_2 \equiv 1, \qquad v_1 = v_2 \equiv m, \qquad w_1(q_1) = \tfrac{1}{2}kq_1^2, \\ &w_2(q_2) = \tfrac{1}{2}kq_2^2 + F(2q_2). \end{aligned}$$

Thus, the system of Fig. 16.4.1 is a Liouville system in the coordinates q_1 and q_2. When the energies (d) are substituted in Lagrange's equations, one finds the uncoupled equations

$$\begin{aligned} m\ddot{q}_1 + kq_1 &= 0, \\ m\ddot{q}_2 + kq_2 + 2F'(2q_2) &= 0, \end{aligned} \tag{e}$$

where the prime denotes $d/d(2q_2)$. Thus, in the coordinates q_1, q_2, the variables are separated.

16.5. Solution in Quadratures

A strictly Newtonian system (SN) with n generalized coordinates has n second-order Lagrange equations [as was shown in (16.1.3) and (16.1.4)]. Therefore, the complete integration requires $2n$ integrals. When these can be written down in the form of $2n$ indefinite integrals, the problem is "solved in quadratures." If these integrals can be evaluated in terms of elementary or tabulated functions the solution is said to be in "closed form." For practical purposes, there is little difference between these two, and we regard a problem as completely solved if it can be solved in quadratures.

Frequently a problem can be solved in quadratures after some first integrals have been found by some of the methods described earlier in this chapter. This is best illustrated by examples.

Example 16.5.1. Consider the classical example of the spherical pendulum. This is a particle under the force of gravity, constrained to move on the surface of a perfectly smooth sphere of radius r. Therefore, its kinetic energy is found from (15.2.16) by setting the time derivative of the radius equal to zero, and

$$\begin{aligned} T &= \tfrac{1}{2}mr^2(\sin^2\theta\,\dot{\varphi}^2 + \dot{\theta}^2), \\ V &= mgr(1 - \cos\theta). \end{aligned} \tag{16.5.1}$$

The energy integral exists, or

$$mr[\tfrac{1}{2}r(\sin^2\theta\dot{\varphi}^2 + \dot{\theta}^2) + g(1 - \cos\theta)] = h = \text{const.} \tag{16.5.2}$$

The Lagrangean function is

$$L = mr[\tfrac{1}{2}r(\sin^2\theta\dot{\varphi}^2 + \dot{\theta}^2) - g(1 - \cos\theta)] \tag{16.5.3}$$

and φ is seen to be an ignorable coordinate. Therefore,

$$p_\varphi = \frac{\partial L}{\partial\dot{\varphi}} = mr^2\sin^2\theta\dot{\varphi} = c_\varphi = \text{const.} \tag{16.5.4}$$

Because of the existence of an energy integral in this example, no benefit is derived from constructing the Routhian function. We proceed by forming from (16.5.4)

$$\dot{\varphi}^2 = \frac{c_\varphi^{\,2}}{m^2r^4\sin^4\theta} \tag{16.5.5}$$

and substitute it in the energy integral (16.5.3). That equation can then be rewritten in the form

$$\dot{\theta} = G(\theta),$$

resulting in the quadrature

$$t = \int d\theta/G(\theta) = t(\theta).$$

If a Routhian function had been constructed, the equation in θ would have been of second order rather than first, and it would have had a first integral $\dot{\theta} = G(\theta)$. Writing the inverse of the last equation

$$\theta = \theta(t)$$

(assuming it exists) and substituting it in (16.5.4) gives the relation for φ and t as the quadrature

$$\varphi = \int \frac{c_\varphi\,dt}{mr^2\sin^2\theta(t)}.$$

We now show the details of this basic procedure. We begin by rewriting the energy integral as

$$\tfrac{1}{2}mr^2(\sin^2\theta\dot{\varphi}^2 + \dot{\theta}^2) - mgr\cos\theta = mgrH, \tag{16.5.6}$$

where we have put

$$h - mgr = mgrH = \text{const.}$$

The momentum integral is rewritten as

$$\sin^2\theta\dot{\varphi} = (2g\alpha/r)^{1/2} \tag{16.5.7}$$

where we have put

$$c_\varphi/mr^2 = (2g\alpha/r)^{1/2} = \text{const.}$$

Then, solving (16.5.7) for $\dot{\varphi}^2$, we find

$$\dot{\varphi}^2 = 2g\alpha/r \sin^4\theta, \tag{16.5.8}$$

and substituting this in (16.5.6) gives

$$\dot{\theta}^2 = \frac{2g}{r}\left(H + \cos\theta - \frac{\alpha}{\sin^2\theta}\right). \tag{16.5.9}$$

Under the transformation

$$z = \cos\theta \tag{16.5.10}$$

one finds

$$\dot{\theta}^2 = \frac{\dot{z}^2}{1 - z^2},$$

so that (16.5.9) becomes

$$\dot{z}^2 = \frac{2g}{r} f(z), \tag{16.5.11}$$

where

$$f(z) = (1 - z^2)(H + z) - \alpha.$$

Thus, $f(z)$ is a cubic polynomial in z. Then, from (16.5.11) we obtain

$$\left(\frac{2g}{r}\right)^{1/2} dt = \frac{dz}{[f(z)]^{1/2}} \tag{16.5.12}$$

and the relation between z and t is the quadrature

$$\left(\frac{2g}{r}\right)^{1/2} t = \int \frac{dz}{[f(z)]^{1/2}}. \tag{16.5.13}$$

Integrals of the form (16.5.13) in which $f(z)$ is a polynomial of third or fourth degree are called elliptic integrals of the first kind; they are tabulated.

Now from (16.5.7) and (16.5.10) we obtain

$$d\varphi = \left(\frac{2g}{r}\right)^{1/2} \frac{dt}{1 - z^2} \alpha^{1/2}$$

or, in view of (16.5.12),

$$\varphi = \alpha^{1/2} \int \frac{dz}{(1 - z^2)\,[f(z)]^{1/2}}. \tag{16.5.14}$$

This is a quadrature relating φ and z. Then by means of (16.5.13), the relation between φ and t is also known. The integral in (16.5.14) is an elliptic integral of the third kind, which is also tabulated.

As a second example we consider a holonomic system having one generalized coordinate.

Example 16.5.2. When a particle having one degree of freedom is acted on by a bounded force which depends on position only, the energy integral always exists and the problem can always be solved in quadratures.

The kinetic energy is

$$T = \tfrac{1}{2}m\dot{x}^2$$

and the generalized force is of the form

$$Q_x = X = X(x)$$

Lagrange's equation of motion is

$$m\ddot{x} = X(x)$$

and the energy integral is

$$\tfrac{1}{2}m\dot{x}^2 + F(x) = h = \text{const},$$

where

$$F(x) = -\int_0^x X(u)\,du.$$

It follows that

$$\dot{x}^2 = \frac{2}{m}[h - F(x)]$$

and the problem is solved by the quadrature

$$\left(\frac{2}{m}\right)^{1/2} t = \int \frac{dx}{[h - F(x)]^{1/2}}.$$

As a final example we consider the Liouville system.

Example 16.5.3. The Liouville system can be solved in quadratures. In consequence, the same is true for the system (16.4.1) because it is the special case of the Liouville system when one puts $\sum_{\beta=1}^{n}\tilde{u}_\beta = 1$ in (16.4.4).

To show this, solve each of the equations (16.4.11) for $u^2/2$ and equate them. This gives

$$\frac{dq_1}{[hu_1(q_1) - w_1(q_1) + c_1]^{1/2}} = \frac{dq_2}{[hu_2(q_2) - w_2(q_2) + c_2]^{1/2}} = \cdots$$

$$= \cdots = \frac{dq_n}{[hu_n(q_n) - w_n(q_n) + c_n)^{1/2}}.$$

Obviously, the variables in these are separated, and the equations can be integrated.

16.6. Qualitative Integration

Sometimes no analytical solution to a dynamics problem can be found (even when existence and uniqueness under specified initial conditions are assured), sometimes an analytical solution is so complicated that the behavior of the system is very difficult to deduce from it, and sometimes the skill required to find a solution exceeds that of the person seeking it. In these cases, a "discussion" by qualitative methods may give the required information. No general rules for these "qualitative integrations" can be given, but an illustrative example may be useful.

Example 16.6.1. We consider the force-free motion of two particles in the plane, partially treated in Example 16.3.1. It was shown there that the energy and momentum integrals exist. They are, respectively,

$$m_1\dot{r}_1^2 + m_2\dot{r}_2^2 + (m_1r_1^2 + m_2r_2^2)\dot{\theta}^2 = h, \tag{16.6.1}$$

and

$$(m_1r_1^2 + m_2r_2^2)\dot{\theta} = C, \tag{16.6.2}$$

where h and C are positive constants. [It is obvious from (16.6.1) that $h > 0$; putting $C > 0$ merely defines the positive direction of $\dot{\theta}$.]

In Example 15.5.1, the equations of motion were derived in (15.5.15). Under force-free motion, they become

$$\begin{gathered} \ddot{r}_1 - r_1\dot{\theta}^2 = 0, \\ \ddot{r}_2 - r_2\dot{\theta}^2 = 0, \\ (m_1r_1^2 + m_2r_2^2)\ddot{\theta} + 2(m_1r_1\dot{r}_1 + m_2r_2\dot{r}_2)\dot{\theta} = 0. \end{gathered} \tag{16.6.3}$$

These are the equations having (16.6.1) and (16.6.2) as first integrals. In fact, (16.6.1) could be found by multiplying the first equation of (16.6.3) by $m_1\dot{r}_1$, the second by $m_2\dot{r}_2$, the third by $\dot{\theta}$, and adding them; the resulting equation is the differential of (16.6.1). The momentum integral could be found by noting that the third equation of (16.6.3) is an exact differential.

Even though two first integrals have been found, the variables cannot be separated, and the problem cannot be reduced to quadratures. We shall use a qualitative integration to deduce the motion. As this is our first example of a qualitative integration, we expose all details.

We deduce from (16.6.3) that

$$\ddot{r}_i/r_i \geq 0 \qquad (i = 1, 2)$$

because $\dot{\theta}$ is real. Therefore, the radial displacements and accelerations have the same sign, or

$$\ddot{r}_i \geq 0 \qquad (i = 1, 2) \tag{16.6.4}$$

for all time because, by definition, the radial positions r_i are positive. Now, we may write

$$\ddot{r}_i = \dot{r}_i \frac{d\dot{r}_i}{dr_i} \geq 0 \tag{16.6.5}$$

because of (16.6.4), i.e., the trajectories in the r_i, $\dot{r}_i$ planes have positive slopes in the upper half-plane ($\dot{r}_i > 0$) and negative slopes in the lower ($\dot{r}_i < 0$). This is shown in Fig. 16.6.1.

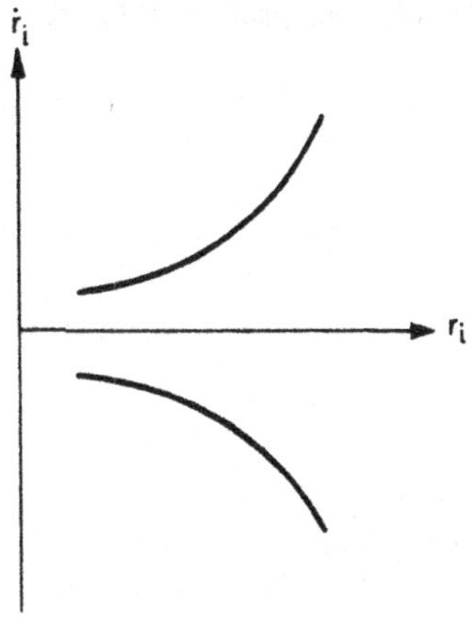

Fig. 16.6.1. State-space trajectories of Example 16.6.1.

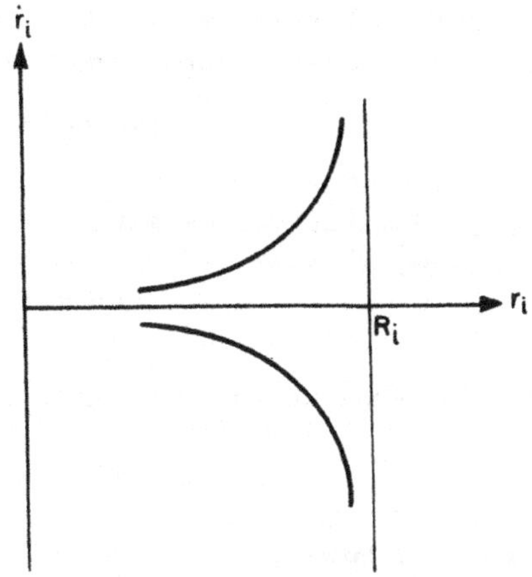

Fig. 16.6.2. State-space trajectories of Example 16.6.1. The r_i approach an upper bound.

We now show that, necessarily,

$$\lim_{t\to\infty} |r_i| = \infty. \tag{16.6.6}$$

Assume to the contrary that either one of the r_i tends to a bounded value R_i, as indicated in Fig. 16.6.2. Then, as is evident from this diagram, the velocities $\dot{r}_i \to \pm\infty$ and $d\dot{r}_i/dr_i \to \pm\infty$ as $t \to \infty$. It follows from (16.6.5) that, under this assumption,

$$r_i \to \infty \qquad (i = 1, 2).$$

But, by this same assumption, the r_i are bounded by R_i. Then, one sees from the first two equations of (16.6.3) that one must have

$$\dot{\theta} \to \pm\infty.$$

This contradicts the momentum integral (16.6.2); therefore, the assumption was false, and (16.6.6) is established. Then, we have as a direct corollary to (16.6.6) that

$$\lim_{t\to\infty} \dot{\theta} = 0 \tag{16.6.7}$$

because, with the radii going to infinity, the momentum (16.6.2) can only be bounded if $\dot{\theta}$ tends to zero.

Next, we demonstrate that the terminal velocities are

$$\lim_{t\to\infty} \dot{r}_i = v_i = \text{const} \qquad (i = 1, 2). \tag{16.6.8}$$

To prove it we note, by combining (16.6.2) and (16.6.7), that

$$\lim_{t\to\infty} (m_1 r_1^2 + m_2 r_2^2)\dot{\theta}^2 = \lim_{t\to\infty} C\dot{\theta} = 0.$$

But this is the limit of the second term in the energy integral (16.6.1). Therefore, the limit of the energy integral itself is

$$\lim_{t\to\infty} (m_1\dot{r}_1^2 + m_2\dot{r}_2^2) = h.$$

This requires that (16.6.8) be true.

Finally, we prove

$$\lim_{t\to\infty} \theta = \theta_0 = \text{const.} \tag{16.6.9}$$

From the momentum integral we have

$$\theta(\infty) - \theta(0) = \lim_{t\to\infty} \int_0^t \frac{C\,dt}{m_1r_1^2 + m_2r_2^2},$$

or, for some large A,

$$\theta(\infty) - \theta(0) = \int_0^A \frac{C\,dt}{m_1r_1^2 + m_2r_2^2} + \int_A^\infty \frac{C\,dt}{m_1r_1^2 + m_2r_2^2}. \tag{16.6.10}$$

Now, the first of these integrals is bounded. We also find, from (16.6.8), that

$$\lim_{t\to\infty} r_i = \lim_{t\to\infty} v_it \qquad (i = 1, 2),$$

i.e., the r_i go to infinity linearly with t. Then, the second integral in (16.6.10) is

$$\int_A^\infty \frac{C\,dt}{(m_1v_1^2 + m_2v_2^2)}$$

and we have in the limit

$$\lim_{A\to\infty} \int_A^\infty \frac{C\,dt}{(m_1v_1^2 + m_2v_2^2)t^2} = \lim_{A\to\infty} \frac{C}{(m_1v_1^2 + m_2v_2^2)A} = 0.$$

Hence, the second integral of (16.6.10) is also bounded, which proves that θ approaches a finite value as time goes to infinity. One also sees from (16.6.2) that $\dot{\theta}$ cannot change sign, or θ always increases (under our assumption that $C > 0$). Therefore, θ must behave as shown in Fig. 16.6.3. We deduce that, regardless of initial conditions, the line on which the masses lie rotates in the same direction and ap-

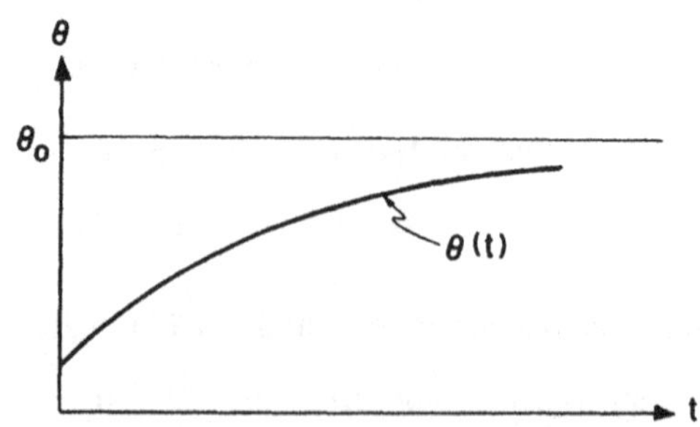

Fig. 16.6.3. Rotation of the system of Example 16.6.1.

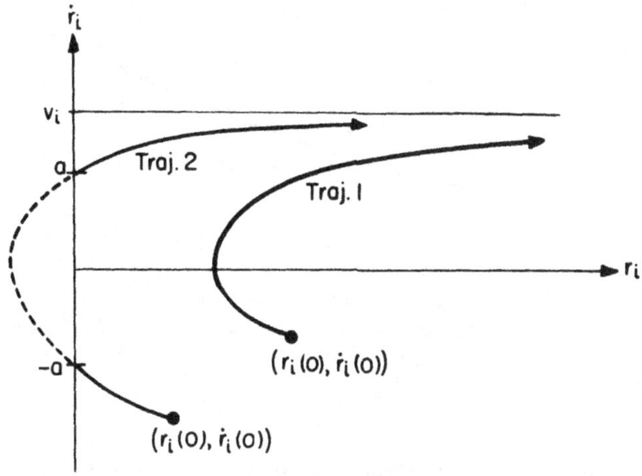

Fig. 16.6.4. State-space trajectories of Example 16.6.1.

proaches a limiting angle as time tends to infinity. It remains to discuss the behavior of the radial positions.

Since the r_i are positive and increase with time for sufficiently large values of t, the terminal velocities v_i are also positive. However, the initial velocity in the r direction of one or both of the m_i might be negative. Therefore, it is of interest to inquire into the behavior of the r_i when this happens.

We may rewrite the first and second equations of (16.6.3) as

$$\frac{d\dot{r}_i}{dr_i} = \frac{r_i\dot{\theta}^2}{\dot{r}_i} \qquad (i = 1, 2); \tag{16.6.11}$$

this shows that if the trajectories in the r_i, $\dot{r}_i$ plane cross the r_i axis (defined by $\dot{r}_i = 0$) at a point where $\dot{\theta}$ and r_i are not zero, they do so with infinite slope. Now every trajectory must end up in the upper half-plane $\dot{r}_i > 0$. Therefore, those with initial velocites $\dot{r}_i(0) < 0$ must either cross the r_i axis, or else they are discontinuous. Both cases arise. If they cross the r_i axis in the right-hand half-plane $r_i > 0$, they are continuous. However, if they have reached a position $r_i = 0$ while the corresponding $\dot{r}_i < 0$, say $\dot{r}_i = -a$, they jump instantaneously to $\dot{r}_i = +a$, as shown in Fig. 16.6.4. In this diagram we show two trajectories with initial points $(r_i(0), \dot{r}_i(0))$. Trajectory 1 is continuous. Even though the initial velocity was negative, the centrifugal force caused the mass to come to rest on its way toward the origin and then move away from it. In trajectory 2, the mass moved initially toward the origin with such large velocity that the centrifugal force was unable to make it come to rest before it reached the origin. Thus, it passed right through that position, emerging instantaneously on the other side with a velocity such that the radius increases; therefore, $\dot{r}_i > 0$ thereafter.

The behavior of the r_i as functions of time is shown in two trajectories in Fig. 16.6.5. The portion of trajectory 2 beyond the corner is the reflection of the dotted portion shown in the lower half-plane.

This discussion has qualitatively described the behavior of the three generalized coordinates as a function of time and concludes our treatment of this problem.

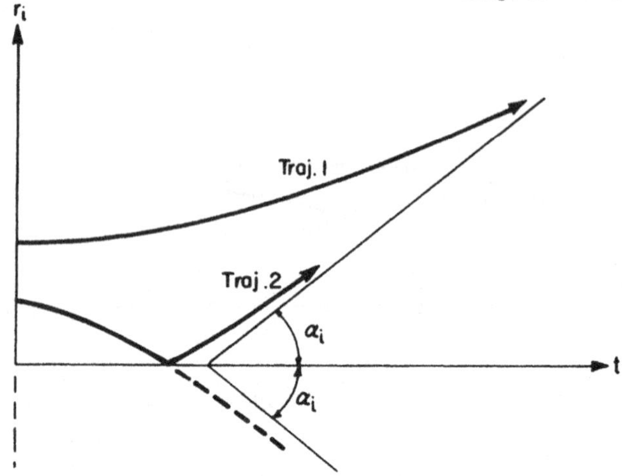

Fig. 16.6.5. Event-space trajectories of Example 16.6.1.

16.7. Problems

16.1. In Example 16.5.1 it was shown that the simple spherical pendulum with fixed support and constant length possess an energy and a momentum integral. Examine the same question when the point of support is moved in a prescribed fashion $f(t)$ along the horizontal x axis.

16.2. Examine the same question as in Problem 16.1 when the point of support is moved in a prescribed fashion $f(t)$ along the vertical z axis.

16.3. Examine the same question as in Problem 16.1 when the point of support is fixed, but the length of the pendulum changes in the prescribed fashion $f(t)$.

16.4. Examine the system of Problem 13.1 for the existence of constants of the motion.

16.5. Two heavy particles P_1 and P_2 of mass m_1 and m_2, respectively, are interconnected by a massless, inextensible string of length l, as shown. P_1 is constrained to move on the surface of an inverted smooth rigid circular cone. The other particle is constrained to move on a vertical line. (The string passes through a smooth hole in the apex of the cone, as shown.) Examine this system for the existence of an energy integral and of momentum integrals.

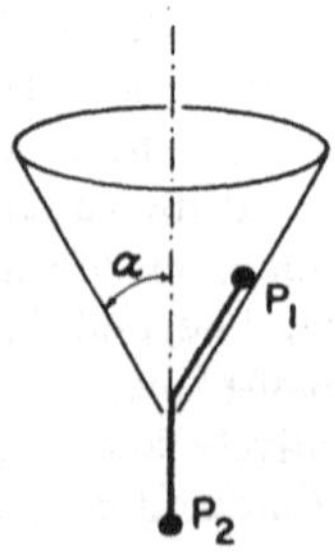

16.6. Examine the system of Problem 16.5 for the existence of constants of the motion when, in addition to the force of gravity, the particles are under mutual Newtonian mass attraction.

16.7. A particle of mass m is attracted to each of two fixed centers with a force inversely proportional to the square of its distance from each. Let the force centers be a distance $2c$ apart, and let the x axis lie in the line connecting them, with origin at the midpoint, as shown. The motion is assumed to take place in the plane defined by the particle position (x, y) and the force

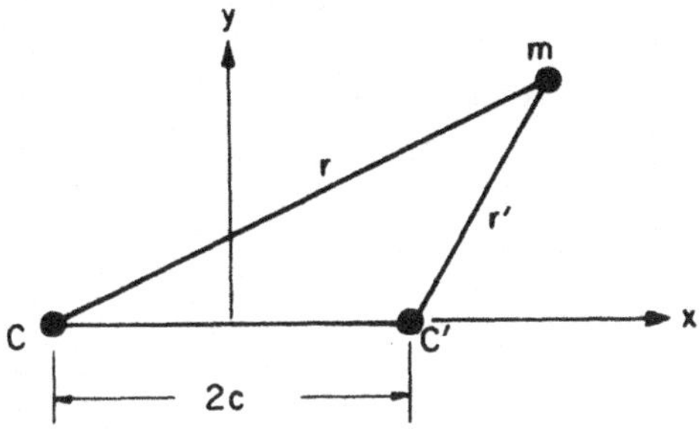

centers. If the distances of the particle from the force centers are r and r', respectively, as shown, the potential energy is $A/r + A'/r'$, where A and A' are positive constants, and the kinetic energy is $T = \frac{1}{2}m(\dot{x}^2 + \dot{y}^2)$. Show that this system is a Liouville system in the coordinates $q_1 = \frac{1}{2}(r + r')$ and $q_2 = \frac{1}{2}(r - r')$.

16.8. A heavy, uniform rod is constrained to move in the vertical plane, and one of its extremities is constrained to move on a horizontal line so that its distance from a fixed point on that line is a prescrived function $f(t)$. Discuss the existence of integrals.

16.9. Let each of a system of $N/3$ particles in 3-space be connected by (linear or nonlinear) springs to all others and let them also be under their mutual Newtonian mass attractions. Show that there exists seven constants of the motion and that three additional constants may be found by quadratures.

16.10. Examine Problem 15.7 for the existence of constants of the motion. Reduce the complete solution to quadratures.

16.11. Consider an unconstrained system of $n = N/3$ particles in x_1, x_2, x_3 space. Let the x_i component of the given force X_i^k acting on the kth particle be a function of x_i^k only, where x_i^k is the x_i component of the position of the kth particle. Show that the solution of the problem can be reduced to quadratures.

16.12. Apply the results of Problem 16.11 to the case where $n = 3$, $m_1 = m_2 = m_3 = 1$, and $X_i^k = (x_i^k)^m$, m and odd integer.

17

Stability

17.1. Introductory Remarks

Having discussed the integration of the Lagrangean equations of motion, it is natural to inquire into the stability of the motions predicted by them.

In the linguistic sense, a process (or a person) is considered unstable when small causes lead to large consequences. This is also the technical meaning of the concept of instability, and an intuitive approach to the idea of stability was briefly touched on in Section 3.11.

An examination of the stability of a given motion seeks an answer to the question: If the motion of a dynamical system is disturbed in any manner at a given instant of time, what will be the effect of this disturbance on the subsequent motion? Disturbances may be changes in initial conditions, or in the parameters of the differential equations of motion, or changes in the equations of motion by the addition of forces, perhaps considered too small to be retained in the original problem formulation.

A convenient way to study stability is offered by the representation of the motion as a T trajectory in state–time space because every point of such a trajectory associates a given state (coordinates and velocity components) with the time at which this state is attained. We may, then, examine the stability of the motion by asking a number of different questions:

(i) Does a disturbance at a time t_0 of a T trajectory produce subsequent changes which remain *permanently* bounded? If the answer is affirmative, the motion is stable in the sense of Liapunov.

(ii) Does an infinitesimal disturbance (in the sense of the calculus) at t_0 of a T trajectory produce infinitesimal changes which do not grow in magnitude *initially*? If the answer is affirmative, the motion possesses short-term, first-order stability.

(iii) Does a disturbance at the time t_0 of an S trajectory in state space produce subsequent changes which remain permanently bounded? If the answer is affirmative, the motion is stable in the sense of Poincaré.

(iv) Does an infinitesimal change in one or more of the parameters of the equations of motion produce at most infinitesimal changes in the T trajectory? If the answer is affirmative, the system is structurally stable.

17.2. Definition of Stability

To define stability it is convenient to utilize the observation (see Section 16.1) that a system of n second-order Lagrangean equations can always be transformed into a system of $2n$ first-order equations. Therefore, we shall deal with the system

$$\frac{dy_s}{dt} = Y_s(y_1, y_2, \ldots, y_{2n}, t) \qquad (s = 1, 2, \ldots, 2n), \tag{17.2.1}$$

in which the first n dependent variables are generalized coordinates, and the remaining ones are generalized velocities.

Let us consider a particular solution $f_s(t)$ $(s = 1, 2, \ldots, 2n)$ of (17.2.1) which we call the *undisturbed motion*. The difference between the values $y_s(t)$ of a disturbed motion and $f_s(t)$ for a given t we call the *disturbance* at t. Then we define:

An undisturbed motion $f_s(t)$ $(s = 1, 2, \ldots, 2n)$ *of* (17.2.1) *is called* stable in the sense of Liapunov (L-stable) *with respect to the* y_s *if for each* $\varepsilon > 0$, *one can find another number* $\eta(\varepsilon) > 0$ *such that for all disturbed motions* $y_s(t)$, *with initial disturbances*

$$| y_s(t_0) - f_s(t_0) | \leq \eta(\varepsilon),$$

the following inequalities hold:

$$| y_s(t) - f_s(t) | < \varepsilon$$

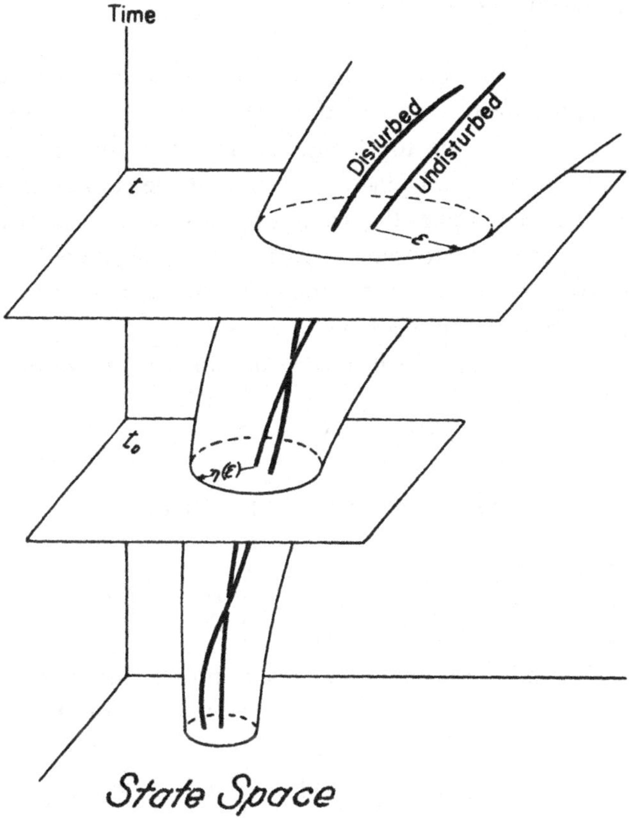

Fig. 17.2.1. State–time space trajectories of Liapunov-stable motion.

for all $t > t_0$ *and for all* $s = 1, 2, \ldots, 2n$. *If* $\varepsilon \to 0$ *as* $t \to \infty$, *the undisturbed motion is* asymptotically stable.

The meaning of this definition is illustrated in Fig. 17.2.1, which shows an undisturbed motion that is L-stable. The definition states essentially that every bounded disturbance at the time t must have been preceded by an earlier disturbance which was also bounded.

Poincaré stability (P-stability) is defined in a similar way. Let the trajectory of the undisturbed motion in state space be given by $f_s(\sigma)$ ($s = 1, 2, \ldots, 2n$) where σ is the arc length measured from some point $f_s(\sigma_0)$. Moreover, let $y_s(\sigma)$ be the S trajectory of the disturbed motion where $y_s(\sigma_0)$ is that point on the disturbed S trajectory which is nearest the point $f_s(\sigma_0)$ on the undisturbed one. Then, the definition of P-stability is obtained by replacing t in the definition of L-stability by σ.

17.3. The Variational Equations

If the motion of a system is examined for stability, one must specify the causes which have produced the disturbance. In this book, we only consider disturbances due to changes in initial conditions. Disturbances which arise from changes in the parameters of the equations, or from the addition of disturbing forces that cannot be ascribed to changed parameters are better treated in monographs devoted to stability theory. The reader who wishes to learn more about stability theory is referred to the excellent treatises by Leipholz† and Malkin.‡

In studying stability, the disturbed motion $y_s(t)$ is taken to be of the form

$$y_s(t) = f_s(t) + \eta_s(t) \qquad (s = 1, 2, \ldots, 2n). \tag{17.3.1}$$

The substitution of this equation in (17.1.1) gives

$$\frac{df_s}{dt} + \frac{d\eta_s}{dt} = Y_s(f_1 + \eta_1, f_2 + \eta_2, \ldots, f_{2n} + \eta_{2n}, t).$$

Under the assumption that the functions Y_s possess first partial derivatives with respect to the y_s, the above equations may be written as

$$\frac{df_s}{dt} + \frac{d\eta_s}{dt} = Y_s(f_1, f_2, \ldots, f_{2n}, t) + \sum_{r=1}^{2n} a_{sr}\eta_r + \Phi_s(\eta_1, \eta_2, \ldots, \eta_{2n}, t), \tag{17.3.2}$$

where the a_{sr} are the partial derivatives $\partial Y_s/\partial y_r$ evaluated at $y_r = f_r$ $(r = 1, 2, \ldots, 2n)$. The functions Φ_s are *not* linear in the η_r, and they satisfy $\Phi_s(0, 0, \ldots, 0, t) = 0$ for all s because (17.3.2) must reduce to

$$\frac{df_s}{dt} = Y_s(f_1, f_2, \ldots, f_{2n}, t) \tag{17.3.3}$$

when the $\eta_r \equiv 0$.

Then, subtracting (17.3.3) from (17.3.2), one finds

$$\frac{d\eta_s}{dt} = \sum_{r=1}^{2n} a_{sr}\eta_r + \Phi_s(\eta_1, \eta_2, \ldots, \eta_{2n}, t) \qquad (s = 1, 2, \ldots, 2n). \tag{17.3.4}$$

† Leipholz, H., *Stability Theory*, Academic Press, New York and London (1970).

‡ Malkin, I. G., *Theory of Stability of Motion*, United States Atomic Energy Commission translation (AEC-tr-3352) from the Russian. (This translation is not satisfactory.) *Theorie der Stabilität einer Bewegung* (German translation), R. Oldenbourg, Munich (1959).

These are called the variational equations of (17.1.1) with respect to the solutions $f_s(t)$, and all stability investigations are based on them. In fact, the stability of the undisturbed solution is identified with the stability of the zero solution $\eta_s = 0$ $(s = 1, 2, \ldots, 2n)$ of (17.3.4). Evidently, that solution satisfies (17.3.4), and it is identified with the undisturbed solution because of (17.3.1). Thus, if the zero solution of (17.3.4) is stable, the undisturbed solution is stable.

Two different methods are used. The so-called "indirect" methods seek to integrate, or approximate the integrals of (17.3.4) and thereby study the behavior of the disturbance with time. Liapunov's "direct" method studies the stability of the zero solution with the aid of a test function called the Liapunov function *without* integrating (17.3.4) (exactly or by approximation).

17.4. Some Remarks on Indirect Methods

The most common indirect method begins with "linearizing" (17.3.4) by deleting all but the linear terms. The rationale for this procedure is that the functions Φ_s are of higher order than first in the η_r; therefore, they tend to zero with the $\eta_r \to 0$ more rapidly than the linear terms. Then, since a *stable* zero solution ensures that small $|\eta_r|$ remain permanently small, it is argued that for sufficiently small $|\eta_r|$, the functions Φ_s may be ignored in the stability analysis of the zero solution of (17.3.4). This heuristic argument is not reliable and leads occasionally to wrong conclusions. If the nonlinear system (17.3.4) is identified by (N), and the linearized one by (L), some results are known as to when stability of (L) implies stability of (N).[†]

When the nonlinear terms are deleted in (17.3.4), one obtains the linear variational equations

$$\frac{d\eta_s}{dt} = \sum_{r=1}^{2n} \left.\frac{\partial Y_s}{\partial y_r}\right|_f \eta_r \qquad (s = 1, 2, \ldots, 2n), \tag{17.4.1}$$

where the symbol

$$\left.\frac{\partial Y_s}{\partial y_r}\right|_f$$

means that these derivatives are evaluated on the undisturbed solution

[†] See, for instance, Coddington, E. A., and Levinson, N., *Theory of Ordinary Differential Equations*, McGraw-Hill Book Co., Inc., New York (1955), pp. 375–401.

$f_s(t)$ ($s = 1, 2, \ldots, 2n$). Equations (17.4.1) are linear equations with variable coefficients, and a study of the stability of their zero solutions is not simple. However, *if we choose the undisturbed solution for which all* $\dot{f}_s = 0$, *and if the functions* Y_s *do not depend explicitly on time*, the equations (17.4.1) become a set of linear equations with constant coefficients, i.e., they are

$$\frac{d\eta_s}{dt} = \sum_{r=1}^{2n} a_{sr}\eta_r \qquad (s = 1, 2, \ldots, 2n). \tag{17.4.2}$$

It should be remembered that the first n derivatives $\dot{f}_s$ are velocities while the remaining ones are accelerations. Therefore, the condition $\dot{f}_s = 0$ for all s means that all velocities are zero (a rest point) and all accelerations are zero (no forces act). It follows that this particular undisturbed "stationary" solution is an equilibrium point of the system or the motions under study are motions in the neighborhood of an equilibrium point.

It is well known that the characteristic values λ_s of (17.4.2) are obtained as the roots of the characteristic equation

$$| a_{rs} - \delta_{rs}\lambda | = 0, \tag{17.4.3}$$

where δ_{rs} is Kronecker's delta [see (6.3.10)], and one has the theorems:

(i) *When all roots* λ_s *of the characteristic equation have negative real parts, the stationary solution is asymptotically stable.*

(ii) *When one or more than one root* λ_s *has positive real part, the stationary solution is unstable.*

(iii) *When all roots* λ_s *are distinct and some roots have zero real part while the remaining ones, if any, have negative real part, the solution is stable, but not asymptotically stable. This is called the critical case.*

The reason for calling the last case "critical" is that a change in the equations of motion, no matter how small, may change the behavior to asymptotic stability or to instability depending on the change. This is illustrated in Example 17.4.3.

There are other theorems known for the system (17.4.2); for additional results, the reader is referred to the above-mentioned monographs.

The essential difficulty in the use of these theorems is that of finding the roots of high-degree polynomials. The relations which exist between the roots of a polynomial and its coefficients can be utilized to derive algebraic criteria which permit the determination of the stability. These are known as the Routh–Hurwitz criteria. The roots of a polynomial may be represented by points in the complex plane and, for stability, none of these points may lie in the right-hand half of that plane. The so-called Nyquist diagram is a method for examining this criterion.

We shall now illustrate the use of the linearized variational equations by some examples.

Example 17.4.1. In Example 16.4.1, the Lagrange equations after separation of the variables were found to be

$$\begin{aligned} m\ddot{q}_1 + kq_1 &= 0, \\ m\ddot{q}_2 + kq_2 + 2F'(2q_2) &= 0. \end{aligned} \tag{a}$$

We assume that $F'(0) = 0$, which is equivalent to the observation that the coupling spring in Fig. 16.4.1 has zero force in the equilibrium position. If we write

$$F'(2q_2) = Kq_2 + \Phi(2q_2),$$

the second equation of (a) is

$$m\ddot{q}_2 + (k + K)q_2 + \Phi(2q_2) = 0, \tag{b}$$

where Φ contains no linear terms. Then, with the transformations

$$q_1 = \eta_1, \qquad \dot{q}_1 = \eta_2, \qquad q_2 = \eta_3, \qquad \dot{q}_2 = \eta_4,$$

the equations of motion become

$$\begin{aligned} \dot{\eta}_1 &= \eta_2, \\ \dot{\eta}_2 &= -\frac{k}{m}\eta_1, \\ \dot{\eta}_3 &= \eta_4, \\ \dot{\eta}_4 &= -\left(\frac{k}{m} + \frac{K}{m}\right)\eta_3 + \frac{2}{m}\Phi(2\eta_3). \end{aligned} \tag{c}$$

These are a special case of (17.3.4), and when the nonlinear term is deleted one has

$$\dot{\eta}_1 = \eta_2, \qquad \dot{\eta}_2 = -\frac{k}{m}\eta_1, \qquad \dot{\eta}_3 = \eta_4, \qquad \dot{\eta}_4 = -\left(\frac{k}{m} + \frac{K}{m}\right)\eta_3, \tag{d}$$

which is a special case of (17.4.2). The characteristic equation of this system is

$$\lambda^4 + \left(\frac{k}{m} + \frac{k+K}{m}\right)\lambda^2 + \frac{k}{m}\cdot\frac{k+K}{m} = 0 \tag{e}$$

and, for stability, no root may have a real positive part. This requires that λ_1^2 and λ_2^2 be real and negative. From the relations between the roots and coefficients of a second-degree polynomial it is known that

$$\begin{aligned} -\lambda_1^2 - \lambda_2^2 &= \frac{2k+K}{m}, \\ \lambda_1^2 \cdot \lambda_2^2 &= \frac{k}{m}\cdot\frac{k+K}{m}. \end{aligned} \tag{f}$$

For negative, real $\lambda_{1,2}^2$ one must have

$$\frac{2k+K}{m} > 0, \qquad \frac{k}{m}\cdot\frac{k+K}{m} > 0. \tag{g}$$

These are, therefore, the requirements for stable motion. The second of these can be satisfied if $k < 0$ and $k + K < 0$, which implies that one must have $k < K < 0$; but in that case, the first equation of (g) is violated. Hence, one must have, from the second equation of (g), $k > 0$ and $k + K > 0$, and in that case the first relation of (g) is automatically satisfied. Thus, stability of small motions about equilibrium requires, under the approximations made, that $k > -K > 0$.

As a second example consider:

Example 17.4.2. Use the linearized equations to determine the stability of motion of a plane double pendulum about all equilibrium positions. For simplicity, let the pendulums be identical.

The equations of motion of the double pendulum were derived in (15.5.23) and (15.5.24). When the pendulums are both of length l and mass m, these equations become

$$\begin{aligned} &2\ddot{\theta}_1 + \cos(\theta_1 - \theta_2)\ddot{\theta}_2 + \dot{\theta}_2^2 \sin(\theta_1 - \theta_2) + 2\frac{g}{l}\sin\theta_1 = 0, \\ &\ddot{\theta}_2 + \cos(\theta_1 - \theta_2)\ddot{\theta}_1 - \dot{\theta}_1^2 \sin(\theta_1 - \theta_2) + \frac{g}{l}\sin\theta_2 = 0. \end{aligned} \tag{a}$$

The double pendulum has four equilibrium positions: (i) Both pendulums hang down; (ii) two positions in which one pendulum points up and the other down; (iii) both pendulums stand on end. Small displacements from these are shown in Figs. 17.4.1, (a)–(d). If we write

$$\begin{aligned} \cos(\theta_1 - \theta_2) &= 1 + \Phi_1(\theta_1 - \theta_2), \\ \sin\theta_1 &= \theta_1 + \Phi_2(\theta_1), \end{aligned}$$

and introduce the transformations

$$\theta_1 = \eta_1, \qquad \dot{\theta}_1 = \eta_2, \qquad \theta_2 = \eta_3, \qquad \dot{\theta}_2 = \eta_4,$$

we find, after deleting the nonlinear terms,

$$\begin{aligned} \dot{\eta}_1 &= \eta_2, \\ 2\dot{\eta}_2 + \dot{\eta}_4 &= -2\frac{g}{l}\eta_1, \\ \dot{\eta}_3 &= \eta_4, \\ \dot{\eta}_4 + \dot{\eta}_2 &= -\frac{g}{l}\eta_3. \end{aligned} \tag{b}$$

The characteristic equation is easily computed to be

$$\lambda_{1,2}^2 = (-2 \pm 2^{1/2})g/l. \tag{c}$$

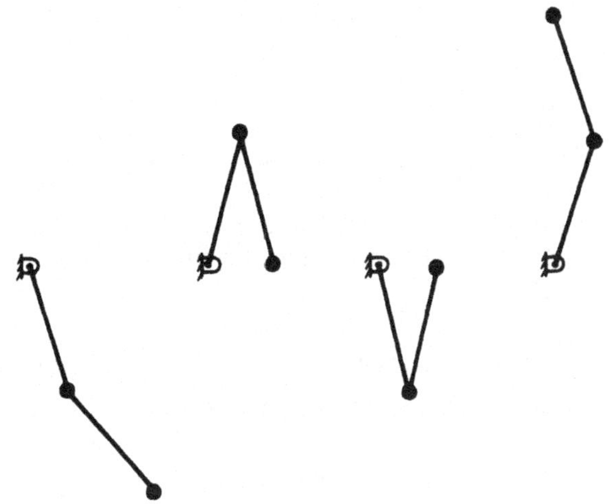

(a) (b) (c) (d)

Fig. 17.4.1. Motions of plane double pendulum of Example 17.4.2 about four equilibrium positions.

If no λ is to have positive real part we require that $\lambda_{1,2}^2$ be real and negative. Since $g/l > 0$, that condition is fulfilled, i.e., motion about the equilibrium position $\theta_1 = \theta_2 = 0$ is stable.

To examine the second equilibrium position we put

$$\theta_1 = \pi + \varphi_1. \tag{d}$$

Then, proceeding as before, we let

$$\varphi_1 = \eta_1, \qquad \dot{\varphi}_1 = \eta_2, \qquad \theta_2 = \eta_3, \qquad \dot{\theta}_2 = \eta_4,$$

and the linearized perturbation equations become

$$\begin{aligned} \dot{\eta}_1 &= \eta_2 \\ 2\dot{\eta}_2 + \dot{\eta}_4 &= 2\frac{g}{l}\eta_1, \\ \dot{\eta}_3 &= \eta_4, \\ \dot{\eta}_4 + \dot{\eta}_2 &= \frac{g}{l}\eta_3. \end{aligned} \tag{e}$$

These become identical with (b) when g/l is replaced by $-g/l$. Hence, here,

$$\lambda_{1,2}^2 = -(-2 \pm 2^{1/2})g/l, \tag{f}$$

so that both λ_1^2 and λ_2^2 are positive real. Therefore, the motion about this equilibrium position is unstable.

The third case is examined by putting

$$\theta_2 = \pi + \varphi_2,$$

and the fourth one, by putting

$$\theta_1 = \pi + \varphi_1, \qquad \theta_2 = \pi + \varphi_2$$

in (a), and proceeding as before. It is left to the reader to demonstrate that motions about all equilibrium positions except the first are unstable under the approximations made.

The final example is one where the linearization of (17.3.4) leads to erroneous conclusions. This is an example of a critical case.

Example 17.4.3. Consider a Lagrangean system for which

$$T = \tfrac{1}{2}m\dot{x}^2, \qquad V = -\tfrac{1}{2}kx^2, \qquad D = \tfrac{1}{4}c\dot{x}^4,$$

where D is a Rayleigh dissipation function [see (13.7.8) and (13.7.9)], and m, k, and c are all positive constants. Discuss the stability of the zero solution if it exists.

Substitution in Lagrange's equation (13.7.9) gives

$$m\ddot{x} + kx - c\dot{x}^3 = 0 \tag{a}$$

or, with $x = \eta_1$, $\dot{x} = \eta_2$, one has

$$\begin{aligned} \dot{\eta}_1 &= \eta_2, \\ \dot{\eta}_2 &= -\frac{k}{m}\eta_1 + \frac{c}{m}\eta_2^3. \end{aligned} \tag{b}$$

This is a special case of (17.3.4) and linearization gives

$$\dot{\eta}_1 = \eta_2, \qquad \dot{\eta}_2 = -\frac{k}{m}\eta_1, \tag{c}$$

which is a special case of (17.4.2). The characteristic values of (c) are

$$\lambda_{1,2} = \pm\left(-\frac{k}{m}\right)^{1/2}$$

and, since k and m are positive, the characteristic values have no positive real part, indicating stability. As all roots have zero real part, we deal here with a critical case. It is easy to show that the conclusion of stability of the solution of (c) does not imply stability of those of (b). We may write

$$m\eta_2\frac{d\eta_2}{d\eta_1} + k\eta_1 = c\eta_2^3, \tag{d}$$

or

$$m\eta_2\, d\eta_2 + k\eta_1\, d\eta_1 = c\eta_2^3\, d\eta_1 = c\eta_2^4\, dt,$$

which gives upon integration

$$\frac{1}{2}m\eta_2^2 + \frac{1}{2}k\eta_1^2 = h + c\int_{t_0}^{t} \eta_2^4\,dt. \tag{e}$$

This is the equation of a curve in the η_1, η_2 plane. Let t_0 be the time of an η_2 axis intercept of this curve (i.e., $\eta_1 = 0$ at $t = t_0$), and let the curve intercept the η_2 axis at $\eta_{2,0}$. Then, from (e),

$$\eta_{2,0}^2 = 2\frac{h}{m}. \tag{f}$$

If the next η_2 axis intercept occurs at t_1 and is given by $\eta_{2,1}$ one has, again from (e),

$$\eta_{2,1}^2 = 2\frac{h}{m} + 2\frac{c}{m}\int_{t_0}^{t_1} \eta_2^4\,dt. \tag{g}$$

But $c > 0$ and $\eta_2^4 > 0$. Therefore,

$$\eta_{2,1}^2 > \eta_{2,0}^2.$$

By induction, if the rth η_2 axis intercept is denoted by $\eta_{2,r}$, and the time at which it occurs is t_r, one has

$$\eta_{2,r}^2 > \eta_{2,r-1}^2.$$

Hence, the zero solution grows indefinitely with time, and the solution is unstable.

This result might have been expected on physical grounds alone because the nonlinear force $-c\dot{x}^3$ is a "negative damping" force and thus pumps energy into the system.

17.5. Some Remarks on Liapunov's Direct Method

It was already mentioned that Liapunov's direct method has the purpose of avoiding integration of (17.3.4) when determining the stability of its zero solution.

Liapunov's celebrated method is not recent; it appeared first in 1892 in a Russian monograph called *General Problem of Stability*, and it became first available in a Western language in 1907. Today, there exists a large body of literature dealing with the Liapunov method and its application to technical problems. Here, we can only present the basic features; the interested reader is referred for further information to the monographs mentioned above.

In Liapunov's direct method, there is no advantage in separating the terms linear in the η_r from the others, as was done in (17.3.4). Therefore, we rewrite that equation as

$$\frac{d\eta_s}{dt} = g_s(\eta_1, \eta_2, \ldots, \eta_{2n}, t) \qquad (s = 1, 2, \ldots, 2n), \tag{17.5.1}$$

where

$$g_s(0, 0, \ldots, 0, t) = 0$$

and our concern is the stability of the zero solution $\eta_s = 0$ $(s = 1, 2, \ldots, 2n)$ of (17.5.1). The stability definition of Liapunov, when applied to (17.5.1), states that the zero solution is stable when the variations satisfy

$$|\eta_s(t)| < \varepsilon \quad \text{for } t > t_0 \quad \text{if } |\eta_s(t_0)| \leq \eta(\varepsilon) \qquad (s = 1, 2, \ldots, n).$$

As mentioned already, when these are satisfied, the undisturbed solution $f_s(t)$ is stable.

We shall distinguish between the "autonomous" case, when the right-hand sides of (17.5.1) are not explicit functions of the independent variable t, and the "nonautonomous" case, when one or more than one of the g_s contains t explicitly.

(a) The Autonomous Case

We consider the autonomous case

$$\frac{d\eta_s}{dt} = g_s(\eta_1, \eta_2, \ldots, \eta_{2n}) \qquad (s = 1, 2, \ldots, 2n) \tag{17.5.2}$$

with

$$g_s(0, 0, \ldots, 0) = 0.$$

We begin by defining a *function, definite in sign.*

If $V(\eta_1, \eta_2, \ldots, \eta_{2n})$ is of class C^1 in an open region Ω containing the origin, defined by $|\eta_r| < A_r$, if $V(0, 0, \ldots, 0) = 0$, and if $V(\eta_1, \eta_2, \ldots, \eta_{2n})$ has the same sign everywhere in Ω except at the origin, the function $V(\eta_1, \eta_2, \ldots, \eta_{2n})$ is called definite in sign *in Ω.*

If $V(\eta_1, \eta_2, \ldots, \eta_{2n})$ is positive everywhere except at the origin, it is called positive-definite. *In the contrary case, it is* negative-definite.

If $V(\eta_1, \eta_2, \ldots, \eta_{2n})$ has the same sign everywhere in Ω where it is not zero, but it can be zero without all of its arguments being zero, it is called semidefinite in sign.

The meaning of these definitions is illustrated in Figs. 17.5.1(a), (b), and (c).

Suppose, we replace the η_s in V by solutions of (17.5.2). Then, we find a curve C lying in the V surface, and its projection down the V axis

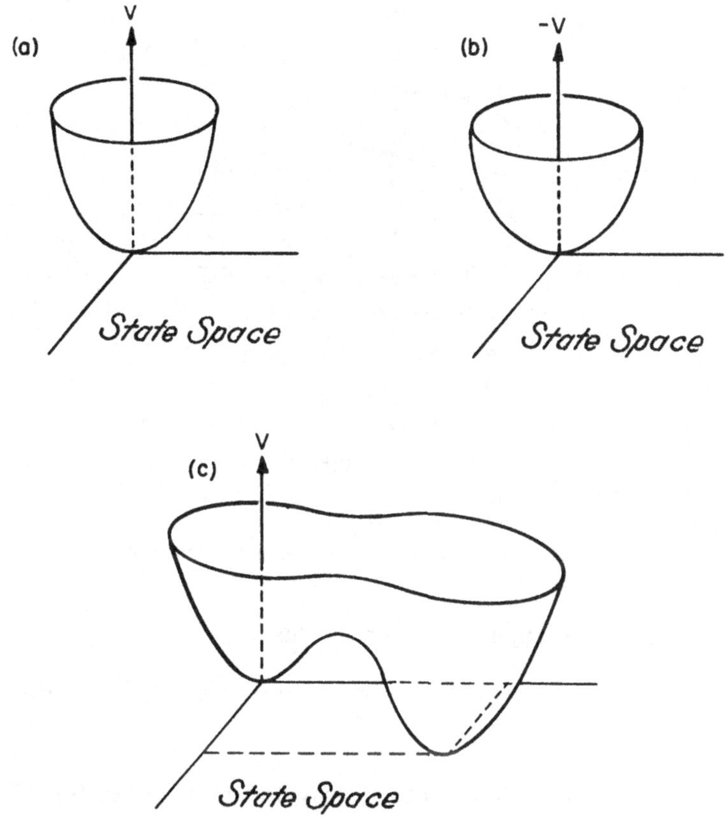

Fig. 17.5.1. (a) Positive-definite function V; (b) negative-definite function V; (c) positive-semidefinite function V.

is an S trajectory of (17.5.2), as shown in Fig. 17.5.2. One may now study the change of V along the C curve, and one finds

$$\frac{dV}{dt} = \sum_{s=1}^{2n} \frac{\partial V}{\partial \eta_s} \dot{\eta}_s = \sum_{s=1}^{2n} \frac{\partial V}{\partial \eta_s} g_s. \tag{17.5.3}$$

This is the rate of change of V with increasing t as one proceeds along the trajectory. We now define:

A function $V(\eta_1, \eta_2, \ldots, \eta_{2n})$ in Ω definite in sign, for which

$$\frac{dV}{dt} = \sum_{s=1}^{2n} \frac{\partial V}{\partial \eta_s} g_s(\eta_1, \eta_2, \ldots, \eta_{2n})$$

is semidefinite and opposite in sign to V, or $dV/dt \equiv 0$, is called a Liapunov function.

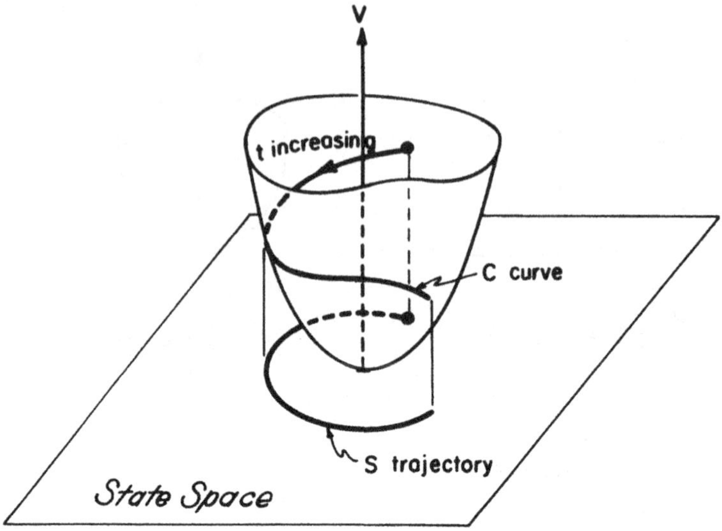

Fig. 17.5.2. Liapunov function V for the autonomous case.

Then, the celebrated Liapunov theorem is this:

> *If one can construct a Liapunov function for* (17.5.2), *the undisturbed solution is stable.*

The idea of the proof of this theorem is easily understood from Fig. 17.5.2. If, in proceeding along the C curve on the V surface in the direction of increasing t, the value of V decreases (this is the case shown), or at least never increases, the C curve will "descend down," or at least will not "climb up" the V surface. Thus, the S trajectory, which is the projection of the C curve on the state space, remains in a bounded domain containing the origin. Therefore, the undisturbed solution is stable.

A famous special case of the Liapunov theorem is Dirichlet's stability theorem. We have already noted that the equilibrium points of a dynamical system are characterized by zero velocities and accelerations. Now, in the equilibrium state, Lagrange's equations become

$$-\frac{\partial U}{\partial q_r} = 0 \qquad (r = 1, 2, \ldots, n), \tag{17.5.4}$$

where we have used the potential function U, the negative of the potential energy, in order to avoid confusion between the symbol for the potential energy and V, which is used here to denote the Liapunov function. Then, if the transformation from Cartesian to Lagrangean coordinates does not contain t explicitly, and if all given forces are potential, we have the energy integral

$$E = T - U = h, \tag{17.5.5}$$

where E is the total energy, and h is a constant.

We deduce from (17.5.4) that, in these systems, U has a stationary value at an equilibrium position.

Dirichlet's stability theorem states:

An equilibrium position is stable if the stationary value of the potential energy at that equilibrium position is a minimum relative to all neighboring points.

The proof of this theorem is simple. Let an equilibrium position be given by q_{r0} ($r = 1, 2, \ldots, n$). Then, we make the coordinate transformations

$$\eta_r = q_r - q_{r0}$$

so that

$$U = U(\eta_1, \eta_2, \ldots, \eta_n), \tag{17.5.6}$$

and we chose the datum of U so that

$$U(0, 0, \ldots, 0) = 0. \tag{17.5.7}$$

If the potential energy $-U$ has the *minimum* value zero at $\eta_r = 0$ ($r = 1, 2, \ldots, n$), then it is positive at every other point in the neighborhood of this equilibrium position. Hence, $-U$ is positive-definite in some neighborhood of the equilibrium position.

Moreover, we have already shown in Section 12.1 that T is always positive-definite. Then, if we define

$$\dot{q}_r = \eta_{n+r} \qquad (r = 1, 2, \ldots, n),$$

the total energy becomes

$$E(\eta_1, \eta_2, \ldots, \eta_{2n}) = T - U, \tag{17.5.8}$$

and it is positive-definite. Moreover, since

$$E(\eta_1, \eta_2, \ldots, \eta_{2n}) = h = \text{const},$$

its time derivative vanishes identically. Therefore, E is a Liapunov function, i.e., small motions about the equilibrium position are stable if the equilibrium position corresponds to a minimum in the potential energy.

It is not always easy to determine where a function of many variables has a minimum. However, it is known that, if there exists a point (x_0, y_0) of a function of two variables $W(x, y)$ such that

$$\frac{\partial W}{\partial x} = 0, \qquad \frac{\partial W}{\partial y} = 0,$$
$$\frac{\partial^2 W}{\partial x^2} \cdot \frac{\partial^2 W}{\partial y^2} - \left(\frac{\partial^2 W}{\partial x\, \partial y}\right)^2 > 0 \tag{17.5.9}$$

at that point, the function W has a minimum at that point.

Example 17.5.1. We return to Example 17.4.2, but here we do not utilize the linearized variational equations.

From (15.5.22), the potential energy of the double pendulum with identical component pendulums is

$$-U = 2mgl(1 - \cos\theta_1) + mgl(1 - \cos\theta_2). \tag{a}$$

This can be seen by inspection to be zero for $\theta_1 = \theta_2 = 0$, and positive for all $\theta_{1,2}$ in $0 < \theta_{1,2} < \pi$. Hence, it is positive-definite. However, we may proceed formally by substituting in (17.5.9),

$$-\frac{\partial U}{\partial \theta_1} = 2mgl \sin\theta_1, \qquad -\frac{\partial U}{\partial \theta_2} = mgl \sin\theta_2, \tag{b}$$

and this vanishes at $\theta_1 = \theta_2 = 0$. Thus $-U$ has a stationary value at that point. Moreover,

$$\frac{\partial^2 U}{\partial \theta_1^2} \cdot \frac{\partial^2 U}{\partial \theta_2^2} - \left(\frac{\partial^2 U}{\partial \theta_1\, \partial \theta_2}\right)^2 = 2\mathrm{m}^2 g^2 l^2 \cos\theta_1 \cos\theta_2, \tag{c}$$

and that quantity is positive in $-\pi/2 < \theta_{1,2} < \pi/2$. Therefore, the criteria (17.5.9) are satisfied.

It is worth noting that in this case, where energy is conserved in the nonlinear and the linearized equations, the deductions of Example 17.4.2 remained unchanged.

The essential difficulty in utilizing Liapunov's direct method is that of constructing a Liapunov function. Sometimes, it is possible to find first integrals which are not in themselves definite in sign, but which can be combined so as to give a Liapunov function. Such a case is quoted by Malkin† and is due to Chetaev. It concerns the stability of motion of a spinning projectile.

† Malkin, I. G., *Theory of Stability of Motion*, United States Atomic Energy Commission translation, pp. 31–33; see also pp. 65–66. *Theorie der Stabilität einer Bewegung*, R. Oldenbourg, Munich (1959), pp. 26–28; see also pp. 56–57.

Example 17.5.2. A spinning projectile whose mass center travels with constant speed along a straight, horizontal path has equations of motion given by Malkin as

$$\begin{aligned} I\ddot{\beta} + I\dot{\alpha}^2 \sin\beta\cos\beta - Jn\dot{\alpha}\cos\beta &= eR\sin\beta\cos\alpha, \\ I\ddot{\alpha}\cos\beta - 2I\dot{\alpha}\dot{\beta}\sin\beta + Jn\dot{\beta} &= eR\sin\alpha, \end{aligned} \tag{a}$$

where

- α is the angle which the projection of the projectile axis on a vertical plane makes with the horizontal path,
- β is the angle between the projectile axis and the vertical plane,
- J is the mass moment of inertia about the projectile axis,
- I is the mass moment of inertia about a transverse axis through the mass center,
- R is the constant air resistance,
- e is the distance from the mass center to the center of pressure,
- n is the rate of spin.

Equations (a) have the steady-state solutions

$$\alpha = \dot{\alpha} = \beta = \dot{\beta} \equiv 0,$$

and we are interested in the stability of that motion.

It is not difficult to find two first integrals. Multiplying the first of (a) by $\dot{\beta}$ and the second by $\dot{\alpha}\cos\beta$ and adding, one finds an equation which is the exact differential of a function F_1 given by

$$F_1(\alpha, \dot{\alpha}, \beta, \dot{\beta}) = \tfrac{1}{2}I(\dot{\beta}^2 + \dot{\alpha}^2\cos^2\beta) + eR(\cos\alpha\cos\beta - 1).$$

Hence, $F_1 =$ const is a first integral; this is, in fact, the energy integral.

Similarly, multiplying the first by $\sin\alpha$, the second by $\cos\alpha\sin\beta$, and subtracting, one finds an equation which is an exact differential of a function F_2 given by

$$F_2(\alpha, \dot{\alpha}, \beta, \dot{\beta}) = I(\dot{\beta}\sin\alpha - \dot{\alpha}\sin\beta\cos\beta\cos\alpha) + Jn(\cos\alpha\cos\beta - 1). \tag{c}$$

Hence, $F_2 =$ const is also a first integral. It is, in fact, the angular momentum integral. We now choose

$$V = F_1 - \lambda F_2 \tag{d}$$

and we note that $dV/dt \equiv 0$. Hence, V is a Liapunov function if we can show that V is definite in sign.

To do this we state without proof the following theorem:

If $V = V_m + V_n$, where V_m is a polynomial definite in sign, and V_n is a polynomial of higher order than V_m, then V is definite in sign in some neighborhood of the origin.

We now expand (d) in a power series and find

$$\begin{aligned} V &= V_m + V_n \\ &= \tfrac{1}{2}[I\dot{\alpha}^2 + 2I\lambda\dot{\alpha}\beta + (Jn\lambda - eR)\beta^2] + \tfrac{1}{2}[I\dot{\beta}^2 - 2I\lambda\alpha\dot{\beta} + (Jn\lambda - eR)\alpha^2] + V_n, \end{aligned} \tag{e}$$

where V_n is a polynomial in α, $\dot{\alpha}$, β, $\dot{\beta}$ of degree higher than second. Therefore, we must examine V_m as to definiteness of sign.

Now, V_m will be positive-definite if each of the forms

$$F(x, y) = Ix^2 \pm 2I\lambda xy + (Jn\lambda - eR)y^2 \tag{f}$$

is positive-definite. Instead of (f) we may write

$$F(x, y) = (x \quad y)\begin{pmatrix} I & \pm I\lambda \\ \pm I\lambda & Jn\lambda - eR \end{pmatrix}\begin{pmatrix} x \\ y \end{pmatrix}, \tag{g}$$

and both functions $F(x, y)$ will be positive-definite if the characteristic values of the 2×2 matrices are all positive. But, the characteristic values of a matrix $\left(\begin{smallmatrix} a & b \\ c & d \end{smallmatrix}\right)$ satisfy the characteristic equation

$$s^2 - (a + d)s + (ad - bc) = 0$$

and are given by

$$2s_{1,2} = a + d \pm [(a + d)^2 - 4(ad - bc)]^{1/2}. \tag{h}$$

For the smaller of these to be positive, one sees from (h) that it is necessary and sufficient that

$$a + d > 0, \qquad ad - bc > 0$$

or, returning to (g), one must have

$$\begin{aligned} I + Jn\lambda - eR &> 0, \\ I(Jn\lambda - eR) - I^2\lambda^2 &> 0. \end{aligned} \tag{i}$$

We will now show that, if the second of these is satisfied, so is the first. If the second is satisfied, one has

$$Jn\lambda - eR > I\lambda^2$$

or, adding I to both sides,

$$I + Jn\lambda - eR > I\lambda^2 + I.$$

Now, the right-hand side of this equation is positive, and the left-hand side is the same as that of the first equation of (i), which we wanted to show.

For the second equation of (i) to be satisfied, one must have

$$f(\lambda) = I\lambda^2 - Jn\lambda + eR < 0. \tag{j}$$

For large λ, $f(\lambda) > 0$. Therefore, for $f(\lambda)$ to become negative, it must cross the λ axis. This is ensured if the roots of $f(\lambda) = 0$ are real, or if

$$J^2n^2 - 4IeR > 0. \tag{k}$$

When (k) is satisfied, V as given in (e) is a Liapunov function, and the steady-state motion is stable. This can always be achieved if the velocity of spin is

$$n > 2\left(\frac{IeR}{J^2}\right)^{1/2}.$$

It is noteworthy that (k) is always satisfied if the resistance R is zero, or if the center of mass and the center of pressure coincide.

(b) The Nonautonomous Case

In the nonautonomous case, the variational equations are (17.5.1), where one or more than one of the g_s ($s = 1, 2, \ldots, 2n$) is an explicit function of time.

For this case, we consider two functions: $V(\eta_1, \eta_2, \ldots, \eta_{2n}, t)$ is an explicit function of time, and $W(\eta_1, \eta_2, \ldots, \eta_{2n})$ is not. Both vanish at $\eta_s = 0$ ($s = 1, 2, \ldots, 2n$), both are single-valued and of class C^1 on a domain Ω defined by $|\eta_s| < A_s$. Moreover, $W(\eta_1, \eta_2, \ldots, \eta_{2n})$ is positive definite. Then, we call $V(\eta_1, \eta_2, \ldots, \eta_{2n}, t)$ positive- or negative-definite according as

$$V(\eta_1, \eta_2, \ldots, \eta_{2n}, t) \geq W(\eta_1, \eta_2, \ldots, \eta_{2n})$$

or (17.5.10)

$$V(\eta_1, \eta_2, \ldots, \eta_{2n}, t) \leq -W(\eta_1, \eta_2, \ldots, \eta_{2n})$$

at every point $\eta = (\eta_1, \eta_2, \ldots, \eta_{2n}) \in \Omega$.

As in the autonomous case, we may compute the change in V as we move along a trajectory on the V surface in the direction of increasing t, but in the nonautonomous case

$$\frac{dV}{dt} = \sum_{s=1}^{2n} \frac{\partial V}{\partial \eta_s} g_s + \frac{\partial V}{\partial t}. \tag{17.5.11}$$

We now define:

A function $V(\eta_1, \eta_2, \ldots, \eta_{2n}, t)$ *definite in sign in accordance with* (17.5.10) *for which* dV/dt *as given in* (17.5.11) *is semidefinite with the opposite sign of* V *is called a* Liapunov function *for* (17.5.1).

Then, the Liapunov theorem for the nonautonomous case states:

If one can construct a Liapunov function for (17.5.1), *the undisturbed solution is stable.*

The proof of this theorem proceeds along similar lines as in the autonomous case. In Fig. 17.5.3 we illustrate the case where V is positive-definite, and dV/dt is negative-definite.

The construction of Liapunov functions for the case of nonautonomous systems of variational equations is quite difficult. Therefore, most stability investigations of nonautonomous systems deal with the linearized equations (17.3.4).

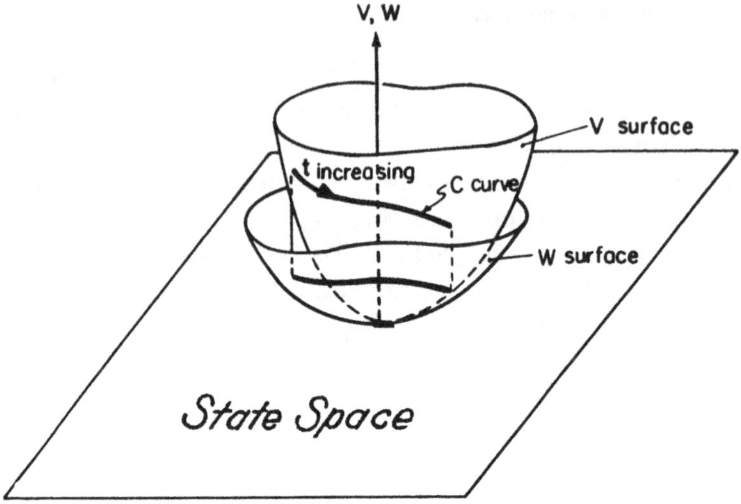

Fig. 17.5.3. Positive-definite Liapunov function in the nonautonomous case.

17.6. Problems

17.1. It was shown in (7.4.9) that the torque-force motion of a rigid body about a point is given by

$$I_x\dot{\omega}_x - (I_y - I_z)\omega_y\omega_z = 0,$$
$$I_y\dot{\omega}_y - (I_z - I_x)\omega_z\omega_x = 0,$$
$$I_z\dot{\omega}_z - (I_z - I_y)\omega_x\omega_y = 0.$$

Use the linearized variational equations to examine the stability of the steady-state rotation $\omega_x = \Omega = \text{const}$, $\omega_y = \omega_z = 0$. In particular, show that the motion is unstable if I_x is intermediate in magnitude between I_y and I_z.

17.2. Derive the stability criterion (k) in Example 17.5.2 by using the linearized variational equations.

17.3. A heavy, inverted pendulum is restrained by identical linear springs, as shown. Examine the stability of small motions about the inverted, vertical position by means of the linearized variational equations.

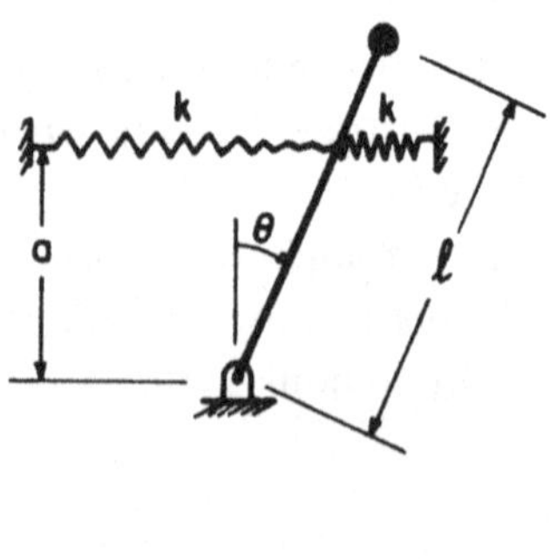

17.4. A heavy pendulum rotates with constant angular velocity about the vertical, as shown. Show that there exists a steady-state motion for which the pendulum angle with the vertical is constant, and examine the stability of this steady state by means of the linearized variational equations.

17.5. In any one or more of the preceding problems, construct a Liapunov function and show that the results from the linearized equations are correct.

17.6. If the equation of a particle motion is given by

$$\ddot{q} + a\dot{q} + 2bq + 3q^2 = 0, \qquad a, b > 0,$$

find a Liapunov function and show that the zero solution is stable.

17.7. The motion of a particle is given by

$$\ddot{x} + f(x)\dot{x} + g(x) = 0$$

and

$$\left.\begin{aligned} xg(x) &> 0 \\ f(x) &> 0 \end{aligned}\right\} \quad \text{for all } x \neq 0,$$

$$\lim_{x\to\infty} G(x) = \lim_{x\to\infty} \int_0^x g(u)\, du = \infty;$$

show that, with $\dot{x} = y$,

$$V(x, y) = \tfrac{1}{2}y^2 + G(x)$$

is a Liapunov function.

18

Applications

18.1. Introductory Remarks

In preceding chapters we used Lagrange's equations principally to formulate problems, not to determine the motion of physical systems. In that formulation the generalized forces occurred merely as the symbols Q_s ($s = 1, 2, \ldots, n$); they were, in general, functions of all coordinates, of their time derivatives, and of time or perhaps they arose as partial derivatives of a potential function, but the precise functional form of the Q_s or of the potential energy V was of no importance in the formulation. However, when one wants to solve the equations of motion, the forces must be specified; in fact, the problem is not fully described until the forces are given.

In this chapter we give examples of the application of Lagrange's equations to problems that are more fully or completely described, and we carry the solutions as far as is possible. Examples were selected, in part, to complete problems formulated in earlier chapters, in part because of their intrinsic interest, in part for the sake of variety, and in part because they either illustrate an important point or they try to make a new one. Among them, there are classical problems treated in many books, and less classical ones, which can, perhaps, also be found elsewhere. In a field which has been so often and so thoroughly plowed as that of the application of Lagrange's equations, such repetition is unavoidable. Where problems were taken from a source which is known to be the original one for that problem, reference is made to that source.

18.2. The Single Particle

We begin by considering a holonomic, scleronomic, conservative system.

Example 18.2.1. Discuss the motion of a bead of mass m that slides without friction on a wire of parabolic shape having the equation

$$y = \pm 2a(x + a^2)^{1/2}, \tag{18.2.1}$$

where x, y are Cartesian coordinates and a is a constant. The force acting on the bead has the Cartesian components

$$X = C\left[\frac{x + (x^2 + y^2)^{1/2}}{2(x^2 + y^2)}\right]^{1/2}, \qquad Y = C\left[\frac{-x + (x^2 + y^2)^{1/2}}{2(x^2 + y^2)}\right]^{1/2}, \tag{18.2.2}$$

where C is a positive constant. The curve of (18.2.1) is sketched in Fig. 18.2.1.

It is reasonable (because of the form of the constraint) to suppose that parabolic u, v coordinates are the most suitable ones for this problem. They are connected with the Cartesian ones by

$$x = u^2 - v^2, \qquad y = 2uv. \tag{18.2.3}$$

One sees that the parabola of Fig. 18.2.1 is the one defined by

$$v = a, \tag{18.2.4}$$

for, if one substitutes (18.2.4) in (18.2.3) and eliminates u between them, (18.2.1) results.

It is now necessary to calculate the generalized forces Q_u and Q_v, which correspond to the u and v coordinates, respectively. By substituting (18.2.3) into

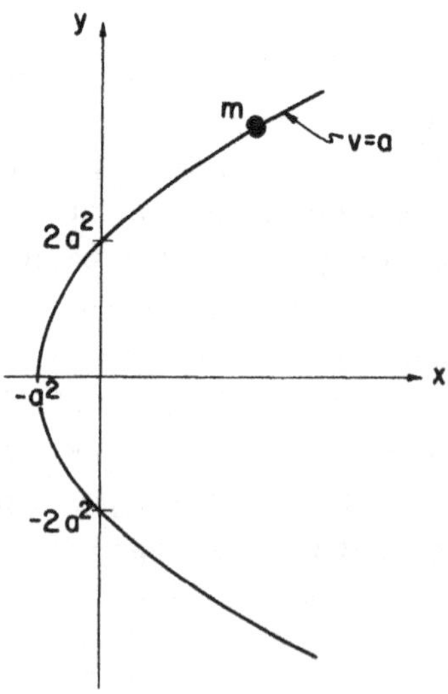

Fig. 18.2.1. Bead on wire of Example 18.2.1.

both of (18.2.2), the Cartesian force components are found in terms of the parabolic coordinates as

$$X = \frac{Cu}{u^2 + v^2}, \qquad Y = \frac{Cv}{u^2 + v^2}. \tag{18.2.5}$$

Then, the generalized forces are, by applying (12.3.8),

$$\begin{aligned} Q_u &= X\frac{\partial x}{\partial u} + Y\frac{\partial y}{\partial u} = 2C, \\ Q_v &= X\frac{\partial x}{\partial v} + Y\frac{\partial y}{\partial v} = 0. \end{aligned} \tag{18.2.6}$$

The velocity components are, from (18.2.3),

$$\dot{x} = 2u\dot{u}, \qquad \dot{y} = 2a\dot{u}, \tag{18.2.7}$$

and these may be directly introduced into the kinetic energy because the constraint is holonomic. Therefore

$$T^{\dagger} = \tfrac{1}{2}m(\dot{x}^2 + \dot{y}^2) = 2m(a^2 + u^2)\dot{u}^2. \tag{18.2.8}$$

It is apparent from (18.2.6) that the force components are derivable from the potential energy

$$V = -2Cu. \tag{18.2.9}$$

If we had calculated this problem in Cartesian coordinates, we would have had to consider the force components (18.2.2). It is not at all evident from inspection that these are derivable from a potential energy $V(x, y)$.

The energy integral exists and is

$$2m(a^2 + u^2)\dot{u}^2 - 2Cu = h, \tag{18.2.10}$$

where h is a constant which is defined from the initial conditions.

Equation (18.2.10) is reducible to the quadrature

$$t - t_0 = \int_{u_0}^{u} (2m)^{1/2}\left(\frac{a^2 + u^2}{h + 2Cu}\right)^{1/2} du, \tag{18.2.11}$$

where

$$\begin{aligned} h &= 2m(a^2 + u_0^2)\dot{u}_0^2 - 2Cu_0, \\ u_0 &= u(t_0), \qquad \dot{u}_0 = \dot{u}(t_0). \end{aligned}$$

To discuss the motion consider the energy integral (18.2.10), which defines a family of curves in the u, $\dot{u}$ plane. They are symmetric with respect to the u axis because (18.2.10) is unchanged when $\dot{u}$ is replaced by $-\dot{u}$. They cross the u axis at $u = -h/2C$, which is negative, zero, or positive according as

$$Cu_0 \lesseqgtr m(a^2 + u_0^2)\dot{u}_0^2.$$

They cross with vertical slope because

$$\frac{d\dot{u}}{du} = \frac{C - 2mu\dot{u}^2}{2m(a^2 + u^2)\dot{u}},$$

and that quantity is infinite for $\dot{u} = 0$, e.g., on the u axis. Stationary values of $\dot{u}$ occur when $d\dot{u}/du = 0$, and this occurs where

$$\dot{u} = \pm\left(\frac{C}{2mu}\right)^{1/2}.$$

It follows from (18.2.10) that, at that value of $\dot{u}$, the value of u becomes

$$u = -\frac{h}{2c} + \left(\frac{h^2}{4c^2} + a^2\right)^{1/2}.$$

(Actually, there is the familiar sign ambiguity in front of the radical. We leave it as an exercise to show why the negative sign is excluded.) From

$$\dot{u}^2 = \frac{h + 2Cu}{2m(a^2 + u^2)}$$

it is seen that all curves in the u, $\dot{u}$ plane must lie to the right of their u axis intercepts, and that $\dot{u} \to 0$ as $u \to \infty$. Curves having these properties are shown in Fig. 18.2.2. The direction of the arrow heads indicates the behavior of the system for t increasing, e.g., when the velocity $\dot{u}$ is negative (in the lower half-plane), u decreases; when $\dot{u} > 0$, u increases.

It is seen that all curves tend to infinity along the u axis as $t \to \infty$ and they do so with positive velocity that tends to zero as $t \to \infty$. Those that have negative initial velocity pass through $\dot{u} = 0$; those that have positive initial velocity do not have a rest point for finite u. Trajectories in the u, t plane having these properties are shown in Fig. 18.2.3.

Let us return now to the parabola in the xy plane shown in Fig. 18.2.1. The variable u is a parameter along the parabola. One sees from

$$y = 2au$$

that $u = 0$ at the vertex of the parabola, in the upper half-plane $u > 0$, while in the lower half-plane $u < 0$. Therefore, in the case of curve 1 in Fig. 18.2.3, the bead is initially on the upper branch, but its initial velocity is directed toward the vertex. It passes the vertex and goes onto the lower branch until $y = -ah/C$. Then it reverses its velocity and its x, y components go to infinity as $t \to \infty$.

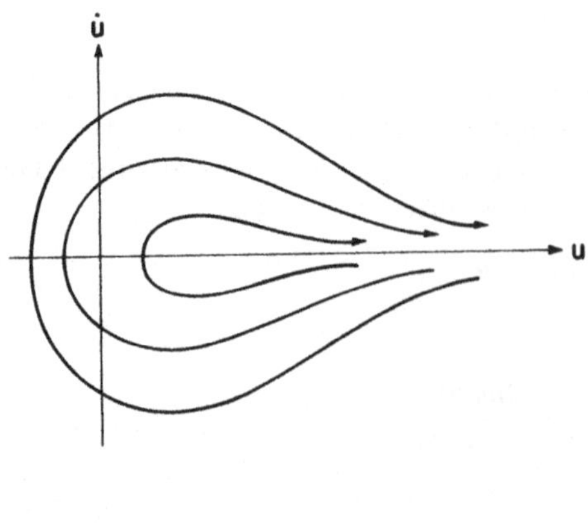

Fig. 18.2.2. State-space trajectories of Example 18.2.1.

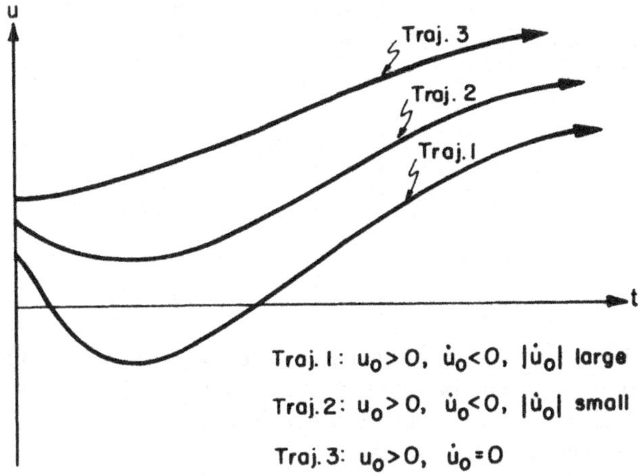

Fig. 18.2.3. Event-space trajectories of Example 18.2.1.

Next we consider the rheonomic system already discussed in Example 15.3.2.

Example 18.2.2. A particle of unit mass moves in the surface of a right circular cylinder of expanding radius $\varrho(t)$. We use cylindrical coordinates ϱ, φ, and z to define a point on the cylinder. The given force is necessarily in the tangent plane of the cylinder (why?). Let its z component be denoted by Z, and the component normal to Z by Φ.

(i) Discuss the integration when Z is a function of z alone, and Φ is a linear function of z, $\dot{z}$, φ, and $\dot{\varphi}$ which is zero when these arguments are zero.

(ii) In the time interval $0 < t_0 \leq t < \infty$, discuss the particular case of (i) when

$$\begin{aligned} \varrho(t) &= Kt, \\ \Phi &= -At^2\varphi + 2K^2t\dot{\varphi} + a_1(t)z + a_2(t)\dot{z}, \end{aligned} \tag{18.2.12}$$

where A and K are constants, and $a_1(t)$ and $a_2(t)$ are given, continuous, smooth and bounded functions.

Under the assumptions of the problem

$$\varrho = \varrho(t) > 0, \qquad \dot{\varrho}(t) > 0.$$

Lagrange's equations are [see (15.3.21)]

$$\begin{aligned} \varrho^2\ddot{\varphi} + 2\varrho\dot{\varrho}\dot{\varphi} - \Phi &= 0, \\ \ddot{z} - Z &= 0. \end{aligned} \tag{18.2.13}$$

The forces are, under assumption (i),

$$\begin{aligned} Z &= Z(z), \\ \Phi &= a_1(t)z + a_2(t)\dot{z} + a_3(t)\varphi + a_4(t)\dot{\varphi}. \end{aligned} \tag{18.2.14}$$

It follows from Example 16.5.2 that the second equation of (18.2.13) is integrable in quadratures. Let this quadrature furnish

$$z = f(t). \tag{18.2.15}$$

Then, the first equation of (18.2.13) becomes

$$\ddot{\varphi} + \frac{2\varrho\dot{\varrho} - a_4(t)}{\varrho^2(t)}\,\dot{\varphi} - \frac{a_3(t)}{\varrho^2(t)}\,\varphi = F(t), \tag{18.2.16}$$

where

$$F(t) = \varrho^{-2}(t)[a_1(t)f(t) + a_2(t)\dot{f}(t)].$$

Therefore, under (i), the equations uncouple; one is integrable in quadratures, and the other becomes a linear, inhomogeneous equation with variable coefficients. In general, (18.2.16) is not integrable in quadratures.

Under assumptions (ii), we obtain

$$\dot{\varrho} = K, \qquad a_3(t) = -At^2, \qquad A_4(t) = 2K^2t. \tag{18.2.17}$$

Substitution of these values in (17.3.16) gives

$$\ddot{\varphi} + \frac{A}{K^2}\,\varphi = F(t). \tag{18.2.18}$$

For the special case that $Z = Bz$, where B is a negative constant, the z equation is the equation of free vibrations of the linear oscillator, i.e., the altitude z is a simple harmonic function of time. If, in addition, $A > 0$ and $a_1(t)$ and $a_2(t)$ are either constant of commensurable† periodic functions, (18.2.18) is the equation of the forced linear oscillator. In that case, the particle executes Lissajous figures on the surface of an expanding cylinder.

A nonholonomic system was completely treated in Example 10.3.1.

18.3. Systems of Particles

Our first example is taken directly from the famous problem collection of Whittaker's *Analytical Dynamics* (p. 69, Example 3).

Example 18.3.1. (see also Example 16.3.3). "In a dynamical system with two degrees of freedom, the kinetic energy is

$$T = \frac{1}{2}\,\frac{\dot{q}_1^2}{(a + bq_2)} + \frac{1}{2}\,q_2^2\dot{q}_2^2,$$

and the potential energy is

$$V = c + dq_2,$$

† Periodic functions are said to be commensurable if the ratios of their periods are rational numbers.

where a, b, c, d are constants. Shew that the value of q_2 in terms of the time is given by an equation of the form

$$(q_2 - k)(q_2 + 2k)^2 = h(t - t_0)^2,$$

where h, k, and t_0 are constants."

While it is not clear to what physical problem reference is made here, we shall suppose that we deal with the rectilinear motion of two particles having positions q_1 and q_2.

From the form of the energies, it is clear that q_1 is an ignorable coordinate so that a momentum integral exists, and that the energy integral exists as well. Thus

$$\begin{aligned} \frac{\partial T}{\partial \dot{q}_1} &= \frac{\dot{q}_1}{a + bq_2} = A = \text{const}, \\ T + V &= \frac{1}{2}\left(\frac{\dot{q}_1^2}{a + bq_2} + q_2^2\dot{q}_2^2\right) + (c + dq_2) = B = \text{const}. \end{aligned} \tag{18.3.1}$$

If the first of these is solved for $\dot{q}_1$ and substituted in the second, one has

$$A^2(a + bq_2) + q_2^2\dot{q}_2^2 + 2(c + dq_2) = 2B.$$

When this equation is solved for $\dot{q}_2$ there results

$$\dot{q}_2 = \pm \frac{1}{q_2}(C - Dq_2)^{1/2}, \tag{18.3.2}$$

where

$$C = 2B - A^2a - 2c,$$
$$D = A^2b + 2d.$$

This last equation integrates to

$$t - t_0 = \pm \frac{2}{3D^2}(C - Dq_2)^{1/2}(-2C - Dq_2). \tag{18.3.3}$$

Then, on squaring, we have

$$(t - t_0)^2 = -\frac{4}{9D}\left(q_2 - \frac{C}{D}\right)\left(q_2 + 2\frac{C}{D}\right)^2. \tag{18.3.4}$$

This is the required expression if one puts

$$-h = 9D/4, \qquad k = C/D.$$

Our next example illustrates the effect of velocity-dependent forces on the stability of a motion.

Example 18.3.2. (Hamel, in part, pp. 275–278). Consider a system with kinetic energy

$$T_0 = \tfrac{1}{2}(\dot{q}_1^2 + \dot{q}_2^2) \tag{18.3.5}$$

and potential energy

$$V = \tfrac{1}{2}(aq_1^2 + bq_2^2), \tag{18.3.6}$$

where a and b are constants. It is assumed that this system is unstable. Can one add either viscous or gyroscopic forces, or both, and thereby make the system stable?

Let us add damping forces derivable from the dissipation function

$$D = \tfrac{1}{2}(\alpha\dot{q}_1^2 + 2\beta\dot{q}_1\dot{q}_2 + \gamma\dot{q}_2^2), \tag{18.3.7}$$

where α, β, and γ are constants which are such that D is negative-definite; for simplicity, let α, β, and γ be negative.

We show next that gyroscopic forces are added by coupling to the system another system having a single degree of freedom q_3, and by operating the added system in such a way that the velocity

$$\dot{q}_3 = B = \text{const.} \tag{18.3.8}$$

The force required to produce this constant velocity is Q_3. The coupling is done such that the kinetic energy of the entire system becomes

$$T = \tfrac{1}{2}(\dot{q}_1^2 + \dot{q}_2^2) + \tfrac{1}{2}\dot{q}_3^2 + Aq_2\dot{q}_1\dot{q}_3, \tag{18.3.9}$$

where A is a constant. In physical terms, we may think of the added system as a flywheel having constant angular velocity $\dot{q}_3$, where it takes a torque Q_3 to drive the flywheel. The applicable form of Lagrange's equation is (13.7.9) with $Q_s = -\partial V/\partial q_s$ for $s = 1, 2$. Direct substitution gives

$$\begin{aligned} \frac{d}{dt}(\dot{q}_1 + Aq_2\dot{q}_3) - \alpha\dot{q}_1 - \beta\dot{q}_2 + aq_1 &= 0, \\ \ddot{q}_2 - A\dot{q}_1\dot{q}_3 - \beta\dot{q}_1 - \gamma\dot{q}_2 + bq_2 &= 0, \\ \frac{d}{dt}(\dot{q}_3 + Aq_2\dot{q}_1) &= Q_3. \end{aligned} \tag{18.3.10}$$

Under the constraint (18.3.8), the last becomes

$$Q_3 = \frac{d}{dt}(Aq_2\dot{q}_1).$$

This is the value of Q_3 needed to produce constant velocity $\dot{q}_3$. When the constraint is introduced into the first two relations of (18.3.10), they become

$$\begin{aligned} \ddot{q}_1 + n\dot{q}_2 - \alpha\dot{q}_1 - \beta\dot{q}_2 + aq_1 &= 0, \\ \ddot{q}_2 - n\dot{q}_1 - \beta\dot{q}_1 - \gamma\dot{q}_2 + bq_2 &= 0, \end{aligned} \tag{18.3.11}$$

where we have put $AB = n$. The terms $n\dot{q}_1$ and $n\dot{q}_2$ are not derivable from a dissipation function; thus they are gyroscopic forces (see Section 13.7). To see this, multiply the first by $\dot{q}_1$, the second by $\dot{q}_2$, and add them. One finds

$$\frac{d}{dt}(T_0 + V) - 2D = 0, \tag{18.3.12}$$

which is independent of n. Equation (18.3.12) shows the effect of the damping terms; as D is negative-definite by supposition, (18.3.12) implies that the time rate of change of total energy is nonpositive, i.e., the total energy decreases with time when damping forces are present. Therefore, the system will eventually come to rest.

Instability of the motion in the absence of damping and gyroscopic forces means that either $a < 0$, or $b < 0$, or both are negative. To see this, substitute in the usual way

$$q_1 = C_1 e^{\lambda t}, \qquad q_2 = C_2 e^{\lambda t} \tag{18.3.13}$$

into the equations when n, α, β, and γ are equated to zero. Then, one finds the characteristic values (see Section 17.4).

$$\lambda_1 = \pm (-a)^{1/2}, \qquad \lambda_2 = \pm (-b)^{1/2}.$$

Let us consider the presence of gyroscopic forces, and the absence of damping forces, or

$$\begin{aligned} \ddot{q}_1 + n\dot{q}_2 + aq_1 &= 0, \\ \ddot{q}_2 - n\dot{q}_1 + bq_2 &= 0. \end{aligned} \tag{18.3.14}$$

The substitution of (18.3.13) in these shows that λ must satisfy the characteristic equation

$$\lambda^4 + (a + b + n^2)\lambda^2 + ab = 0. \tag{18.3.15}$$

For stability we require that the λ^2 be real and negative. They will be real if

$$(a - b)^2 + n^2[n^2 + 2(a + b)] > 0, \tag{18.3.16}$$

and they will be negative if their sum is negative and their product positive, or if

$$\begin{aligned} a + b + n^2 &> 0, \\ ab &> 0. \end{aligned} \tag{18.3.17}$$

In the absence of gyroscopic forces $n = 0$, and the conditions of stability reduce to

$$a + b > 0, \qquad ab > 0.$$

By supposition, these are not both satisfied. We see that if a and b are of opposite sign, the motion is always unstable and cannot be stabilized by gyroscopic forces. Let us assume that a and b are both negative. Then, for stability one must have

$$n^2 > -(a + b) \tag{18.3.18}$$

in order to satisfy the first equation of (18.3.17). However, to satisfy (18.3.16) as well, one must have

$$n^2 \geq -(a + b) \pm 2(ab)^{1/2}.$$

The negative sign gives a weaker, and the positive sign a stronger condition than (18.3.18). Therefore, gyroscopic forces can stabilize the system provided $a < 0$, $b < 0$, and

$$n^2 > -(a + b) + 2(ab)^{1/2}. \tag{18.3.19}$$

Since

$$n = AB = A\dot{q}_3,$$

this requires that $|\dot{q}_3|$ be sufficiently large.

To study the effect of damping and gyroscopic forces, we simplify the system so that

$$a = b < 0, \qquad \alpha = \gamma, \qquad \beta = 0.$$

For these values of the parameters, the equations (18.3.11) become

$$\begin{aligned} \ddot{q}_1 - \alpha\dot{q}_1 + n\dot{q}_2 + aq_1 &= 0, \\ \ddot{q}_2 - \alpha\dot{q}_2 - n\dot{q}_1 + aq_2 &= 0. \end{aligned} \tag{18.3.20}$$

It is easy to verify that the characteristic equation of (18.3.20) is

$$\lambda^4 - 2\alpha\lambda^3 + (n^2 + \alpha^2 + 2a)\lambda^2 - 2a\alpha\lambda + a^2 = 0. \tag{18.3.21}$$

Now, it is shown in the elementary theory of equations that the roots of a fourth-degree polynomial with real coefficients

$$\lambda^4 + A\lambda^3 + B\lambda^2 + C\lambda + D = 0$$

have nonpositive real parts if and only if A, B, C, and D are all positive and if

$$C^2 + A^2D - ABC < 0.$$

The last condition becomes, in the case of (18.3.21),

$$-4a\alpha^2(n^2 + \alpha^2) < 0$$

and this is violated when $a < 0$, as is assumed here. Moreover, since α is negative, $-\alpha > 0$, or

$$C = -2a\alpha,$$

and this is negative under our assumptions. Therefore, the addition of viscous damping to a gyroscopically stabilized system destroys the stability. A well-known example of this is the children's toy top. When the top is simply set down on its point it falls over. Spinning it makes this position stable, but the addition of viscous forces will cause the top to fall over eventually.

Our final example shows that the dynamics of deformable systems with infinitely many particles falls within the compass of the preceding theory provided the admitted configurations are describable by a finite number of coordinates.

Example 18.3.3. Explain the crack of the whip.

We idealize the whip as an inextensible uniform string with no bending stiffness. Let it be arranged in a straight line as in Fig. 18.3.1. It is bent double in the manner shown. The distance of one end of the string from some fixed datum is x,

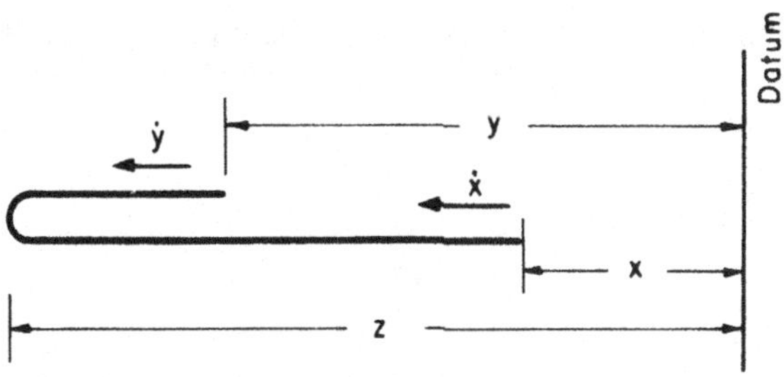

Fig. 18.3.1. Model of the whip of Example 18.3.3.

and its velocity is $\dot{x}$, that of the other end is y, and its velocity is $\dot{y}$. The kink is a distance z from the datum and the rope is of length l. The kinked string is initially in a straight line and the velocities $\dot{x}$ and $\dot{y}$ also lie in this line. We are given

$$x = x_0 - vt, \tag{18.3.22}$$

where $v > 0$ is a constant; in other words, the free end of the lower portion of the string in Fig. 18.3.1 is moved with constant speed v to the right. The entire portion of length $z - x$ will move to the right with speed v because the end moves with that velocity, and the string is inextensible. Similarly, the entire portion of length $z - y$ moves with the velocity $\dot{y}$ because that portion is pulled by the element just above the kink in Fig. 18.3.1.

As the string is inextensible, the coordinates y and z satisfy the holonomic, rheonomic constraint

$$(z - x) + (z - y) - l = 0$$

or

$$z = \tfrac{1}{2}(l + x + y), \tag{18.3.23}$$

where x is a prescribed function of time, given in (18.3.22).

If the mass per unit length of the string is ϱ, the kinetic energy is

$$T = \tfrac{1}{2}\varrho[(z - x)\dot{x}^2 + (z - y)\dot{y}^2]$$

or, introducing (18.3.23) into it,

$$T^\dagger = \tfrac{1}{4}\varrho[l - x + y)\dot{x}^2 + (l + x - y)\dot{y}^2].$$

As there is no potential energy, Lagrange's equation is easily found as

$$(l + x_0 - vt - y)\ddot{y} - \tfrac{1}{2}(v + \dot{y})^2 = 0. \tag{18.3.24}$$

A convenient form of this equation is found by the transformation

$$l + x_0 - vt - y = u. \tag{18.3.25}$$

This transforms (18.3.24) into

$$u\ddot{u} + \tfrac{1}{2}\dot{u}^2 = 0; \tag{18.3.26}$$

this is the differential equation of motion of the kink. By using the identity

$$\ddot{u} = \dot{u}\,\frac{d\dot{u}}{du},$$

one easily finds a first integral

$$u\dot{u}^2 = C = u_0\dot{u}_0^{\,2}, \tag{18.3.27}$$

where

$$u_0 = l + x_0 - y(0), \qquad \dot{u}_0 = -v - \dot{y}(0), \tag{18.3.28}$$

and $y(0)$ and $\dot{y}(0)$ are the initial values of these quantities.

It is perhaps worthwhile to point out that (18.3.27) is not an energy integral. In fact, even though no nonpotential forces exist, and even though the mass of the system remains constant, no energy integral exists because the Pfaffian form of (18.3.23) is

$$-dy + 2dz + v\,dt = 0,$$

i.e., the system is acatastatic.

Let us choose as initial condition

$$\dot{y}(0) = 0. \tag{18.3.29}$$

Then, utilizing (18.3.28) and (18.3.29), we may write (18.3.27) in the form

$$\dot{y} = v[(u_0/u)^{1/2} - 1]. \tag{18.3.30}$$

This is seen to become unbounded as $u \to 0$. Now, in view of (18.3.22) and (18.3.25), the vanishing of u implies $y - x = l$ or, as $u \to 0$, the string stretches out completely, and the velocity of the free end tends to infinity. (This result agrees with Kucharski's.†) In consequence, as the free end whips around it passes the velocity of sound, and a sonic boom is heard.

18.4. Nonholonomic Systems

We discuss three nonholonomic problems. The first is the classical problem of the knife edge, already mentioned in Example 9.4.2.

Example 18.4.1. (Hamel, pp. 465–470). Discuss the force-free motion of a knife edge whose mass center does not coincide with its contact point.

Let the contact point of the knife edge have the coordinates x, y; let its direction relative to the positive x axis be θ, as shown in Fig. 18.4.1, let its mass center be a distance s from the contact point, and let it have mass m and mass moment of inertia I about an axis through the contact point normal to the xy plane.

† W. Kucharski, Zur Kinetik dehnungsloser Seile mit Knickstellen, *Ingenieur-Archiv*, Vol. 12, (1941), p. 109.

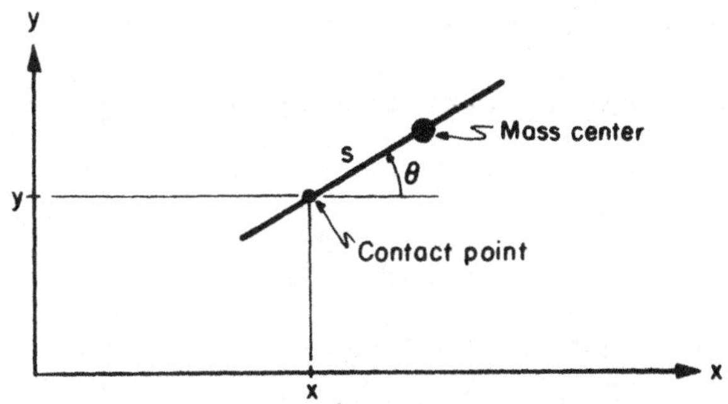

Fig. 18.4.1. Model of the knife edge of Example 18.4.1.

The x and y components of the velocity of the mass center are, respectively,

$$\dot{x} - s\dot{\theta}\sin\theta \quad \text{and} \quad \dot{y} + s\dot{\theta}\cos\theta.$$

Therefore, the kinetic energy of the knife edge is

$$T = \tfrac{1}{2}m(\dot{x}^2 + \dot{y}^2) + ms\dot{\theta}(\dot{y}\cos\theta - \dot{x}\sin\theta) + \tfrac{1}{2}I\dot{\theta}^2. \tag{18.4.1}$$

As the motion is force-free, the potential energy vanishes identically.

The constraint was already derived in Example 9.4.2. We write it as

$$\dot{y}\cos\theta - \dot{x}\sin\theta = 0 \tag{18.4.2}$$

or, in Pfaffian form,

$$\sin\theta\, dx - \cos\theta\, dy = 0. \tag{18.4.3}$$

Introducing the Lagrange multiplier $\mu = m\lambda$, the equations of motion are

$$\frac{d}{dt}(\dot{x} - s\dot{\theta}\sin\theta) + \lambda\sin\theta = 0,$$

$$\frac{d}{dt}(\dot{y} + s\dot{\theta}\cos\theta) - \lambda\cos\theta = 0, \tag{18.4.4}$$

$$\frac{d}{dt}[s(\dot{y}\cos\theta - \dot{x}\sin\theta)] + K^2\ddot{\theta} + s\dot{\theta}(\dot{y}\sin\theta + \dot{x}\cos\theta) = 0,$$

where K is the radius of gyration of the knife edge about an axis through the contact point normal to the xy plane.

To eliminate λ, we multiply the first equation of (18.4.4) by $\cos\theta$, the second by $\sin\theta$, and add them. This gives

$$\ddot{x}\cos\theta + \ddot{y}\sin\theta - s\dot{\theta}^2 = 0. \tag{18.4.5}$$

Then, substituting the kinematic constraint equation (18.4.2) into the third equation of (18.4.4) (which is permitted *after* the equations of motion have been formulated), that equation becomes

$$K^2\ddot{\theta} + s\dot{\theta}(\dot{y}\sin\theta + \dot{x}\cos\theta) = 0. \tag{18.4.6}$$

Equations (18.4.5), (18.4.6) together with (18.4.2) are three equations in the three unknowns x, y, and θ.

If we denote the velocity of the contact point in the direction of the knife edge by v, we have

$$\begin{aligned} \dot{x} &= v\cos\theta, \qquad \dot{y} = v\sin\theta, \\ \ddot{x} &= \dot{v}\cos\theta - v\dot{\theta}\sin\theta, \\ \ddot{y} &= \dot{v}\sin\theta + v\dot{\theta}\cos\theta. \end{aligned} \tag{18.4.7}$$

These transform the equations of motion into

$$\begin{aligned} \dot{v} - s\dot{\theta}^2 &= 0, \\ K^2\ddot{\theta} + s\dot{\theta}v &= 0, \end{aligned} \tag{18.4.8}$$

and the constraint equation is identically satisfied.

As the system is catastatic, the energy integral exists (even though the problem is nonholonomic; see Section 10.3 and Example 10.3.1). Here, this means that the kinetic energy is a constant. If one sets $T =$ const, substitutes the kinematical constraint into it, introduces (18.4.7), and divides by m, the energy integral becomes

$$v^2 + K^2\dot{\theta}^2 = 2h/m = v_0^2.$$

where $h > 0$ is the total energy and v_0 is a real constant. It can easily be verified that this is a first integral of (18.4.8) by multiplying the first relation by v, the second by $\dot{\theta}$, and adding them.

Solving the energy integral for $\dot{\theta}$ and substituting it in the first equation of (18.4.8) gives

$$\dot{v} + \frac{sv^2}{K^2} = \frac{sv_0^2}{K^2}. \tag{18.4.9}$$

This is the equation governing the velocity in the direction of the knife edge.

One may also solve the energy integral for v and substitute this in the second equation of (18.4.8). The result is

$$\dot{\omega} \pm \frac{s}{K^2}(v_0^2 - K^2\omega^2)^{1/2}\,\omega = 0, \tag{18.4.10}$$

where

$$\omega = \dot{\theta}$$

is the angular velocity of the knife edge; thus, the energy integral has permitted separation of the variables.

Equation (18.4.9) reduces to the quadrature

$$t = a\int_{v(0)}^{v} \frac{du}{v_0^2 - u^2}, \tag{18.4.11}$$

where $a = K^2/s$. Now, from the energy integral one has

$$\dot{\theta}^2 = \frac{1}{K^2}(v_0^2 - v^2) \tag{18.4.12}$$

and this shows that one must have $|v_0| > |v|$. With that inequality satisfied, (18.4.11) integrates to

$$v = v_0 \tanh v_0 t/a, \tag{18.4.13}$$

where the constant of integration was so chosen that $v(0) = 0$. Finally, the substitution of this value of v in the energy integral gives

$$\dot{\theta} = \omega = \pm \frac{v_0}{K} (1 - \tanh^2 v_0 t/a)^{1/2}. \tag{18.4.14}$$

Then, the direction of the knife edge as a function of time is

$$\begin{aligned} \theta &= \pm \frac{v_0}{K} \int \operatorname{sech}(v_0 t/a)\, dt \\ &= \pm \frac{a}{K} \tan^{-1}(\sinh v_0 t/a). \end{aligned} \tag{18.4.15}$$

We note from (18.4.11) that v does not depend on the sign of v_0; therefore, we take $v_0 > 0$. Then, v is positive for positive t, negative for negative t, and zero when $t = 0$. The meaning of positive and negative v is that when $v > 0$, and hence $t > 0$, the mass center is ahead of the contact point and when $t < 0$, the mass center trails the contact point. Also, as $t \to \infty$, $v \to v_0$, and as $t \to -\infty$, $v \to -v_0$; these results follow directly from $\tanh x \to \pm 1$ as $x \to \pm \infty$. It follows that v increases from $-v_0$ through $v = 0$ to $+v_0$.

The angular velocity cannot change sign; it is either always positive or always negative, depending on the choice of sign in

$$\dot{\theta} = \pm \frac{1}{K} (v_0^2 - v^2)^{1/2}.$$

One sees from this equation that the angular velocity is not zero when $v = 0$. Therefore, the trajectory has a cusp at $t = 0$.

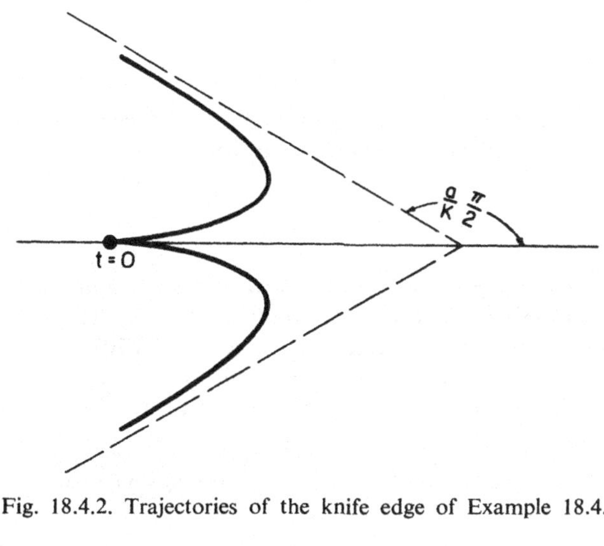

Fig. 18.4.2. Trajectories of the knife edge of Example 18.4.1.

The angle through which the knife edge turns in the interval from $t = 0$ to $t = \infty$ is

$$\Delta\theta = \pm \frac{a}{K} (\tan^{-1}\infty - \tan^{-1}0) = \pm \frac{a}{K} \frac{\pi}{2}.$$

The lines making this angle with the axis are asymptotes of the trajectory because $\dot{\theta} \to 0$ as $t \to \pm \infty$. These trajectories appear as shown in Fig. 18.4.2.

Our second example is no less classic; it is the continuation of the disk problem, formulated in Example 15.5.3.

Example 18.4.2. Discuss the heavy disk which rolls without sliding on a horizontal plane.

The problem has already been formulated. The equations of motion (15.5.30) are repeated here for convenience. They are

$$\frac{d}{dt} [m\dot{x} + mr(-\dot{\theta} \sin\theta \cos\varphi - \dot{\varphi} \cos\theta \sin\varphi)] + \lambda_1 = 0, \tag{18.4.16}$$

$$\frac{d}{dt} [m\dot{z} + mr(\dot{\theta} \sin\theta \sin\varphi - \dot{\varphi} \cos\theta \cos\varphi)] + \lambda_2 = 0, \tag{18.4.17}$$

$$\begin{aligned} &\frac{d}{dt} [mr^2\dot{\theta} + mr(-\dot{x} \sin\theta \cos\varphi + \dot{z} \sin\theta \sin\varphi) + I\dot{\theta}] + mr^2\dot{\varphi}^2 \cos\theta \sin\theta \\ &\quad + mr(\dot{x}\dot{\theta} \cos\theta \cos\varphi - \dot{x}\dot{\varphi} \sin\theta \sin\varphi - \dot{z}\dot{\theta} \cos\theta \sin\varphi - \dot{z}\dot{\varphi} \sin\theta \cos\varphi) \\ &\quad - I\dot{\varphi}^2 \sin\theta \cos\theta + J(\dot{\psi} + \dot{\varphi} \cos\theta)\dot{\varphi} \sin\theta + mgr \cos\theta = 0, \end{aligned} \tag{18.4.18}$$

$$\begin{aligned} &\frac{d}{dt} [mr^2\dot{\varphi} \cos^2\theta + mr(-\dot{x} \cos\theta \sin\varphi - \dot{z} \cos\theta \cos\varphi) + I\dot{\varphi} \sin^2\theta \\ &\quad + J(\dot{\psi} + \dot{\varphi} \cos\theta) \cos\theta] + mr(-\dot{x}\dot{\theta} \sin\theta \sin\varphi + \dot{x}\dot{\varphi} \cos\theta \cos\varphi \\ &\quad - \dot{z}\dot{\theta} \sin\theta \cos\varphi - \dot{z}\dot{\varphi} \cos\theta \sin\varphi) = 0, \end{aligned} \tag{18.4.19}$$

$$\frac{d}{dt} [J(\dot{\psi} + \dot{\varphi} \cos\theta)] + \lambda_1 r \sin\varphi + \lambda_2 r \cos\varphi = 0. \tag{18.4.20}$$

In these equations λ_1 and λ_2 are Lagrange multipliers, and the translational velocities satisfy the following kinematical equations of constraint:

$$\begin{aligned} \dot{x} + r\dot{\psi} \sin\varphi &= 0, \\ \dot{z} + r\dot{\psi} \cos\varphi &= 0. \end{aligned} \tag{18.4.21}$$

As a first step in the integration, (18.4.16), (18.4.17), and (18.4.20) are combined into a single equation. The procedure is to multiply the first by $r \sin\varphi$, the second by $r \cos\varphi$, and to substitute their sum in (18.4.20); this eliminates the Lagrange multipliers. Next, $\dot{x}$ and $\dot{z}$ are replaced by their equivalents from (18.4.21), and some of the differentiations are carried out. In this way, one finds easily the equation

$$(J + mr^2) \frac{d}{dt} (\dot{\psi} + \dot{\varphi} \cos\theta) - mr^2 \dot{\theta}\dot{\varphi} \sin\theta = 0. \tag{18.4.22}$$

Next, one substitutes (18.4.21) in (18.4.19). Then, a number of terms combine, and (18.4.19) becomes

$$\frac{d}{dt}\,[(J + mr^2)(\dot\psi + \dot\varphi\cos\theta)\cos\theta + I\dot\varphi\sin^2\theta] + mr^2\dot\theta\dot\psi\sin\theta = 0. \qquad (18.4.23)$$

Another simplification can be achieved by replacing d/dt by $\dot\theta d/d\theta$. Then the last two equations become, respectively,

$$\begin{aligned}&(J + mr^2)\frac{d}{d\theta}(\dot\psi + \dot\varphi\cos\theta) - mr^2\dot\varphi\sin\theta = 0,\\ &\frac{d}{d\theta}\,[(J + mr^2)(\dot\psi + \dot\varphi\cos\theta)\cos\theta + I\dot\varphi\sin^2\theta] + mr^2\dot\psi\sin\theta = 0.\end{aligned} \qquad (18.4.24)$$

These are two equations which determine the angular velocities $\dot\varphi$ and $\dot\psi$ as functions of θ.

If we write

$$\dot\psi + \dot\varphi\cos\theta = \varrho, \qquad (18.4.25)$$

multiply the first equation of (18.4.24) by $\cos\theta$ and subtract it from the second, then, after carrying out some of the indicated differentiations, the resulting equation is

$$-J\varrho\sin\theta + I\frac{d}{d\theta}(\dot\varphi\sin^2\theta) = 0. \qquad (18.4.26)$$

Next, we solve the first equation of (18.4.24) for $\dot\varphi$ and substitute the results in (18.4.26); we find the linear equation

$$\frac{d^2\varrho}{d\theta^2} + \cot\theta\frac{d\varrho}{d\theta} - \frac{mJr^2}{I(mr^2 + J)}\,\varrho = 0. \qquad (18.4.27)$$

This equation was (probably) first given by Appell[†] in his famous *Traité de Mecanique Rationelle* and his *Les Mouvements de Roulement en Dynamique* although he used a totally different method to find it.

The equation can be transformed into one closely related to the Legendre equation. This is done by introducing the new independent variable

$$\xi = \cos^2\theta.$$

With it, (18.4.27) becomes

$$2\xi(1 - \xi)\frac{d^2\varrho}{d\xi^2} + (1 - 3\xi)\frac{d\varrho}{d\xi} - \frac{mJr^2}{2I(mr^2 + J)}\,\varrho = 0. \qquad (18.4.28)$$

There is a great deal of literature[‡] on this equation. It is satisfied by a linear combination of hypergeometric series, or by Legendre polynomials. Hence, two integrals are now available because the general solution of (18.4.28) contains two constants of integration.

[†] Appell, P., *Traité de Mécanique Rationelle*, Vol. 2, 2nd ed., Gauthier-Villars (1904), pp. 368–374; *Les Mouvements de Roulement en Dynamique*, Collection Scientia (1899), Chapter 3.

[‡] See Kamke, E., *Differentialgleichungen, Lösungsmethoden und Lösungen*, Chelsea Publishing Co. (1948), p. 419 (2.68) and p. 455 (2.240).

Another integral is the energy integral. When the constraint equations (18.4.21) are substituted in the kinetic energy, given in (15.5.28), the energy integral becomes

$$\tfrac{1}{2}(J + mr^2)(\dot{\psi} + \dot{\varphi}\cos\theta)^2 + \tfrac{1}{2}I(\dot{\theta}^2 + \dot{\varphi}^2\sin^2\theta) + \tfrac{1}{2}mr^2\dot{\theta}^2 + mgr\sin\theta = h, \quad (18.4.29)$$

where $h > 0$ is a constant equal to the total energy.

From the integration of the Legendre equation (18.4.28) we find ϱ as a function of θ. Now, the first equation of (18.4.24) is

$$(J + mr^2)\frac{d\varrho}{d\theta} = mr^2\dot{\varphi}\sin\theta.$$

Therefore, the substitution of ϱ in this last equation furnishes $\dot{\varphi}$ as a function of θ. Substituting this value into (18.4.25) gives

$$\dot{\psi} = \varrho(\theta) - \dot{\varphi}(\theta)\cos\theta.$$

Therefore, $\dot{\psi}$ is now also known as a function of θ. When $\dot{\psi}(\theta)$ and $\dot{\varphi}(\theta)$ are substituted into the energy integral (18.4.29), that equation becomes a first-order, second-degree equation in θ which is solvable in quadratures. Its solution $\theta(t)$ is now substituted in $\dot{\psi}(\theta)$ and $\dot{\varphi}(\theta)$. These are then first-order equations that may be integrated in simple quadratures to yield $\psi(t)$ and $\varphi(t)$. The substitution of $\dot{\psi}(\theta(t))$ and $\varphi(t)$ in (18.4.21) gives first-order equations in x and z that can also be integrated in quadratures. This gives the locus of the contact point as a function of time and completes the integration.

While it has been shown how the problem of the rolling disk may be integrated, it is also evident that closed-form solutions cannot be expected except, perhaps, in special cases such as rolling in a straight line or in a circle. However, it is remarkable that, except for quadratures, the problem can be solved exactly by integrating a single Legendre equation.

Our third nonholonomic problem deals with the motion of a two-wheeled cart.

Example 18.4.3. (Hamel, in part, p. 471). Discuss the motion of the street vendor's cart. The bed of the cart remains permanently horizontal.

Consider the cart shown in Fig. 18.4.3. The distance between the wheels is $2b$, the radius of each wheel is r, and the inertia parameters are as follows: m_c is the mass of the cart without wheels; the mass of each wheel is m_w; the mass moment of inertia of the cart (exclusive of the wheels) about a vertical axis through the intersection of the axis of symmetry with the wheel axis is I_c; the mass moment of inertia of each wheel about the wheel axis is C, and the mass moment of inertia of each wheel about a wheel diameter is A.

Let the coordinates describing the configuration of the system be as follows: The coordinates of the intersection of the axis of symmetry with the wheel axle are x, y; the axis of symmetry is inclined by θ to the positive x axis; finally, the angular positions of left and right wheels are, respectively φ_l and φ_r. The center of mass G is on the axis of symmetry a distance s from the point (x, y). All this is shown in Fig. 18.4.3.

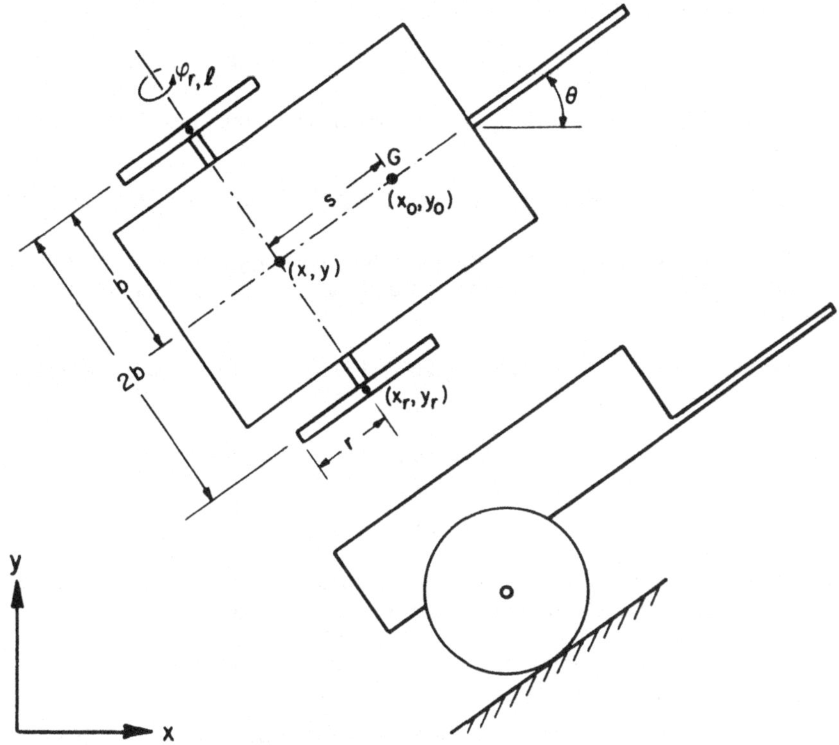

Fig. 18.4.3. Street vendor's car of Example 18.4.3.

There are three constraints on the coordinates. One is the knife edge constraint, i.e., the cart must move in the direction of the axis of symmetry, or

$$\cos\theta\dot{y} - \sin\theta\dot{x} = 0. \tag{18.4.30}$$

In addition, there are the rolling constraints, which state that the tangential velocity of each wheel rim must be equal to the linear velocity of the wheel rim with respect to the ground. One sees readily from Fig. 18.4.3 that this condition leads to the two equations

$$\begin{aligned} \dot{x}\cos\theta + \dot{y}\sin\theta + b\dot{\theta} &= r\dot{\varphi}_r, \\ \dot{x}\cos\theta + \dot{y}\sin\theta - b\dot{\theta} &= r\dot{\varphi}_l. \end{aligned} \tag{18.4.31}$$

All the constraints are nonholonomic.

The coordinates of the mass center of the cart are

$$\begin{aligned} x_0 &= x + s\cos\theta, \\ y_0 &= y + s\sin\theta. \end{aligned}$$

From this, one finds the kinetic energy of the cart (without wheels) as

$$T_c = \tfrac{1}{2}m_c[\dot{x}^2 + \dot{y}^2 - 2s\dot{\theta}(\dot{x}\sin\theta - \dot{y}\cos\theta) + s^2\dot{\theta}^2] + \tfrac{1}{2}I_0\dot{\theta}^2, \tag{18.4.32}$$

where I_0 is the yawing moment of inertia of the cart about the mass center.

The coordinates of the right wheel axis are

$$x_r = x + b\sin\theta, \qquad y_r = y - b\cos\theta,$$

and there are similar expressions for the left wheel. Then, the kinetic energy of the right wheel is

$$T_r = \tfrac{1}{2}m_w[\dot{x}^2 + \dot{y}^2 + 2b\dot{\theta}(\dot{x}\cos\theta + \dot{y}\sin\theta) + b^2\dot{\theta}^2] + \tfrac{1}{2}A\dot{\theta}^2 + \tfrac{1}{2}C\dot{\varphi}_r^2, \qquad (18.4.33)$$

and a similar expression is found for the kinetic energy of the left wheel.

Adding the kinetic energies of the cart and the wheels, one finds for the total kinetic energy

$$T = \tfrac{1}{2}m(\dot{x}^2 + \dot{y}^2) + m_c s\dot{\theta}(\dot{y}\cos\theta - \dot{x}\sin\theta) + \tfrac{1}{2}I\dot{\theta}^2 + \tfrac{1}{2}C(\dot{\varphi}_r^2 + \dot{\varphi}_l^2), \qquad (18.4.34)$$

where

$$\begin{aligned} m &= m_c + 2m_w, \\ I &= I_c + 2m_w b^2 + 2A, \\ I_c &= I_0 + m_c s^2. \end{aligned}$$

When the three constraints are adjoined by means of the Lagrange multipliers λ_1, λ_2, and λ_3, the Lagrange equations of motion are found to be

$$\begin{aligned} m\ddot{x} - m_c s(\ddot{\theta}\sin\theta + \dot{\theta}^2\cos\theta) - \lambda_1\sin\theta - (\lambda_2 + \lambda_3)\cos\theta &= X, \\ m\ddot{y} + m_c s(\ddot{\theta}\cos\theta - \dot{\theta}^2\sin\theta) + \lambda_1\cos\theta - (\lambda_2 + \lambda_3)\sin\theta &= Y, \\ -m_c s(\ddot{x}\sin\theta - \ddot{y}\cos\theta) + I\ddot{\theta} + b(\lambda_3 - \lambda_2) &= M, \qquad (18.4.35) \\ C\ddot{\varphi}_r + \lambda_2 r &= 0, \\ C\ddot{\varphi}_l + \lambda_3 r &= 0. \end{aligned}$$

These five equations together with the three kinematical constraint equations are sufficient to determine the five coordinates and the three Lagrange multipliers.

Now, the first equation of (18.4.35) is multiplied by $\cos\theta$, the second by $\sin\theta$, and then they are added together. Into the resulting equation one substitutes the last two relations of (18.4.35). Then the reader may readily verify that the result of these operations is

$$m\dot{v} - m_c s\dot{\theta}^2 + \frac{C}{r}(\ddot{\varphi}_r + \ddot{\varphi}_l) = F, \qquad (18.4.36)$$

where the tangential acceleration is

$$a_T = \dot{v} = \ddot{x}\cos\theta + \ddot{y}\sin\theta,$$

and the tangential force is

$$F = X\cos\theta + Y\sin\theta.$$

When the last two equations of (18.4.35) are substituted in the third, one finds

$$-m_c s a_N + I\ddot{\theta} + \frac{C}{r}(\ddot{\varphi}_r - \ddot{\varphi}_l) = M, \qquad (18.4.37)$$

where the normal acceleration is

$$a_N = \ddot{x}\sin\theta - \ddot{y}\cos\theta.$$

This system of equations can be further simplified. Subtracting the second relation of (18.4.31) from the first, one has

$$2b\dot{\theta} - r(\dot{\varphi}_r - \dot{\varphi}_l) = 0$$

or, on differentiating that equation,

$$\ddot{\varphi}_r - \ddot{\varphi}_l = \frac{2b}{r}\ddot{\theta}. \tag{18.4.38}$$

Adding the two equations (18.4.31) gives

$$\dot{\varphi}_r + \dot{\varphi}_l = \frac{2}{r}(\dot{x}\cos\theta + \dot{y}\sin\theta),$$

and when that equation is differentiated the result is

$$\begin{aligned}\ddot{\varphi}_r + \ddot{\varphi}_l &= \frac{2}{r}[\ddot{x}\cos\theta + \ddot{y}\sin\theta - \dot{\theta}(\dot{x}\sin\theta - \dot{y}\cos\theta)]\\ &= \frac{2}{r}(\ddot{x}\cos\theta + \ddot{y}\sin\theta)\end{aligned}$$

because one has from (18.4.30)

$$\dot{x}\sin\theta - \dot{y}\cos\theta = 0.$$

It follows that

$$\ddot{\varphi}_r + \ddot{\varphi}_l = \frac{2}{r}\dot{v}. \tag{18.4.39}$$

Finally, on differentiating (18.4.30), one finds

$$\ddot{x}\sin\theta - \ddot{y}\cos\theta + \dot{\theta}(\dot{x}\cos\theta + \dot{y}\sin\theta) = 0$$

or,

$$a_N + \dot{\theta}v = 0, \tag{18.4.40}$$

where the tangential velocity is

$$v = \dot{x}\cos\theta + \dot{y}\sin\theta.$$

When (18.4.38), (18.4.39), and (18.4.40) are substituted in (18.4.36) and (18.4.37), the equations of motion become

$$\begin{aligned}D\dot{v} - m_c s\dot{\theta}^2 &= F,\\ J\ddot{\theta} + m_c s\dot{\theta}v &= M,\end{aligned} \tag{18.4.41}$$

where

$$D = m + \frac{2C}{r^2},$$

$$J = I + \frac{2bC}{r^2}.$$

These equations are more easily interpreted when they are written in terms of the path length u, so that

$$\dot{u} = v.$$

Then, they become

$$\begin{aligned} D\ddot{u} - m_c s\dot{\theta}^2 &= F(u, \theta, t), \\ J\ddot{\theta} + m_c s\dot{\theta}\dot{u} &= M(u, \theta, t). \end{aligned} \tag{18.4.42}$$

Now, if the cart is drawn at sensibly constant speed, one has approximately

$$F = 0. \tag{18.4.43}$$

The moment is applied to turn the cart when needed, such as when there is a curve in the path. Therefore, the moment is a function of the location of the cart along the path, or

$$M = M(u). \tag{18.4.44}$$

Then

$$\begin{aligned} D\ddot{u} - m_c s\dot{\theta}^2 &= 0, \\ J\ddot{\theta} + m_c s\dot{\theta}\dot{u} &= M(u). \end{aligned} \tag{18.4.45}$$

If the first of these is multiplied by $\dot{u}$, the second by $\dot{\theta}$, and they are added together, one finds

$$\frac{1}{2}(D\dot{u}^2 + J\dot{\theta}^2) = \int M(u)\, d\theta + h, \tag{18.4.46}$$

where h is a constant of integration.

Now, the angle θ is controlled by the street vendor who may be presumed to know his route. In other words, he will turn the cart when his route calls for it, i.e., the function

$$\theta = \theta(u) \tag{18.4.47}$$

may be presumed known. Hence,

$$\int M(u)\, d\theta = \int M(u)\theta'(u)\, du = G(u), \tag{18.4.48}$$

i.e., this integral can be evaluated under our assumptions. When (18.4.47) and (18.4.48) are substituted in (18.4.46) one finds the quadrature

$$t - t_0 = \int \left\{\frac{D + J\theta'^2}{2[G(u) + h]}\right\}^{1/2} du, \tag{18.4.49}$$

in which θ' and G are known functions of u.

Perhaps, the assumptions (18.4.43) is considered to be too stringent. However, it is reasonable to suppose that, like M, the tangential force is a function of position along the path, or

$$F = F(u). \tag{18.4.50}$$

Then, if one writes

$$\int F(u)\, du = H(u), \tag{18.4.51}$$

one finds instead of (18.4.49)

$$t - t_0 = \int \left\{ \frac{D + J\theta'^2}{2[G(u) + H(u) + h]} \right\}^{1/2} du. \tag{18.4.52}$$

It may be possible to integrate the equations of motion by iteration even when the assumption (18.4.47) is relaxed.

A two-wheeled cart is generally so designed and loaded that the center of mass is very nearly above the wheel axle, i.e., $|s|$ is a small quantity. Then, retaining the assumptions (18.4.44) and (18.4.50), one may write as a first approximation, in place of (18.4.45),

$$\begin{aligned} D\ddot{u}_0 &= F(u_0), \\ J\ddot{\theta}_0 &= M(u_0). \end{aligned} \tag{18.4.53}$$

The first of these integrates to

$$t - t_0 = \int \left\{ \frac{2}{D} [\Phi(u_0) + h] \right\}^{-1/2} du_0 = \Theta(u_0), \tag{18.4.54}$$

where

$$\Phi(u_0) = \int F(u_0)\, du_0.$$

If (18.4.54) is solved for u_0 (assuming the inversion exists) one has

$$u_0 = u_0(t),$$

and the substitution of this in the second equation of (18.4.53) results in

$$J\ddot{\theta}_0 = M(u_0(t))$$

or, integrated,

$$\theta_0(t) = \frac{1}{J} \left[\iint M(u_0(t))\, dt\, dt + \dot{\theta}(0)t + \theta(0) \right]. \tag{18.4.55}$$

Then, as a second approximation, one may integrate directly:

$$\begin{aligned} D\ddot{u}_1 &= m_c s[\dot{\theta}_0(t)]^2 + F(u_0(t)), \\ J\ddot{\theta}_1 &= -m_c s\dot{\theta}_0(t)\dot{u}_0(t) + M(u_0(t)). \end{aligned} \tag{18.4.56}$$

This process may be continued by iteration in the form

$$\begin{aligned} D\ddot{u}_{i+1} &= m_c s[\dot{\theta}_i(t)]^2 + F(u_i(t)), \\ J\ddot{\theta}_{i+1} &= -m_c s\dot{\theta}_i(t)\dot{u}_i(t) + M(u_i(t)). \end{aligned} \tag{18.4.57}$$

Whether or not this process converges to the solution requires a special investigation as well as knowledge of the specific form of the functions $F(u)$ and $M(u)$. Usually, when the iteration procedure converges, the convergence is very rapid.

18.5. Problems

18.1. Let a particle of unit mass slide without friction in the gutter described in Example 4.2.2.

(a) Show that the motion is restricted to a domain of the xy plane and define its boundary.

(b) Show that there is a particular solution for which $y \equiv 0$, and find $x(t)$ for that solution.

(c) Examine small motions about $y \equiv 0$.

(d) Give these motions and their stability for $z_0(x) = -bx$, where b is a positive constant.

18.2. Discuss the motion of the mechanism described in Problem 12.6.

18.3. Discuss the motion of the system described in Problem 13.2. In particular:

(a) Find two constants of integration.

(b) Find the equation governing θ.

(c) Are there initial conditions such that the system can rotate about the z axis with $\theta \equiv$ const? If so, is that motion stable?

18.4. Answer the same questions as in Problem 18.3 if the mechanism is that of Problem 13.3.

18.5. A heavy hoop of mass M and radius R is constrained to remain in a vertical plane; it can rotate without friction about one of its points, as shown. A second heavy hoop of mass m and radius $r < R$ can roll without slipping inside the first hoop. It remains in the same vertical plane as the first hoop.

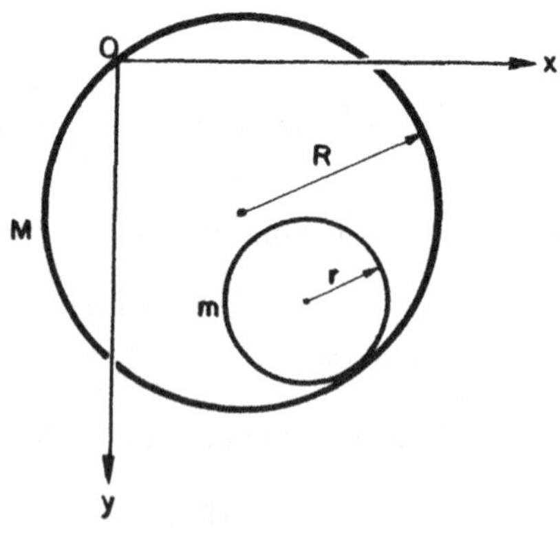

(a) Show that this is a Liouville system.

(b) Determine the equilibrium positions and their stability.

(c) Study small motions about stable equilibrium.

18.6. The given forces acting on a particle of unit mass are derivable from the potential energy

$$V(r, \theta, \varphi) = f(r) + \frac{g(\theta)}{r^2} + \frac{h(\varphi)}{r^2 \sin^2 \theta}$$

where r, θ, and φ are spherical coordinates. Show that Lagrange's equations can be integrated in quadratures.

18.7. A particle of mass m slides under no forces on an elliptical wire. Discuss the motion. (Hint: See Problem 12.10.)

18.8. Discuss the same problem when the particle is heavy.

18.9. Discuss Problem 18.7 when the elliptic wire rotates about one of its axes with constant angular velocity.

18.10. A uniform heavy rope is thrown in the air bent double, as shown; it satisfies the same conditions as the string of Example 18.3.3. Let the initial velocity of one extremity be u_0, that of the other v_0, both are vertically up, and $u_0 \neq v_0$. Discuss the motion; in particular, explain what happens when the rope straightens out.

18.11. Treat the same problem when a heavy particle of mass m moves along the right-hand side of the rope with constant velocity v relative to the rope, as shown. This is the problem of the "Indian rope trick," in which the mass is that of a little boy who climbs up a rope which has been thrown in the air.

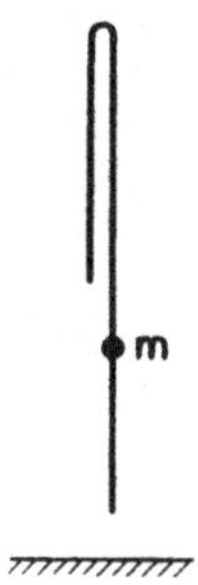

18.12. Find two constants of the motion of the mechanism described in Problem 13.9 and reduce the solution of the equation of motion to a quadrature.

18.13. Discuss the motion of the rod described in Problem 13.10.

19

About Celestial Problems

19.1. Historical Notes[†]

In all of the history of mechanics there is no problem which has had a more profound influence on its development than the Kepler problem.

Accurate information about the motion of the planets did not exist prior to the painstaking and admirable astronomical observations by Tycho Brahé (1546-1601), a Danish noble. While Tycho also searched for the laws governing planetary motion, his mathematical skills were not equal to this task.

It was Kepler (1571-1630), contemporary and sometimes collaborator of Tycho, who succeeded in discovering the three laws named after him. The first:

Planets move in ellipses with the sun in one focus

and the second:

The line joining sun and planet sweeps out equal areas in equal times

were published in 1609. The third:

The square of the time of revolution of each planet is proportional to the cube of its mean distance from the sun

was not announced until 1618. The book in which it was published (1620)

† These are largely gleaned from *Pioneers of Science* by Sir Oliver Lodge, Macmillan Co., Ltd. (1893), reprinted by Dover Publications, New York (July, 1960).

Table 19.1.1. Experimental Values of the Quantities Needed to Verify Kepler's Third Law

Planet	(Mean distance)^3	(Time of revolution in earth years)^2
Mercury	0.05801	0.05801
Venus	0.37845	0.37846
Earth	1.00000	1.00000
Mars	3.5375	3.5375
Jupiter	140.83	140.70
Saturn	867.92	867.70

bore the title *On Celestial Harmonies.* It is interesting to compare Kepler's third law with observation (see Table 19.1.1). Considering the accuracy with which Kepler's laws agreed with astronomically observed data, their correctness was acknowledged then and has remained beyond doubt.

What was still lacking was a mechanism, some basic law of nature, which would "explain" Kepler's laws. It was Sir Isaac Newton (1642-1727) who found that basic law—his law of gravitation.

Newton's genius was discovered early; he became the second Lucasian professor of mathematics at the age of twenty-six when the first, Isaac Barrow, resigned to yield his place to the young Newton. However, what established Newton's fame in his own lifetime as one of the greatest scientists of all time was his discovery of the mechanism of the solar system.

At the age of twenty-three, in the year of the Great Plague, Newton had been on the verge of discovering the inverse square law, but in applying it to the earth–moon problem an error in the then-imagined size of the earth seemed to invalidate the theory. It was nineteen years later that he communicated his result to Halley.

In January 1684, Christopher Wren, English architect and amateur scientist, Charter Member of the Royal Society, offered as prize "a book worth forty shillings" to his fellow members Robert Hooke (of Hooke's law fame) and Edmund Halley (illustrious astronomer and discoverer of the comet named after him) if either could prove that an inverse square law produces elliptic orbits. Whereupon Halley asked Newton what path a body would describe if it were attracted to a center by a force varying as the inverse square of the distance. Newton answered unhesitatingly that the path would be an ellipse with the force center at a focus. To substantiate his answer he produced a portion of the manuscript of the *Principia.* In this way Halley discovered the existence of this manuscript and became eventually responsible (and even paid personally) for its publication. While Halley made many first-class discoveries in his own right, perhaps his greatest is that of Newton's manuscript.

Newton's law of gravitation states:

Every particle attracts every other particle with a force proportional to the mass of each and to the inverse square of the distance between them and acts along the line joining them.

It lies at the basis of all celestial mechanics because in the computation of their orbits the heavenly bodies may be considered as particles.

19.2. The Central Force Problem

The central force problem arises as a generalization of the two-body problem with one body fixed in inertial space.† The moving body is then attracted to a fixed point called the force center with a force which is inversely proportional to the square of the distance from that point, and its trajectory is a conic section. Therefore, every trajectory that does not go to infinity is an ellipse. These results are familiar from a first course in mechanics.

The so-called *central force problem* is that in which a particle is acted on by a force whose line of action passes for all time through a fixed point, and whose magnitude depends only on the distance between the particle and that point. It was shown by Bertrand‡ that among all central forces which tend to zero as the distance tends to infinity, Newton's gravitational force is the only one for which every trajectory not going to infinity is closed. Moreover, there is only one other central force for which every finite trajectory is closed; this is the central force, which is directly proportional to the distance of the particle from the force center.

It is interesting to speculate that, had Newton been in possession of the mathematical tools available to Bertrand, he would not have developed doubts about his gravitational law and might have announced it twenty years sooner.

In Chapter 15, Lagrange's equations were used to obtain the equations of motion of a particle in spherical coordinates. The kinetic energy was

† Kepler's third law does not hold when neither body is fixed, as shown later.

‡ M. J. Bertrand, Theorème relatif au mouvement d'un point attiré vers un centre fixe, *Comptes Rendus, Academie des Sciences*, Vol. 77, p. 149, Paris (1873).

given in (15.2.16) as

$$T = \tfrac{1}{2}m(\dot{r}^2 + r^2 \sin^2 \theta \dot{\varphi}^2 + r^2\dot{\theta}^2), \tag{19.2.1}$$

the potential energy is

$$V(r) = -\int F(r)\, dr, \tag{19.2.2}$$

where $F(r)$ is assumed to be single-valued and integrable on any interval, and the equations of motion were found in (15.2.19) as

$$\frac{d}{dt}(m\dot{r}) - mr \sin^2 \theta \dot{\varphi}^2 - mr\dot{\theta}^2 = F(r), \tag{19.2.3}$$

$$\frac{d}{dt}(mr^2\dot{\theta}) - mr^2\dot{\varphi}^2 \sin\theta \cos\theta = 0, \tag{19.2.4}$$

$$\frac{d}{dt}(mr^2 \sin^2 \theta \dot{\varphi}) = 0. \tag{19.2.5}$$

We note from (19.2.1) and (19.2.2) that φ is an ignorable coordinate. Therefore,

$$p_\varphi = \frac{\partial T}{\partial \dot{\varphi}} = mr^2 \sin^2 \theta \dot{\varphi} = k = \text{const}, \tag{19.2.6}$$

as follows directly from (19.2.5). This last equation may be used to eliminate the variable $\dot{\varphi}$ from the remaining equations (this is a simple example of the process called in Section 16.3 "ignoration of coordinates"). But, if r and θ are sufficient to specify the configuration, the motion is two-dimensional and takes place in a plane $\varphi = \text{const}$. This implies $\dot{\varphi} = 0$ and, hence, $k = 0$. Then, the kinetic energy is

$$T = \tfrac{1}{2}m(\dot{r}^2 + r^2\dot{\theta}^2), \tag{19.2.7}$$

and the remaining equations of motion are

$$\begin{aligned} \frac{d}{dt}(m\dot{r}) - mr\dot{\theta}^2 &= F(r), \\ \frac{d}{dt}(mr^2\dot{\theta}) &= 0. \end{aligned} \tag{19.2.8}$$

The last of these gives another momentum integral; it is

$$p_\theta = \frac{\partial T}{\partial \dot{\theta}} = mr^2\dot{\theta} = K = \text{const}, \tag{19.2.9}$$

and is, in fact, equivalent to Kepler's second law. This follows from the observation that the area dA swept out by a radius vector r rotating through an infinitesimal angle $d\theta$ is

$$dA = \tfrac{1}{2}r^2\, d\theta.$$

Then (19.2.9) states that the time rate of change of this quantity is constant.

As the problem is catastatic and only potential forces act, one also has the energy integral

$$\tfrac{1}{2}m(\dot{r}^2 + r^2\dot{\theta}^2) + V(r) = h$$

or, combining this with (19.2.9),

$$\frac{1}{2}\, m\dot{r}^2 + \frac{1}{2}\, \frac{K^2}{mr^2} + V(r) = h. \tag{19.2.10}$$

This is a first-order, second-degree equation in r, which is reducible to a simple quadrature.

It is convenient in the central force problem to utilize the transformation

$$u = 1/r \qquad (r \neq 0). \tag{19.2.11}$$

Then, from (19.2.9), we obtain

$$\dot{\theta} = \frac{K}{m}\, u^2, \qquad r^2\dot{\theta}^2 = \left(\frac{K}{m}\right)^2 u^2,$$

and from these and the chain rule,

$$\dot{r} = -\frac{K}{m}\, \frac{du}{d\theta}, \qquad \ddot{r} = -\left(\frac{K}{m}\right)^2 u^2\, \frac{d^2u}{d\theta^2}.$$

If we discuss the motion of a unit mass (or, which is the same thing, we regard h as the energy per unit mass, K as the momentum per unit mass, and $F(r)$ as the force per unit mass) the energy integral (19.2.10) becomes the trajectory equation

$$\left(\frac{du}{d\theta}\right)^2 + u^2 = \frac{2[h - V(u^{-1})]}{K^2}. \tag{19.2.12}$$

Differentiating this equation with respect to θ gives

$$\frac{d^2u}{d\theta^2} + u = \frac{(d/du)[U(u)]}{K^2u^2} = \frac{f(u)}{K^2u^2} = \frac{g(u)}{K^2}, \tag{19.2.13}$$

where

$$U(u) = V(u^{-1}),$$
$$f(u) = -F(u^{-1}) = -F(r),$$
$$g(u) = f(u)/u^2,$$

and $F(r)$ is the central force of (19.2.2). Equation (19.2.13) is evidently the first of (19.2.8) under the transformation (19.2.11).

We summarize the results obtained so far.

In every central force problem in which the force is single-valued and integrable:

(i) The motion is two-dimensional;
(ii) the motion satisfies Kepler's second law;
(iii) the energy is a constant of the motion;
(iv) the angular momentum is a constant of the motion [which is the same as (ii) above].

19.3. The Central Force Problem Continued—The Apsides

In the central force problem, a point of the trajectory that has stationary distance from the force center is called an *apsis* (plural: apsides) and, by definition, $du/d\theta = 0$ at an apsis. The radius vector from the force center to an apsis is called the apsidal radius, and its length is an apsidal distance.

It is easy to see that every apsidal radius lies in an axis of symmetry of the trajectory because every trajectory satisfies (19.2.12). Thus, let the line defined by $\theta = 0$ be along an apsidal radius, and consider the integral of (19.2.12) under the initial condition $u = 0$ when $\theta = 0$. Then, neither the differential equation nor the initial condition is changed if one replaces θ by $-\theta$; this proves the assertion.

It is a consequence of this observation that the angle subtended by adjacent apsides, called the apsidal angle, is the same for all adjacent pairs and that there can be no more than two apsidal lengths. This is shown in Fig. 19.3.1. Suppose two adjacent apsides lie at $\theta = 0$ and $\theta = \alpha$, and suppose their apsidal distances are different; both apsides are axes of symmetry for the trajectory. Then, folding along the line $\theta = \alpha$ shows that the image of the apse at $\theta = 0$ falls on the line $\theta = 2\alpha$; this proves the above two assertions. It has the additional consequences that every apsidal length is either a maximum or a minimum but not a point of inflection

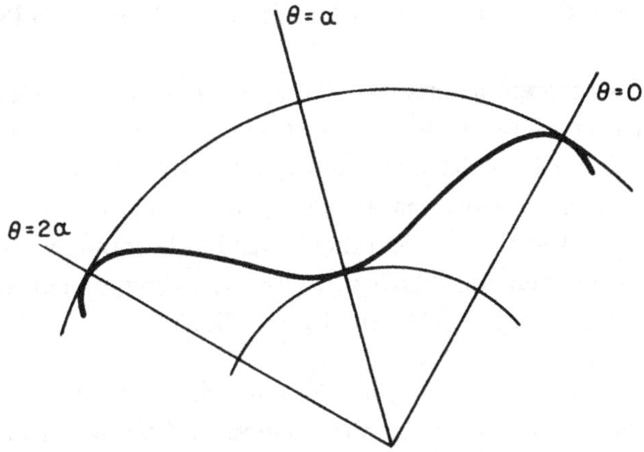

Fig. 19.3.1. Apsides in central force motion.

among neighboring radial distances from the force center to the trajectory, and that maxima and minima alternate. Therefore, the trajectory must lie in the annulus bounded by the two concentric circles about the force center having radii equal to the apsidal lengths, respectively. Moreover, if there is only one apsidal length, the trajectory either goes to infinity, or it is a circle.

It follows from (19.2.12), that if r_i is an apsidal radius and $u_i = 1/r_i$, then the u_i $(i = 1, 2)$ are the roots of the equation

$$\tfrac{1}{2}K^2u^2 + U(u) = h. \tag{19.3.1}$$

From (19.2.12), it follows that

$$\theta = \pm \int \frac{du}{\left\{\frac{2}{K^2}[h - U(u)] - u^2\right\}^{1/2}}, \tag{19.3.2}$$

and the apsidal angle α is

$$\alpha = \pm \int_{u_i}^{u_{i+1}} \frac{du}{\left\{\frac{2}{K^2}[h - U(u)] - u^2\right\}^{1/2}}, \tag{19.3.3}$$

where u_i and u_{i+1} are any two adjacent roots of the equation (19.3.1).

It is evident that (19.3.2) is the integral in quadratures of the trajectory.

19.4. The Central Force Problem Continued—On Bertrand's Theorem

Bertrand's theorem asserts that there are only two central force laws under which every finite trajectory is closed—that in which the force is directly proportional to the distance from the force center, and that in which it is inversely proportional to the square of that distance. By means which are a great deal simpler than those used by Bertrand, we prove here a somewhat weaker theorem, which is, however, strong enough to establish Newton's gravitational law.† We shall prove that:

The only central force law for which the force diminishes in absolute value with distance from the force center, and for which every trajectory not going to infinity is a closed, simple orbit is Newton's gravitational law.

A simple trajectory is one which has no points of self-crossing (see Section 3.1).

It is evident from the symmetry of trajectories about apsidal radii that if a finite, simple trajectory has an apsis on the line $\theta = 0$, it also has one on the line $\theta = \pi$ because, if this were not true, the trajectory would either have a corner or a point of self-crossing at $\theta = \pi$, as shown in Fig. 19.4.1. But nonsimple trajectories are excluded, and corners only occur at rest points or where unbounded forces act, as shown in Section 3.1.

We suppose that the trajectory is not a circle. In that case, there is a finite number of apsides in the interval $0 \leq \theta \leq \pi$, and the angle between any adjacent ones is α. Therefore, $\pi = n\alpha$, where n is an integer, or

$$\alpha = \pi/n. \tag{19.4.1}$$

Let r_0 and r_1 be two adjacent apsidal radii, and let r_0 lie in the line $\theta = 0$. We already saw that one of these has maximum length, the other has minimum length and, in $0 \leq \theta \leq \alpha$, the trajectory lies in the annulus bounded by the circles with radii $r = r_0$ and $r = r_1$, as shown in Fig. 19.3.1. If we use $u_0 = 1/r_0$ and $u_1 = 1/r_1$, the apsidal angle is, from (19.3.3),

$$\alpha = \frac{\pi}{n} = \int_{u_0}^{u_1} \frac{du}{\left\{\dfrac{2}{K^2}[h - U(u)] - u^2\right\}^{1/2}}, \tag{19.4.2}$$

† Rosenberg, R. M., On Newton's Law of Gravitation, *American Journal of Physics* Vol. 40, pp. 975–978 (July, 1972).

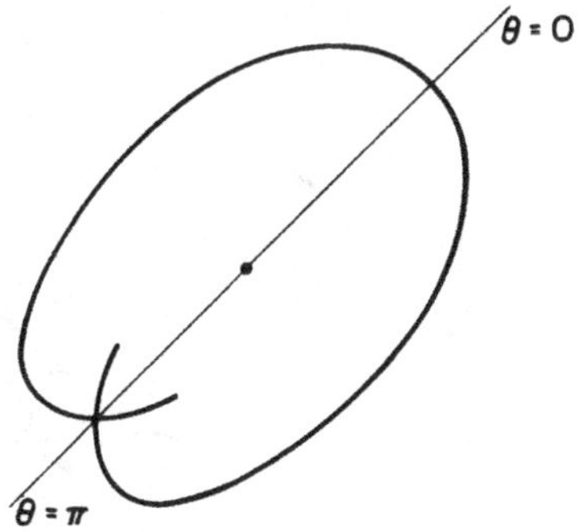

Fig. 19.4.1. A nonsimple trajectory in central force motion.

where the apsidal radii are defined by

$$\begin{aligned} \tfrac{1}{2}K^2u_0^2 + U(u_0) &= h, \\ \tfrac{1}{2}K^2u_1^2 + U(u_1) &= h; \end{aligned} \tag{19.4.3}$$

these follow immediately from (19.3.1). They show that u_0 and u_1 are continuous functions of the constants h and K, and these in turn are continuous functions of the initial conditions because of (19.2.9) and (19.2.10).

We may draw two conclusions: For one, n must be a constant for all initial conditions, because if it were not it would be a continuous function of u_0 and u_1. Then, one could find initial conditions such that n is not an integer, and the corresponding trajectory would then not be a simple closed one. For another, $U(u)$ must be such that (19.4.2) holds for all possible values of u_0 and u_1 for, if it did not, one could again find initial conditions for which the trajectory is not closed.

In particular, let us choose initial conditions such that u_0 and u_1 differ only by an infinitesimal amount, or

$$u_1 = u_0 + \varepsilon \tag{19.4.4}$$

with $|\varepsilon|$ small. But then the trajectory segment in the interval $0 \leq \theta \leq \alpha$ is very nearly a circular arc because $r(\theta)$ is bounded by two circles of nearly equal radii. [In fact, from (19.4.4)

$$\frac{1}{r_1} = \frac{1}{r_0} + \varepsilon$$

or

$$r_1 = r_0/(1 + r_0\varepsilon) = r_0(1 - \varepsilon r_0 + \cdots)$$

within first-order terms in small quantities.] We assume that this trajectory satisfies (19.4.2).

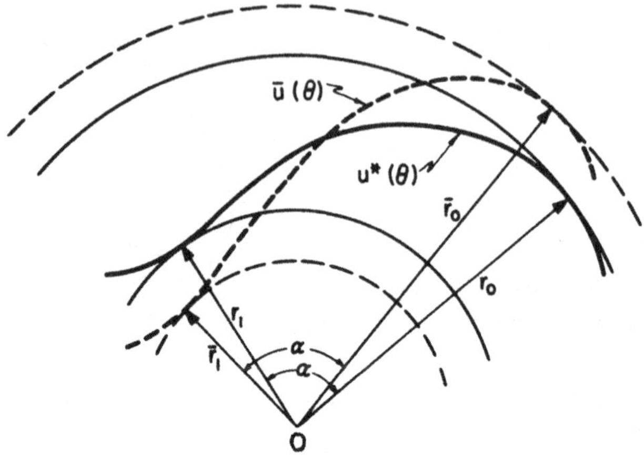

Fig. 19.4.2. Trajectory $u^*(\theta)$ of central force motion, and perturbed trajectory $\bar{u}(\theta)$.

Consider now a perturbed trajectory with slightly different initial conditions. Let it have adjacent apsidal radii $\bar{r}_0$ and $\bar{r}_1$ as shown in Fig. 19.4.2. Since the apsidal radii are continuous functions of the initial conditions, $\bar{r}_0$ must be nearly equal to r_0, and $\bar{r}_1$ to r_1. It follows that the unperturbed and the perturbed trajectories must lie near each other at $\theta = 0$ and at $\theta = \alpha$.

As a matter of fact, the perturbed trajectory must lie near the unperturbed one in the entire interval $0 \leq \theta \leq \alpha$ because, if it did not, there would be another apsis in that interval. This is seen by supposing to the contrary that the perturbed trajectory crosses one of the dotted circles in Fig. 19.4.2. Then it must cross a second time in order to satisfy the conditions at $\theta = 0$ and $\theta = \alpha$, and between the points of crossing there must be another point of stationary distance from 0. This violates the condition satisfied by closed or nonclosed trajectories that the apsidal angle is a constant independent of initial conditions.

Now, the requirement that the perturbed trajectory lie near the unperturbed one over a finite interval is merely another way of saying that the unperturbed trajectory must have first-order stability (see Chapter 17). By supposition, the unperturbed trajectory differs from a circular arc only by first-order terms in infinitesimals. This has the consequence that its stability is the same as that of the circular arc.

To show this, assume that the curve

$$u = u^*(\theta)$$

is a trajectory, i.e., it satisfies (19.2.13), or

$$\frac{d^2u^*}{d\theta^2} + u^* = \frac{g(u^*)}{K^2}. \tag{19.4.5}$$

Consider now a neighboring curve

$$\bar{u}(\theta) = u^*(\theta) + \xi(\theta), \tag{19.4.6}$$

where ξ is a disturbance. Then, substituting (19.4.6) in the trajectory equation (19.2.13) and expanding $g(\bar{u})$ in Taylor series, one has, to the first order in infinitesimals,

$$\frac{d^2u^*}{d\theta^2} + \frac{d^2\xi}{d\theta^2} + u^* + \xi = \frac{1}{K^2}\,[g(u^*) + \xi g'(u^*) + \cdots],$$

where

$$g'(u^*) = \left.\frac{dg}{du}\right|_{u=u^*}.$$

Then, it follows from (19.4.5) that $\bar{u}$ will be a solution if ξ satisfies

$$\frac{d^2\xi}{d\theta^2} + \left[1 - \frac{g'(u^*)}{K^2}\right]\xi = 0. \tag{19.4.7}$$

This is an example of the linearized variational equation (17.4.1) in which the coefficient is variable. Thus, the first-order stability of the solution is determined by (19.4.7).

Let us now suppose that $u^*(\theta)$ *is* stable and ξ *is* the solution of (19.4.7). Then, $\bar{u}$ is a solution of (19.2.13), or

$$\frac{d^2\bar{u}}{d\theta^2} + \bar{u} = \frac{g(\bar{u})}{K^2}. \tag{19.4.8}$$

Let us examine the stability of $\bar{u}$ by considering the neighboring trajectory

$$\tilde{u}(\theta) = \bar{u}(\theta) + \eta(\theta), \tag{19.4.9}$$

where η is an infinitesimal. The substitution of (19.4.9) in (19.2.13), and the identity (19.4.8) show that η satisfies

$$\frac{d^2\eta}{d\theta^2} + \left[1 - \frac{g'(\bar{u})}{K^2}\right]\eta = 0. \tag{19.4.10}$$

But, to first order in small quantities, we obtain

$$g'(\bar{u}) = g'(u^* + \xi) = g'(u^*) + \xi g''(u^*) + \cdots \tag{19.4.11}$$

and the substitution of this in (19.4.10) gives, to first order in infinitesimals,

$$\frac{d^2\eta}{d\theta^2} + \left[1 - \frac{g'(u^*)}{K^2}\right]\eta = 0. \tag{19.4.12}$$

This is identical with (19.4.7); hence the first-order stability of the perturbed trajectory $\bar{u}(\theta)$ is the same as that of the unperturbed trajectory $u^*(\theta)$.

We have now reduced the proof of the above theorem to the problem of finding the conditions on $U(u)$ such that all circular arcs are stable, and n is an integer independent of initial conditions.

On a circular arc the radius is constant, or

$$u = c = \text{const.} \tag{19.4.13}$$

Then, from (19.2.13), on circular arcs, we have

$$K^2 = \frac{f(c)}{c^3}. \tag{19.4.14}$$

Let us now substitute the neighboring arc

$$u = c + \xi, \tag{19.4.15}$$

where ξ is an infinitesimal, into the trajectory equation (19.2.13). One finds

$$\frac{d^2\xi}{d\theta^2} + c + \xi = \frac{f(c + \xi)}{K^2(c + \xi)^2}. \tag{19.4.16}$$

When the right-hand side is expanded in Taylor series one finds, up to linear terms in small quantities,

$$\frac{f(c + \xi)}{K^2(c + \xi)^2} = \frac{f(c)}{K^2c^2}\left[1 + \xi\left(\frac{f'(c)}{f(c)} - \frac{2}{c}\right) + \cdots\right],$$

and the substitution of this in (19.4.16) gives

$$\frac{d^2\xi}{d\theta^2} + \left(3 - \frac{cf'(c)}{f(c)}\right)\xi = 0, \tag{19.4.17}$$

where K^2 was replaced by its equivalent from (19.4.14).

As (19.4.17) is a linear equation with constant coefficients, its solutions are of the form $Ce^{\lambda\theta}$ and, for stability, λ may not have positive real parts. This requires that

$$3 - \frac{cf'(c)}{f(c)} > 0, \tag{19.4.18}$$

i.e., the solution is of the form

$$\xi = A \cos\left(\left[3 - \frac{cf'(c)}{f(c)}\right]^{1/2} \theta + B\right), \tag{19.4.19}$$

where A and B are constants of integration. At an apsis, $du/d\theta = 0$, and this requires, in view of (19.4.15), that $d\xi/d\theta = 0$. Now, choosing the origin of θ such that $\theta = 0$ coincides with an apsis is equivalent to setting $B = 0$ in (19.4.19). Then one sees that $d\xi/d\theta$ vanishes where

$$[3 - cf'(c)/f(c)]^{1/2}\theta = \pi,$$

or the apsidal angle is

$$\theta = \alpha = \pi/[3 - cf'(c)/f(c)]^{1/2}. \tag{19.4.20}$$

Now, the apsidal angle has been shown to be independent of initial conditions, i.e., of c. This requires that

$$\frac{cf'(c)}{f(c)} = p = \text{const.} \tag{19.4.21}$$

This is a differential equation in f whose integral is

$$f = Pc^p, \tag{19.4.22}$$

where P is a constant of integration.

It follows that the central force law must be a power law, or

$$f(u) = Pu^p \quad (= Pr^{-p}). \tag{19.4.23}$$

Now, if the force is to go to zero as the distance r tends to infinity, it is evident that p must be positive. On combining this observation with (19.4.18) and (19.4.21), one has the bounds

$$0 < p < 3. \tag{19.4.24}$$

But, it will be remembered from (19.4.1) that α must be an integer submultiple of π or, from (19.4.20) and (19.4.21), p must also satisfy the relation

$$3 - p = n^2, \tag{19.4.25}$$

where n is an integer. The only value of p which satisfied (19.4.24) and (19.4.25) is $p = 2$. Therefore, the central force law must be Newton's inverse-square law.

It remains to show that when $p = 2$, there are no finite trajectories that are not closed.

The substitution of (19.4.23) with $p = 2$ into (19.2.13) gives

$$\frac{d^2u}{d\theta^2} + u = \frac{P}{K^2}. \tag{19.4.26}$$

The general solution of this equation is

$$u = A\cos(\theta + B) + P/k^2, \tag{19.4.27}$$

where A and B are constants of integration. But, this is the equation of a conic in polar coordinates. Hence, all finite orbits are ellipses; this proves the theorem.

It is interesting to determine the condition under which elliptic orbits occur. Since the central force is an attraction (i.e., it acts in the direction of diminishing radius) it is

$$F(r) = -Pr^{-2},$$

where

$$P > 0. \tag{19.4.28}$$

Then, placing the datum for V at infinity, i.e., $V(\infty) = 0$, the potential energy is

$$V(r) = -\int F(r)\,dr = -P/r. \tag{19.4.29}$$

The substitution of this equation into (19.2.10) with $\dot{r} = 0$ [and $m = 1$ in accordance with the observation preceding (19.2.12)] shows that the apsidal lengths are the positive real roots of

$$hr^2 + Pr - K^2/2 = 0 \tag{19.4.30}$$

with P, $K^2 > 0$, and h is the total energy. By Descartes' "rule of signs," (18.4.30) has only one positive, real root when $h \geq 0$ and two or none when $h < 0$. This demonstrates in a very simple way the well-known result that elliptic orbits can only occur when the total energy is negative, because the ellipse is the only conic having two apsidal distances (except for the circle, which has an infinity).

In the solution (19.4.27), we may always choose the origin of θ so that $B = 0$. Doing this, we have

$$\begin{aligned} u &= A\cos\theta + P/K^2, \\ \frac{du}{d\theta} &= -A\sin\theta, \end{aligned} \tag{19.4.31}$$

and the trajectory equation is, with $V(r)$ as given in (19.4.29),

$$\left(\frac{du}{d\theta}\right)^2 + u^2 = 2\left(\frac{h}{K^2} + \frac{Pu}{K^2}\right). \tag{19.4.32}$$

When the solution (19.4.31) is substituted in the trajectory equation, an easy calculation gives

$$A^2 = \frac{2h}{K^2} + \frac{P^2}{K^4}.$$

In this way, the constant of integration is expressed in terms of the dynamical constraints of the problem, and the solution becomes

$$u = \frac{P}{K^2}\,[1 + (1 + 2hK^2/P^2)^{1/2}\cos\theta], \tag{19.4.33}$$

which is the result of substituting (19.4.33) back into the first of (19.4.31). But the equation of a conic in polar coordinates may be written as

$$u = \frac{1}{l}\,(1 + e\cos\theta) \tag{19.4.34}$$

where l is the semilatus rectum, e is the eccentricity, and θ is zero on the perpendicular dropped from the focus onto the directrix. The conic is an ellipse for $e < 1$, a parabola for $e = 1$, and a hyperbola for $e > 1$. Identifying coefficients between (19.4.33) and (19.4.34), one finds

$$l = K^2/P, \qquad e = (1 + 2hK^2/P^2)^{1/2}. \tag{19.4.35}$$

For an ellipse with major axis $2a$ and minor axis $2b$, one knows from analytical geometry that

$$l = \frac{b^2}{a} = a(1 - e^2),$$

and hence

$$b = a(1 - e^2)^{1/2}. \tag{19.4.36}$$

The period for one orbit is found from Kepler's third law. From (19.2.9), we have

$$dt = \frac{m}{K} r^2 \, d\theta = 2 \frac{m}{K} \, dA,$$

where dA is the area swept out by the radius vector in the time dt. Therefore, the period is

$$T = \frac{2}{K} A \tag{19.4.37}$$

if we take K as the angular momentum per unit mass, as before, and

$$A = \pi ab \tag{19.4.38}$$

is the area of the ellipse. Then, with the aid of (19.4.35), (19.4.36), (19.4.37), and (19.4.38), one finds by some simple algebra

$$T = 2\pi \left(\frac{a^3}{P} \right)^{1/2} = \frac{2\pi P}{(-2h)^{3/2}}. \tag{19.4.39}$$

The first equality shows that

$$T^2 = \frac{4\pi^2}{P} a^3. \tag{19.4.40}$$

Since P is a constant not depending on initial conditions, one sees that the square of the periodic time to complete one orbit is directly proportional to the cube of the major axis of the ellipse; this is Kepler's third law. The second equality shows that the total energy h must be negative if the periodic time is to be real. This result was deduced already from (19.4.30) by very simple means.

19.5. The *n*-Body Problem

The celebrated n-body problem of astronomy is the following:

Each particle in a system of a finite number of particles is subjected to a Newtonian gravitational attraction from all the other particles, and to no other forces. If the initial state of the system is given, how will the particles move?

Inasmuch as the system is catastatic [the only constraint being the inequality constraint from the impenetrability principle (see Section 6.1)] and has constant total mass, and the given forces are derivable from a potential energy, the energy integral exists.

Let the particle P_i have mass m_i and position vector x^i relative to a fixed point in inertial space, and let there be n particles. Then, the total kinetic energy of the system is

$$T = \frac{1}{2} \sum_{i=1}^{n} m_i \dot{x}^i \cdot \dot{x}^i = \frac{1}{2} \sum_{i=1}^{n} m_i (\dot{x}^i)^2. \tag{19.5.1}$$

The potential energy between the particles P_i and P_j is

$$V^{ij} = -\frac{G m_i m_j}{| x^j - x^i |} = -\frac{k_{ij}}{| x^j - x^i |}, \tag{19.5.2}$$

where G is the universal gravitational constant. This expression follows immediately from (19.4.29) applied to the n-body problem. Then, the potential energy from which the force on P_i is derivable is found by summing on j, i.e.,

$$V^i = \sum_{j=1}^{n} V^{ij} = -\sum_{j=1}^{n} \frac{k_{ij}}{| x^j - x^i |} \qquad (j \neq i). \tag{19.5.3}$$

The total potential energy of the system is precisely one-half of the sum of (19.5.3) over i, or

$$V = \frac{1}{2} \sum_{i=1}^{n} V^i = -\frac{1}{2} \sum_{i=1}^{n} \sum_{j=1}^{n} \frac{k_{ij}}{| x^j - x^i |} \qquad (j \neq i). \tag{19.5.4}$$

[A simple way of deriving (19.5.4) is to note that the resultant of all forces on the ith particle is

$$\sum_j F_{ij} = \sum_j k_{ij} \frac{1}{| x^j - x^i |^2} \cdot \frac{x^j - x^i}{| x^j - x^i |} = \sum_j k_{ij} \frac{x^j - x^i}{| x^j - x^i |^3}. \tag{a}$$

But this must be derivable from the potential V, or $\sum_j F_{ij} = -\partial V / \partial x^i$. Then,

$$\sum_i \sum_j k_{ij} \frac{x^j - x^i}{| x^j - x^i |^3} \dot{x}^i = -\sum_i \frac{\partial V}{\partial x^i} \dot{x}^i = -\frac{dV}{dt}. \tag{b}$$

But,

$$\frac{d}{dt} \frac{1}{| x^j - x^i |} = \frac{d}{dt} \left[\sum_{k=1}^{3} (x_k{}^j - x_k{}^i)^2 \right]^{-1/2}$$
$$= \frac{(x^j - x^i)(\dot{x}^j - \dot{x}^i)}{| x^j - x^i |^3}.$$

Then, it follows that

$$\sum_i \sum_j \frac{d}{dt} \frac{k_{ij}}{| x^j - x^i |} = -\sum_i \sum_j k_{ij} \frac{x^j - x^i}{| x^j - x^i |^3} (\dot{x}^j - \dot{x}^i)$$

$$= -\sum_i \sum_j k_{ij} \frac{x^j - x^i}{| x^j - x^i |^3} \dot{x}^j$$

$$+ \sum_i \sum_j k_{ij} \frac{x^j - x^i}{| x^j - x^i |^3} \dot{x}^i. \tag{c}$$

If we exchange i and j in the first term on the right-hand side of (c) we find

$$\sum_i \sum_j \frac{d}{dt} \frac{k_{ij}}{| x^i - x^i |} = 2 \sum_i \sum_j k_{ij} \frac{x^j - x^i}{| x^j - x^i |} \dot{x}^i. \tag{d}$$

The substitution of (d) in (a) gives

$$-\frac{dV}{dt} = \frac{d}{dt} \left[\frac{1}{2} \sum_i \sum_j \frac{k_{ij}}{| x^j - x^i |} \right], \tag{e}$$

and (19.5.4) follows at once.]

Then, the energy integral is

$$T + V = h \tag{19.5.5}$$

with T and V as defined in (19.5.1) and (19.5.4).

The substitution of T and V into Lagrange's equations gives

$$m\ddot{x}^i = \sum_{j=1}^{n} k_{ij} \frac{x^j - x^i}{| x^j - x^i |^3} \qquad (i = 1, 2, \ldots, n;\ j \neq i), \tag{19.5.6}$$

as is also evident from (a) above.

The force exerted by P_j on P_i is

$$F_{ij} = k_{ij} \frac{x^j - x^i}{| x^j - x^i |^3}, \tag{19.5.7}$$

from which it follows that

$$F_{ij} = -F_{ji}, \tag{19.5.8}$$

which is, in fact, Newton's third law. If we substitute (19.5.7) in (19.5.6) and sum these equations over i we find

$$\sum_{i=1}^{n} m_i \ddot{x}^i = \sum_{i=1}^{n} \sum_{j=1}^{n} F_{ij} \qquad (j \neq i). \tag{19.5.9}$$

But the right-hand side of (19.5.9) vanishes because of (19.5.8) and so, one finds

$$\sum_{i=1}^{n} m_i \dot{x}^i = a \sum_{i=1}^{n} m_i, \tag{19.5.10}$$

where a is an arbitrary constant vector. Integrating again, we obtain

$$\sum_{i=1}^{n} m_i x^i \Big/ \sum_{i=1}^{n} m_i = at + b, \tag{19.5.11}$$

where b is also an arbitrary constant vector. If we utilize the mass center position vector

$$x_0 = \sum_{i=1}^{n} m_i x^i \Big/ \sum_{i=1}^{n} m_i, \tag{19.5.12}$$

we find

$$x_0 = at + b \tag{19.5.13}$$

or:

The mass center moves with uniform velocity.

Actually, (19.5.13) has furnished six integrals, because the constant vectors a and b define six constants of integration. It follows from the uniform motion of the mass center that that point may be regarded as a fixed point in inertial space, i.e., the position vectors of all particles may be taken with respect to the mass center.

Additional integrals of the n-body problem are found by writing

$$m_i \ddot{x}^i \times x^i = \sum_{j=1}^{n} F_{ij} \times x^i \qquad (i = 1, 2, \ldots, n;\ j \neq i).$$

When these are summed over i, one finds

$$\sum_{i=1}^{n} m_i \ddot{x}^i \times x^i = \sum_{i=1}^{n} \sum_{j=1}^{n} F_{ij} \times x^i \qquad (j \neq i),$$

where the right-hand side vanishes because the moments of the forces cancel in pairs. Then, since

$$\frac{d}{dt}(\dot{x}^i \times x^i) = \ddot{x}^i \times x^i + \dot{x}^i \times \dot{x}^i = \ddot{x}^i \times x^i$$

one finds

$$\sum_{i=1}^{n} m_i (\dot{x}^i \times x^i) = c, \tag{19.5.14}$$

where c is also an arbitrary, constant vector. Equation (19.5.14) furnishes three additional integrals because the constant vector c defines three constants of integration.

Equation (19.5.10) states that the linear momentum of the system is conserved, and (19.5.14) that the angular momentum is conserved. It should not come as a surprise that these momenta are constant because we know that they can only be changed by external forces and/or moments. But, in the n-body problem all forces are internal; hence, the causes of momentum changes are absent.

The three conservation laws (momenta and energy) furnish, in general, ten of the $6n$ integrals needed to completely integrate the system.

19.6. The Two-Body Problem

Consider two bodies of mass m_1 and m_2, respectively. Let the distance between them be

$$r = r_1 + r_2, \tag{19.6.1}$$

where r_1 and r_2 are the distances of m_1 and m_2, respectively, from their mass center, as shown in Fig. 19.6.1. As we observed above, the mass center may be taken as fixed in inertial space, and hence may be used as the origin of the position vectors. They are

$$r_1 = r\frac{m_2}{m_1 + m_2}, \qquad r_2 = r\frac{m_1}{m_1 + m_2}. \tag{19.6.2}$$

As the line of action of the force on each passes through the mass center, each executes a central force motion relative to the mass center. It follows that Kepler's second law holds for each (see Section 19.2), or each radius vector r_1 or r_2 sweeps out equal areas in equal times. It is simple to show that Kepler's first law also holds for each, or each travels on an ellipse about the mass center. This follows from the fact that the equation of

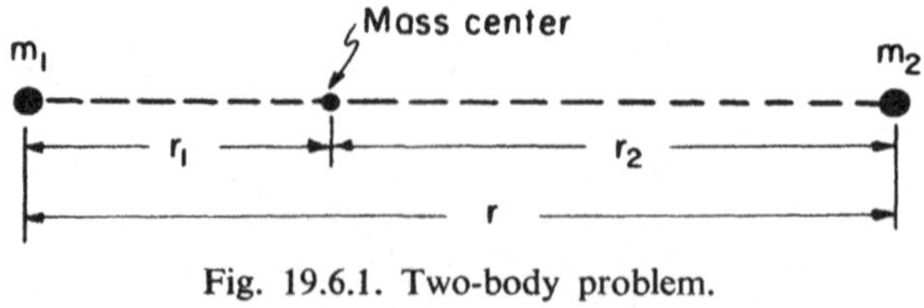

Fig. 19.6.1. Two-body problem.

motion of m_1 (for instance) is

$$m_1\ddot{r}_1 = \frac{Gm_1m_2}{r^2} = \left(\frac{m_2}{m_1 + m_2}\right)^2 \frac{Gm_1m_2}{{r_1}^2}, \tag{19.6.3}$$

where the second equality is the result of computing r from the first relation of (19.6.2). Therefore, the motion of m_1 satisfies the inverse square law with respect to the mass center; the same is, of course, true for m_2.

However, the third Kepler law does not hold, as was pointed out already in a footnote on p. 527. The third Kepler law states that the square of the time for one revolution about the force center is proportional to the cube of the major axis of the ellipse [see (19.4.40)]. Therefore, for $r_1 \neq r_2$ Kepler's third law would imply that the times for one revolution are different for the two masses. But, obviously, these times are necessarily the same because the two masses are at all times (on opposite sides) on a line through the mass center.

Exercise. Evidently, the two bodies of the two-body problem move in a plane; hence, four coordinates specify the configuration. It follows that eight first integrals completely integrate the equations of motion because each coordinate satisfies one second-order equation. But we have shown that the conservation laws furnish ten first integrals. Explain this apparent contradiction.

19.7. Some Information about the Three-Body Problem

The three-body problem is one of the most celebrated of the unsolved problems of dynamics. To the early investigators [largely Laplace (1741–1827) and Lagrange] it presented a challenge because it is the smallest value of n for which the n-body problem is not soluble by elementary means. To solve the three-body problem requires the integration of nine second-order equations, or eighteen first integrals. Ten of these are furnished by the momentum and energy integrals satisfied by all n-body problems. However, the remaining eight have not been found. In fact, Bruns has shown (1887) that no algebraic integrals other than these ten exist.

Two special three-body problems have been intensively investigated: the two-dimensional one, in which all three particles remain permanently in a plane, and the so-called "restricted three-body problem," in which the mass of one particle is so small compared to the other two that its influence on the two large masses is negligible. This latter problem is of interest to astronomers, who wish to calculate the moon's trajectory under the influence of the earth and the sun. It turns out that even these special problems are of great difficulty; they are not reducible to quadratures and cannot be integrated by elementary means.

In recent years, the three-body problem has received renewed and intense attention because of the problems of space flight. Certainly, a space probe wending its way through the solar system is at all times under the influence of the sun and one or more planets, without however influencing their trajectories measurably. The recent activity in investigating the three-body problem has been coupled both with modern attitudes toward solving problems, and with modern means of attacking them. Today, a problem is frequently regarded as solved if it can be put into a form that can be entrusted to a computer. With the incredible capacity of modern high-speed computers, computational problems which were considered totally hopeless only a few years ago are solved routinely today. As a result, a great deal of new information about the three-body problem has come to light. In addition, recent theoretical assaults on the problem have yielded many new results in the past twenty years.

The information gathered in the last few years is not suitable for presentation in a second course in dynamics. The theoretical advances require very highly sophisticated methods of analysis, and the numerical results from computed solutions are quite unsuitable for presentation in a text for self-evident reasons. The reader interested in the three-body problem is referred to an admirable and exhaustive book by Victor Szebehely.[†]

19.8. Problems

19.1. A particle of mass M is connected by an inextensible massless string to a heavy particle of mass m; the string passes through a smooth hole in a horizontal table. The particle of mass M lies on the table, and the other particle is constrained to move on a vertical line. Discuss the motion if the string on the table has some nonzero initial angular velocity.

19.2. Discuss the system of Example 16.6.1 when the two particles are interconnected by a massless, linear spring of rate $k = 1$.

19.3. A particle of unit mass is attached to one extremity of a linear spring, the other extremity being hinged at a fixed point in inertial space. What constants of the motion exist? Find the motion in quadratures and show that finite orbits are closed.

[†] Victor Szebehely, *Theory of Orbits*, Academic Press, New York and London (1967).

19.4. A particle is attracted to a force center by a force which varies inversely as the cube of its distance from the center. Discuss the motion.

19.5. A particle of mass m, lying on a smooth, horizontal table, is connected to a spring under the table by means of a massless inextensible string which passes through a hole in the table. The spring resists a length change u in the amount $f(u)$, where $f(u)$ is of class C^1. When the spring is unstretched and the string is taut, the distance between the particle and the hole is l. If the particle is set into motion in any way which stretches the spring initially, reduce the solution to quadratures.

19.6. Prove that the following particular solution of the three-body problem is correct: Three equal masses rotate as corners of a constant equilateral triangle with constant angular velocity about a fixed line which passes normal by to the plane of the masses through the mass center.

19.7. Lagrange proved that there exists another solution to Problem 19.6, in which the equilateral triangle remains similar to itself. Show that this is also a solution.

19.8. Let the central force be

$$F(r) = \frac{M}{r^2} + \frac{\varepsilon}{r^3}$$

where M and ε are constants. Discuss the motion for small $|\varepsilon|$.

19.9. What is the general solution to the central force problem when the central force is that of Problem 19.8.

19.10. In the central force problem it was implicitly assumed in (19.2.9) that $K^2 \neq 0$. Discuss the planetary motion when $K^2 = 0$.

20

Topics in Gyrodynamics

20.1. Introduction

In a broad sense, gyrodynamics is the term used for the dynamics of rotation of rigid bodies. In a narrower sense, it is usually understood to describe the theory of the gyroscope and of the heavy symmetrical top.

The gyroscope is a fascinating device which exhibits extraordinary properties. Try to move it about, and it rebels; even when it yields, it seems to have a mind of its own and tends to move in ways other than intended. Because of these properties it has been used as the basic element in many toys.

There exists much specialized literature on the gyroscope. The classical work is a four-volume treatise[†] by F. Klein and A. Sommerfeld, which appeared between 1897 and 1910. R. Grammel wrote an excellent two-volume work[‡] dealing both with the theory and with many interesting applications. A number of special books and monographs on the subject have appeared since, and even now papers dealing with gyrodynamics appear frequently in the current professional literature, particularly in the USSR.

The importance of the gyroscope lies both in the intrinsic interest of the subject and in its technical applications. Some of the older applications include the ship stabilizer, the gyroscopic compass, and the artificial horizon. However, renewed interest in gyroscopic devices and instruments was

[†] F. Klein and A. Sommerfeld, *Über die Theorie des Kreisels*, B. G. Teubner, Leipzig (1897–1910).

[‡] R. Grammel, *Der Kreisel*, Springer, Göttingen (1950).

awakened by the technology of rockets and space vehicles. Among the modern applications are the rate gyro, the integrating gyro, and the so-called "stable platform," which maintains an inertial reference triad when mounted on a moving space vehicle.

The subject of gyrodynamics is too basic to be ignored in a second course in dynamics; yet, many of its applications are too specialized to be discussed here. Therefore, we restrict our treatment to some topics pertaining to the gyroscope and the heavy symmetrical top.

20.2. The Heavy Symmetrical Top

Consider a heavy rigid body with one point, *not* the mass center, fixed in inertial space. Let x, y, z be body-fixed principal axes with origin at the fixed point, and let the principal moments of inertia be

$$I_x, \quad I_y, \quad I_z.$$

Then, if the angular velocity vector is

$$\omega = (\omega_x, \omega_y, \omega_z),$$

the kinetic energy of the rotating body is [in an inertial frame; see (7.4.11)]

$$T = \tfrac{1}{2}[I_x\omega_x^2 + I_y\omega_y^2 + I_z\omega_z^2]. \tag{20.2.1}$$

We assume that the body has an axis of rotational (geometric and dynamic) symmetry as, for instance, in a body of revolution having homogeneous density. This is the property which is understood by the term "symmetrical" top. Clearly, such an axis of rotational symmetry is a principal axis because the products of inertia relative to it and either of the other two orthogonal axes are zero. Let the axis of rotational symmetry be the z axis. Then we write

$$I_x = I_y = I, \qquad I_z = J, \tag{20.2.2}$$

and the kinetic energy becomes

$$T = \tfrac{1}{2}[I(\omega_x^2 + \omega_y^2) + J\omega_z^2]. \tag{20.2.3}$$

We shall use Euler angles to describe the configuration. Thus, we substitute (6.8.14) in (20.2.3). We find

$$T = \tfrac{1}{2}[I(\dot\theta^2 + \dot\varphi^2 \sin^2\theta) + J(\dot\psi + \dot\varphi\cos\theta)^2]. \tag{20.2.4}$$

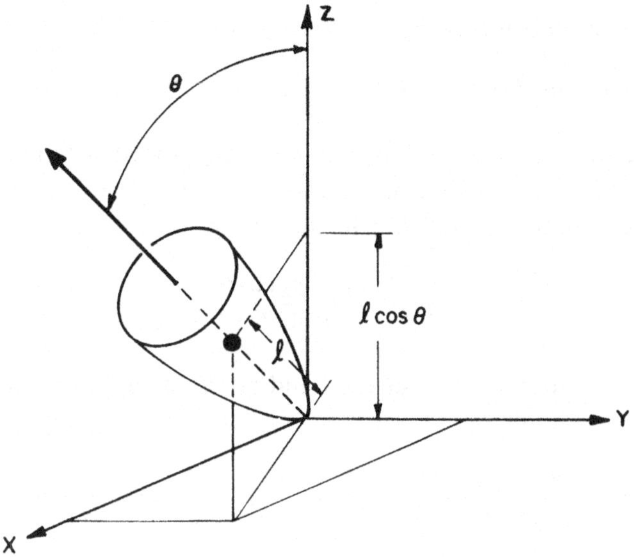

Fig. 20.2.1. Heavy symmetrical top.

If the mass center is a distance l from the fixed point, the potential energy is

$$V = mgl \cos \theta, \tag{20.2.5}$$

where $l \cos \theta$ is the height of the mass center above the fixed point, as shown in Fig. 20.2.1.

The gravitational force is the only given force. As it is a potential force and the problem is catastatic, the energy integral exists, i.e.,

$$T + V = h = \text{const} \tag{20.2.6}$$

with T and V defined in (20.2.4) and (20.2.5), respectively.

It is seen that φ and ψ are ignorable coordinates. Therefore,

$$\begin{aligned} p_\varphi &= \frac{\partial T}{\partial \dot{\varphi}} = I\dot{\varphi} \sin^2 \theta + J(\dot{\psi} + \dot{\varphi} \cos \theta) \cos \theta = p_1, \\ p_\psi &= \frac{\partial T}{\partial \dot{\psi}} = J(\dot{\psi} + \dot{\varphi} \cos \theta) = p_2, \end{aligned} \tag{20.2.7}$$

where p_1 and p_2 are constants. We form the Routhian function (16.3.17), which is here

$$R = \dot{\varphi} p_1 + \dot{\psi} p_2 - (T - V),$$

and the equation satisfied by the Routhian is

$$I\ddot{\theta} - I\dot{\varphi}^2 \sin\theta\cos\theta + J\dot{\varphi}(\dot{\psi} + \dot{\varphi}\cos\theta)\sin\theta = mgl\sin\theta. \tag{20.2.8}$$

Evidently, three of the six integrals have already been found in (20.2.6) and (20.2.7). Because of them, the variables are separable. Substitution of the second equation of (20.2.7) in the first gives

$$\dot{\varphi} = \frac{p_1 - p_2\cos\theta}{I\sin^2\theta}, \tag{20.2.9}$$

and the substitution of this equation and of the second relation of (20.2.7) in (20.2.8) results in

$$I\ddot{\theta} - \frac{p_1 - p_2\cos\theta}{I\sin^3\theta}(p_1\cos\theta - p_2) - mgl\sin\theta = 0, \tag{20.2.10}$$

while substitution in the energy integral gives

$$\frac{1}{2}\left\{I\left[\dot{\theta}^2 + \left(\frac{p_1 - J\omega_z\cos\theta}{I\sin^2\theta}\right)^2\sin^2\theta\right] + J\omega_z{}^2\right\} + mgl\cos\theta = h, \tag{20.2.11}$$

where we have used the relation

$$p_2 = J\omega_z = J(\dot{\psi} + \dot{\varphi}\cos\theta) = \text{const.}$$

It is obvious that (20.2.11) is a first integral of (20.2.10). It is reducible to the quadrature

$$t = \int\left\{\frac{2}{I}\left[h - mgl\cos\theta - \frac{1}{2}J\omega_z{}^2 - \frac{1}{2}\,\frac{(p_1 - J\omega_z\cos\theta)^2}{I\sin^2\theta}\right]\right\}^{-1/2} d\theta. \tag{20.2.12}$$

When the evaluation $t(\theta)$ of this integral is inverted to $\theta(t)$ (assuming the inversion exists), one finds from (20.2.9) as applications of (16.3.21)

$$\varphi = \int \frac{p_1 - p_2\cos\theta(t)}{I\sin^2\theta(t)}\,dt = \varphi(t),$$

and from the second equation of (20.2.7)

$$\psi = \int\left[\frac{p_2}{J} - \dot{\varphi}(t)\cos\theta(t)\right]dt = \psi(t).$$

Therefore, the integration of the equations of motion centers on the discussion of the integral in (20.2.12). This discussion will now be given. If one puts

$$u = \cos \theta \tag{20.2.13}$$

and defines constants

$$\begin{aligned} a &= \frac{2mgl}{I}, \qquad & \alpha &= \frac{2h}{I} - \frac{J\omega_z^2}{I}, \\ \beta &= p_1/I, \qquad & \gamma &= J\omega_z/I = p_2/I, \end{aligned} \tag{20.2.14}$$

(20.2.11) takes on the form

$$\dot{u}^2 = (1 - u^2)(\alpha - au) - (\beta - \gamma u)^2 = f(u), \tag{20.2.15}$$

and (20.2.9) becomes

$$\dot{\varphi} = \frac{\beta - \gamma u}{1 - u^2}. \tag{20.2.16}$$

One sees from (20.2.15) that

$$t = \int \frac{du}{[f(u)]^{1/2}}, \tag{20.2.17}$$

where $f(u)$ is a cubic polynomial.

It was pointed out in Example 16.5.1 in connection with (16.5.13) that the right-hand side of (20.2.17) is an elliptic integral of the first kind and is tabulated. Moreover, inversions of the elliptic integrals exist [see the remark following (20.2.12)] and are also tabulated; they are called elliptic functions and are periodic. In what follows, we discuss some features of the motion of the top without developing the theory of elliptic functions.

We begin by studying the graph of $f(u)$, defined in (20.2.15). Inasmuch as $a > 0$, $f(u) \to \pm\infty$ according as $u \to \pm\infty$. Also, when $u = \pm 1$,

$$f(\pm 1) = -(\beta \mp \gamma)^2 \leq 0.$$

But, if $f(+1) \leq 0$ and $f(\infty) = \infty$, $f(u)$ has a zero in $1 \leq u < \infty$. Moreover, if it has three real zeros, the other two lie necessarily in the interval $-1 \leq u \leq 1$ because $f(-1) \leq 0$, and $f(-\infty) = -\infty$. Therefore, in general, the graph of $f(u)$ looks as shown in Fig. 20.2.2, where it is assumed that three real zeros u_1, u_2, and u_3 exist, which satisfy

$$-1 \leq u_1 \leq u_2 \leq 1 \leq u_3.$$

It is evident from the definition (20.2.13) of u that this variable is bounded

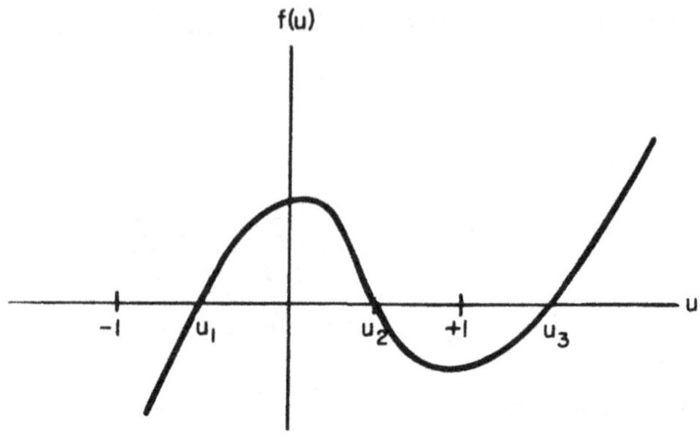

Fig. 20.2.2. The function $f(u)$.

by $-1 \leq u \leq 1$, and from the graph of Fig. 20.2.2 that, in that interval, $f(u)$ is nonnegative only in $u_1 \leq u \leq u_2$. Then, as $\dot{u}$ must be real, one sees from (20.2.15) and Fig. 20.2.2 that the motion must occur in the interval where $f(u)$ is nonnegative.

We shall discuss a few special cases. Let us suppose that u_1 and u_2 coincide. In Fig. 20.2.3 we show

Case (a). The double zero occurs at $u < 1$;

Case (b). The double zero occurs at $u = 1$.

Let

$$u_1 = u_2 = u_0.$$

Then, to that value of u_0 there corresponds an angle θ_0 defined by

$$\theta_0 = \cos^{-1} u_0.$$

This is the only value which θ can take on because for $\theta \neq \theta_0$, the derivative

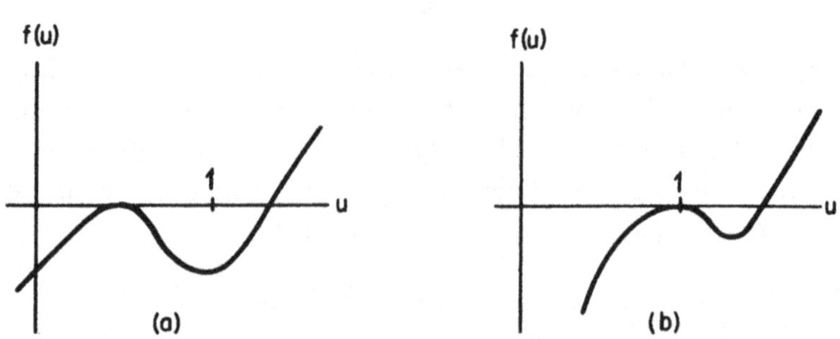

Fig. 20.2.3. Two possible cases of double zeros of $f(u)$.

$\dot{u}$ is imaginary when $u < 1$. Let us rewrite (20.2.8) in terms of the constants (20.2.14); it is

$$\ddot{\theta} - \dot{\varphi}^2 \sin\theta \cos\theta + \gamma\dot{\varphi} \sin\theta = \tfrac{1}{2}a \sin\theta.$$

But, when $\theta = \theta_0 = \text{const}$, this equation becomes

$$\dot{\varphi}^2 \cos\theta_0 - \gamma\dot{\varphi} + a/2 = 0. \tag{20.2.18}$$

The roots of this equation are

$$\dot{\varphi}_{1,2} = \frac{\gamma}{2\cos\theta_0}\left[1 \pm \left(1 - \frac{2a\cos\theta_0}{\gamma^2}\right)^{1/2}\right],$$

and they will be real if

$$\gamma^2 \geq 2a, \tag{20.2.19}$$

or, in physical terms, if

$$n^2 = (\dot{\psi} + \dot{\varphi}\cos\theta_0)^2 \geq \frac{4Imgl}{J^2}.$$

This condition is always satisfied for a sufficiently large spin velocity $\dot{\psi}$. We shall assume throughout that

$$\gamma^2 \geq 2a \qquad \text{or} \qquad n^2 \geq \frac{4Imgl}{J^2}.$$

The φ motion is called "precession," and the case $\theta = \theta_0 = \text{const}$ and $\dot{\varphi} = \text{const}$ is called "steady precession."

When the equality holds in (20.2.19), there is only one precessional speed $\dot{\varphi}_1$, but when the inequality holds, there is a fast and a slow precession. The phenomenon usually seen is the slow precession because it requires less kinetic energy.

The motion corresponding to case (a) is conveniently shown by placing the fixed point of the top at the center of a unit sphere fixed in inertial space, and by imagining the z axis to pierce the sphere, as shown in Fig. 20.2.4. Then the motion may be described by rotation about the z axis and by the curve which this axis traces on the sphere. For the case considered, the inclination θ_0 of the top is a constant, and the curve traced by the z axis on the sphere is a circle. The point tracing this circle moves with angular velocity $\dot{\varphi} = \text{const}$ along it. In case (b), $u_0 = 1$ or $\theta_0 = 0$. Then, the z axis points vertically up. This is called the "sleeping top."

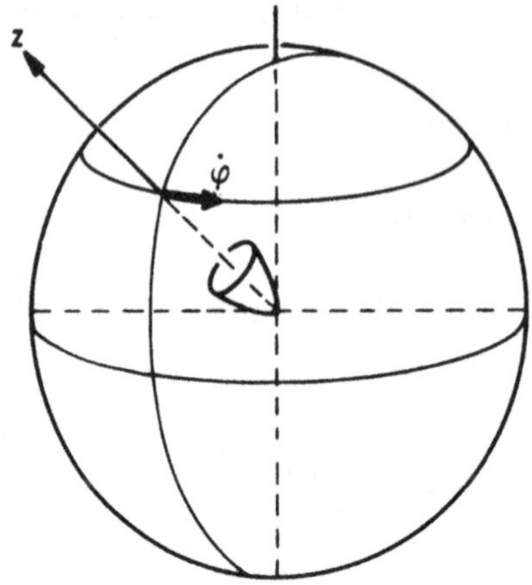

Fig. 20.2.4. Steady precession.

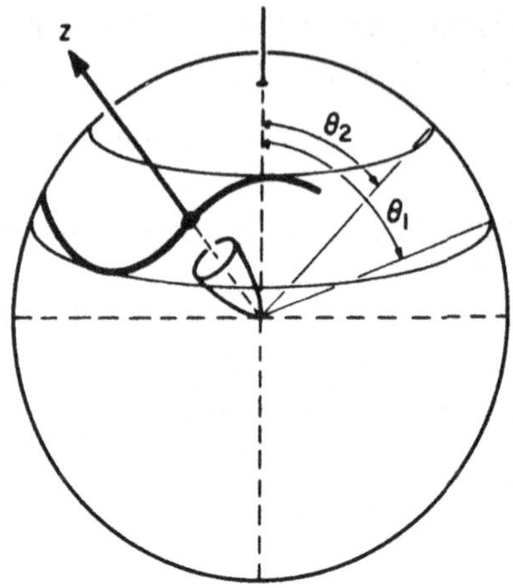

Fig. 20.2.5. Motion of top with precession and nutation.

When $f(u)$ has two real zeros in $-1 < u < 1$, the values of u must lie between u_1 and u_2, i.e., θ must lie in the corresponding interval $\theta_1 \geq \theta \geq \theta_2$, where

$$\theta_1 = \cos^{-1} u_1, \qquad \theta_2 = \cos^{-1} u_2.$$

This is shown in Fig. 20.2.5.

We consider now a special motion called "cuspidal." We release the spinning top from an initial angle of inclination $\theta_1 \neq 0$ without initial θ velocity and φ velocity, i.e.,

$$\theta(0) = \theta_1 \neq 0, \qquad \dot\theta(0) = \dot\varphi(0) = 0.$$

Then, to $\theta = \theta_1$ there corresponds

$$u_1 = \cos^{-1} \theta_1$$

and, since $\dot\varphi(0) = 0$, one finds from (20.2.16)

$$\beta = \gamma u_1. \tag{20.2.20}$$

Moreover, $\dot\theta(0) = 0$ implies $\dot u(0) = 0$. Then one finds from (20.2.15) and (20.2.20) that initially

$$(1 - u_1^2)(\alpha - a u_1) = 0,$$

and, since $\theta_1 \neq 0$, or $u_1 \neq 1$, one also has

$$\alpha = au_1. \tag{20.2.21}$$

When these are substituted back into (20.2.15), there results

$$f(u) = (1 - u^2)a(u_1 - u) - \gamma^2(u_1 - u)^2, \tag{20.2.22}$$

and the zeros of $f(u)$ define the values of θ_1 and θ_2 in Fig. 20.2.5. Obviously, one of these zeros is $u = u_1$. Dividing (20.2.22) by this zero, the other two are the roots of

$$(1 - u^2)a - \gamma^2(u_1 - u) = 0$$

or

$$u_{2,3} = \frac{\gamma^2}{2a} \pm \left(\frac{\gamma^4}{4a^2} - \frac{\gamma^2}{a} u_1 + 1\right)^{1/2}.$$

The larger of these two is known to lie in $u > 1$ and is, therefore, excluded. Hence, the root of physical significance is

$$u_2 = \frac{\gamma^2}{2a} - \left(\frac{\gamma^4}{4a^2} - \frac{\gamma^2}{a} u_1 + 1\right)^{1/2}, \tag{20.2.23}$$

or, in terms of the angles,

$$\cos\theta_2 = \frac{\gamma^2}{2a} - \left(\frac{\gamma^4}{4a^2} - \frac{\gamma^2}{a} \cos\theta_1 + 1\right)^{1/2}. \tag{20.2.24}$$

Therefore, the z axis falls from an inclination θ_1 to an inclination θ_2 and then rises again to θ_1, where it is instantaneously at rest before it falls again.

From (20.2.16) we find

$$\dot{\varphi} = \frac{\gamma(u_1 - u)}{1 - u^2}, \tag{20.2.25}$$

where we have utilized (20.2.20). Now, $u_1 \leq u \leq u_2$, and $u^2 < 1$ for the case considered. Hence, $\dot{\varphi}$ is always positive except that it vanishes when $u = u_1$. The resulting motion is illustrated in Fig. 20.2.6. The rise and fall of the z axis is called the "nutation." Precession is produced by the Coriolis torque, familiar from a first course in mechanics. As the axis of symmetry of the spinning top falls, the moment of the Coriolis force produces the precession. The motion just examined is called cuspidal because the trajectory of the z axis intercept with the sphere has a cusp at the top.

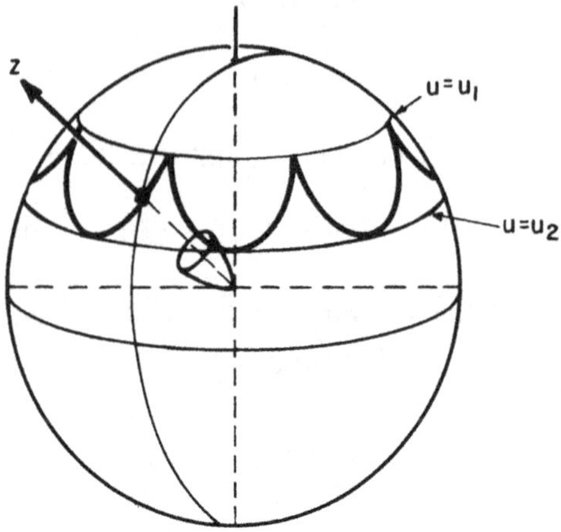

Fig. 20.2.6. Cuspidal motion of top.

The diagram of Fig. 20.2.6 illustrates the utility of describing the motion of the top in terms of the precession, rotation, and spin. The cuspidal motion is a limiting case of the motions "without loop" and "with loop," illustrated in Figs. 20.2.7 and 20.2.8. The motion without loop may be produced by initial conditions $\theta(0) = \theta_1$, $\dot{\theta}(0) = 0$, $\dot{\varphi}(0) = \dot{\varphi}_0 > 0$, and the motion with loop results from the same initial conditions except $\dot{\varphi}_0 < 0$. For a complete discussion of the motion, the reader is referred to the excellent books by A. G. Webster or Osgood, or to other complete treatments of this subject.

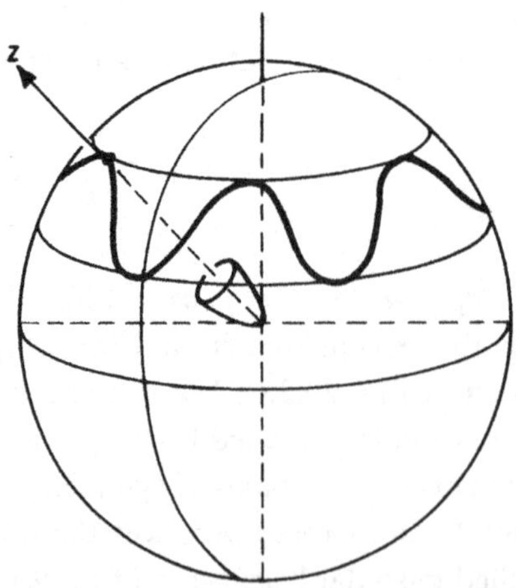

Fig. 20.2.7. Motion without loops.

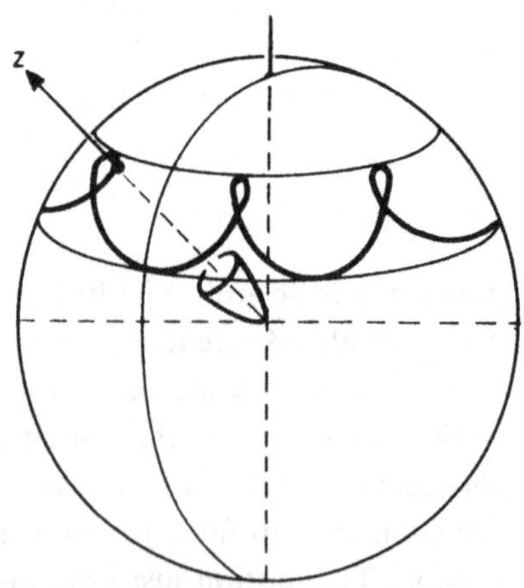

Fig. 20.2.8. Motion with loops.

20.3. The Gyroscope

The gyroscope is shown in Fig. 20.3.1. It is a thin, heavy, circular, homogeneous disk which spins about its axis of rotational symmetry. This axis is mounted in a ring called a gimbal, and the disk can rotate with respect to the gimbal. The gimbal can rotate about an axis in its own plane normal to that of the disk. It, in turn, is mounted in a second gimbal, which is free to rotate about an axis in its own plane normal to that of the first, and this axis is fixed in inertial space. This type of mounting is called a Cardan suspension; it fixes the mass center in inertial space, but permits rotation about any axis through the mass center. If we define, from the last equation of (20.2.7),

$$n = p_2/J, \tag{20.3.1}$$

we find from (20.2.8)

$$I\ddot{\theta} - I\dot{\varphi}^2 \sin\theta \cos\theta + J\dot{\varphi}n \sin\theta = M_\theta, \tag{20.3.2}$$

and from the derivative of the first equation of (20.2.7),

$$I\ddot{\varphi} \sin^2\theta + 2I\dot{\varphi}\dot{\theta} \sin\theta \cos\theta - Jn\dot{\theta} \sin\theta = M_\varphi, \tag{20.3.3}$$

where M_θ and M_φ are the components of some applied torque.

To simplify the treatment let us suppose that the spin $|\dot{\psi}|$ is so large compared to $|\dot{\theta}|$ and $|\dot{\varphi}|$ that second-order terms in $\dot{\theta}$ and $\dot{\varphi}$ may be

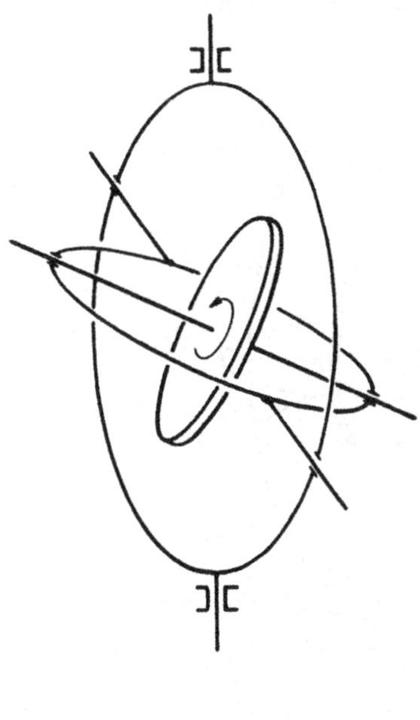

Fig. 20.3.1. Gyroscope.

neglected relative to $\dot{\psi}$. Then, we find in place of (20.3.2) and (20.3.3)

$$\begin{aligned} I\ddot{\theta} + Jn\dot{\varphi}\sin\theta &= M_\theta, \\ J\ddot{\varphi}\sin^2\theta - Jn\dot{\theta}\sin\theta &= M_\varphi. \end{aligned} \tag{20.3.4}$$

Again, as in (18.3.11), the velocity-dependent terms $Jn\dot{\varphi}\sin\theta$ and $-Jn\dot{\theta}\sin\theta$ are not derivable from a dissipating function and, hence, are gyroscopic forces. Here, they are seen to arise naturally in the treatment of the gyroscope, and this has given the name to these forces.

Let us ask what torque must be applied to have steady precession, e.g., $\theta = \theta_0 = \text{const}$, $\dot{\varphi} = \dot{\varphi}_0 = \text{const}$. On substituting these in the above equation, we find

$$M_\theta = Jn\dot{\varphi}_0 \sin\theta_0, \qquad M_\varphi = 0. \tag{20.3.5}$$

Therefore, the torque, such as that produced by the gravity force in the heavy symmetrical top, is a vector lying in the line of nodes [see Section 6.8(a)]. The interpretation of this result is illustrated in Fig. 20.3.2. If one wishes to give the outer gimbal a steady angular velocity $\dot{\varphi}_0$ about the vertical axis, the inner gimbal tends to rotate about its axis. Then it requires a torque given in (20.3.5) to prevent that rotation, and to hold the inner gimbal at the steady angle θ_0.

This effect is utilized in the so-called "stable platform." Consider Fig. 20.3.3. A gyroscope is mounted in a single gimbal. The gimbal in turn is mounted on a platform which can rotate about an axis normal to that of the gimbal. Denote the angular velocity of the platform by $\dot{\varphi}$, and that

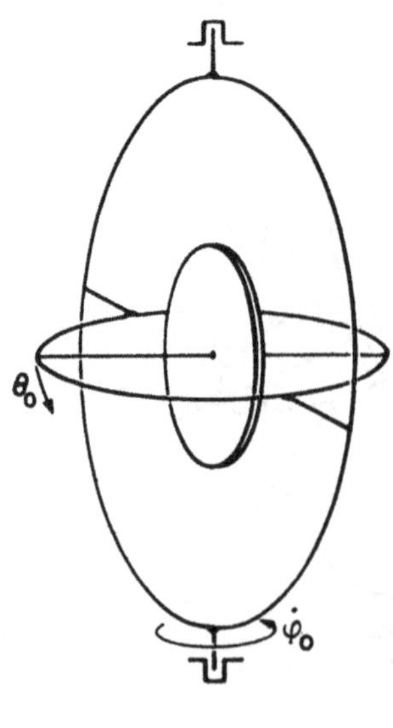

Fig. 20.3.2. Motion of outer gimbal induced by motion of inner gimbal.

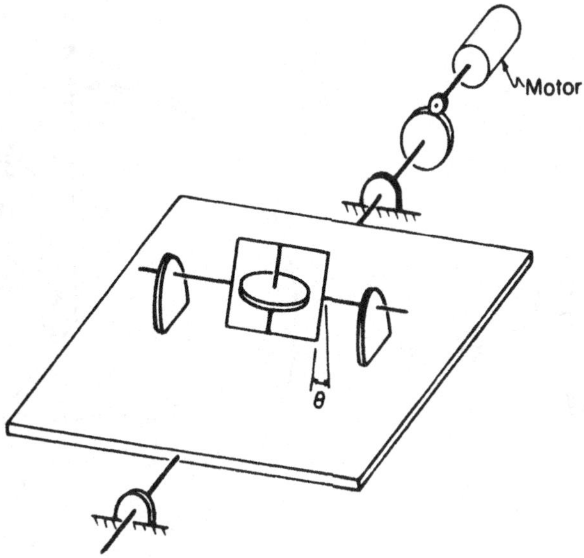

Fig. 20.3.3. One component of stable platform.

of the gimbal by $\dot{\theta}$. An angular velocity $\dot{\varphi}$ produces an angular velocity $\dot{\theta}$; this is the effect that has just been discussed. Now, the gimbal is equipped with an electric sensing device reading changes in θ. It in turn drives a servo motor which produces a torque to counteract that which caused the platform rotation. As one gyroscope can make corrections about one axis, three gyroscopes can maintain an inertia-fixed orientation.

For this and other interesting applications of gyrodynamics the reader is referred to Thomson's book.[†]

20.4. The Gyrocompass

The gyrocompass is a gyroscopic instrument which indicates north without making use of the magnetic properties of the earth. Instead it utilizes the tendency of the gyroscope to align its axis of spin parallel to the axis of rotation of a rotating vehicle on which it is mounted. In the gyrocompass, this rotating vehicle is the earth.

A gyrocompass is a gyroscope arranged in such a way that its axis of spin is free to move in a "horizontal" plane, i.e., a plane tangent to the earth's surface, but the spin axis cannot rise out of this plane, as shown in

[†] Thomson, W. T., *Introduction to Space Dynamics*, John Wiley and Sons, New York and London (1961).

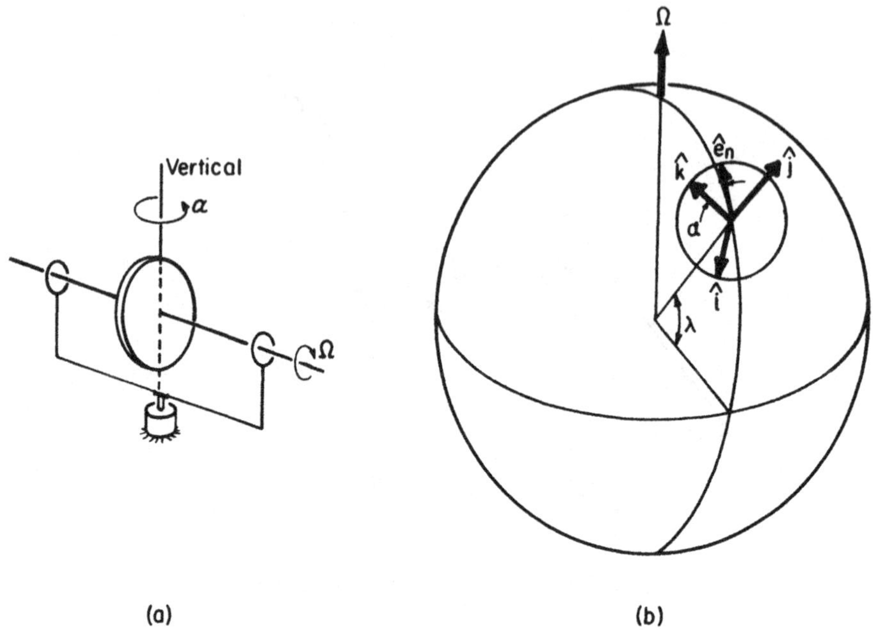

Fig. 20.4.1. (a) Model of gyrocompass; (b) coordinates used in analysis.

Fig. 20.4.1(a). Consider Fig. 20.4.1(b), where the gyroscope is at a latitude of λ degrees north. The $\hat{\imath}$, $\hat{\jmath}$, $\hat{k}$ are unit vectors fixed in the gyroscope such that the axis of spin lies in the $\hat{k}$ direction, and the $\hat{\jmath}$ direction points away from the center of the earth, i.e., it lies in the local vertical, positive upward. The earth's angular velocity is Ω, $\hat{e}_n$ is the unit vector pointing north, and α is the angle between the axis of spin of the gyroscope and the northerly direction.

One sees from this illustration that the component of Ω in the $\hat{\jmath}$ direction is $\Omega \sin \lambda$, and that in the northerly $\hat{e}_n$ direction is $\Omega \cos \lambda$. The component of $\Omega \cos \lambda$ in the $\hat{\imath}$-direction is $-\Omega \cos \lambda \sin \alpha$, and that in the $\hat{k}$ direction is $\Omega \cos \lambda \cos \alpha$.

Therefore, the earth's velocity in components along the body-fixed triad is

$$\Omega = -\Omega \cos \lambda \sin \alpha \hat{\imath} + \Omega \sin \lambda \hat{\jmath} + \Omega \cos \lambda \cos \alpha \, \hat{k}.$$

Then, if the spin axis has the angular velocity $\dot{\alpha}\hat{\jmath}$, the combination of the earth's spin and the α motion produces the angular velocity vector

$$\omega' = -\Omega \cos \lambda \sin \alpha \hat{\imath} + (\dot{\alpha} + \Omega \sin \lambda)\hat{\jmath} + \Omega \cos \lambda \cos \alpha \hat{k}. \qquad (20.4.1)$$

Let s denote the $\hat{k}$ component of the angular velocity of the gyrocompass including the spin velocity and the $\hat{k}$ component of the earth's angular velocity about the polar axis. Then, the angular velocity of the gyroscope is

$$\omega = -\Omega \cos \lambda \sin \alpha \hat{\imath} + (\dot{\alpha} + \Omega \sin \lambda)\hat{\jmath} + s\hat{k}, \tag{20.4.2}$$

and the angular momentum is

$$M = -I\Omega \cos \lambda \sin \alpha \hat{\imath} + I(\dot{\alpha} + \Omega \sin \lambda)\hat{\jmath} + Js\hat{k}. \tag{20.4.3}$$

The moment which prevents the spin axis from rising out of the horizontal plane is a vector in the $\hat{\imath}$ direction; let it be denoted by $N_x\hat{\imath}$. Then, equating the time rate of change of angular momentum to this moment, one has

$$-I\Omega \cos \lambda \cos \alpha\dot{\alpha}\hat{\imath} + I\ddot{\alpha}\hat{\jmath} + J\dot{s}\hat{k} + \omega' \times M = N_x\hat{\imath}.$$

The $\hat{\imath}$, $\hat{\jmath}$ and $\hat{k}$ components of this equation are, respectively,

$$\begin{aligned} (Js - 2I\Omega \cos \lambda \cos \alpha)\dot{\alpha} - I\Omega^2 \sin \lambda \cos \lambda \cos \alpha + Js\Omega \sin \lambda &= N_x, \\ I\ddot{\alpha} - I\Omega^2 \cos^2 \lambda \sin \alpha \cos \alpha + Js\Omega \cos \lambda \sin \alpha &= 0, \\ J\dot{s} &= 0. \end{aligned} \tag{20.4.4}$$

Now, the angular velocity of the earth is 7.29×10^{-5} radians per second while the spin velocity of a gyroscope is on the order of 10^2 radians per second, or approximately one hundred million times as great. Therefore, we shall regard Ω as an infinitesimal and ignore its powers greater than the first. We shall also assume that the angle α is so small that its powers higher than the first may be deleted. This is equivalent to assuming that the spin axis points nearly in a northerly direction at all times. Under these simplifying assumptions equations (20.4.4) become

$$\begin{aligned} (Js - 2I\Omega \cos \lambda)\dot{\alpha} + Js\Omega \sin \lambda &= N_x, \\ I\ddot{\alpha} + Js\Omega \cos \lambda\alpha &= 0, \\ J\dot{s} &= 0. \end{aligned} \tag{20.4.5}$$

We deduce from the last of these that the spin is constant. In consequence, the second may be written as

$$\ddot{\alpha} + \omega^2\alpha = 0, \tag{20.4.6}$$

where

$$\omega^2 = Js\Omega \cos \lambda/I = \text{const.} \tag{20.4.7}$$

Therefore, the spin axis oscillates in simple harmonic motion about the northerly direction, thus indicating true north. Once α has been found from (20.4.6), the moment N_x is given by the first equation of (20.4.5).

Exercise. It appears from (20.4.7) that ω^2 changes sign with s; hence if it is positive for one direction of spin, it is negative when the spin direction is reversed. Therefore, the gyroscopic compass will only work if the spin direction is properly selected. This is not the case. Explain this apparent contradiction.

Experience has shown that the most easily understood treatment of gyrodynamics utilizes Euler's equation of motion of rigid bodies about a point, rather than the Lagrangean formulation. The reader interested in a clear treatment of gyrodynamics is referred to Thomson, W. T., *Introduction to Space Dynamics*† and R. F. Deimel, *Mechanics of the Gyroscope.*‡

20.5. Problems

The problems of this chapter are sometimes formulated most easily by utilizing the Euler equations (7.4.8).

20.1. A homogeneous circular disk is pivoted about its center. It is set spinning with angular velocity ω about a line making an angle α with the normal to the disk. Find the time as a function of ω and α required by the axis of the disk to describe a cone in space.

20.2. When a force-free homogeneous disk is set spinning in a plane which coincides nearly with the plane of the disk, the disk makes two wobbles for every cycle of spin. Prove this result.

20.3. A thin, homogeneous circular disk is rigidly mounted on a rigid shaft so that the normal to the disk makes a least angle α with the shaft. If the shaft is mounted in fixed bearings a distance l apart and is driven at the angular speed ω, what are the bearing reactions if the disk is halfway between the bearings?

20.4. Answer the same question as in Problem 20.3 when the disk is replaced by a homogeneous, rectangular plate of sides a and b and mass m, and when the shaft follows one of the diagonals. [Note that $\alpha = \tan^{-1}(a/b)$.]

20.5. The center of a thin uniform disk of radius r and mass m is attached to one extremity of a thin rod of length l so that the rod is normal to the plane of the disk, as shown. The other end of the rod is smoothly hinged

† John Wiley and Sons, Inc., New York (1961).

‡ The Macmillan Book Co., New York (1929); reprinted by Dover Publications (1950).

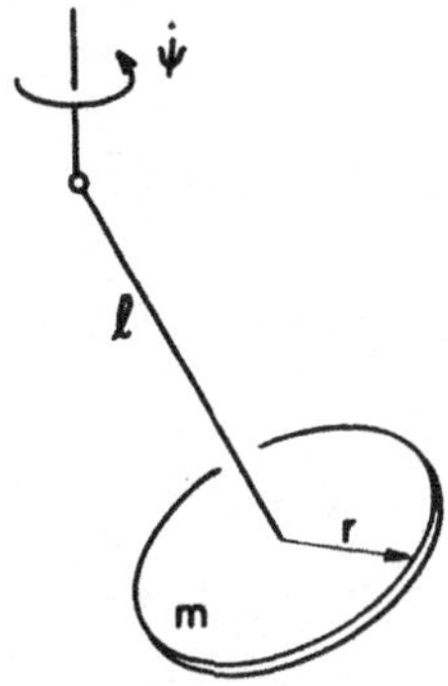

to a vertical shaft which is driven at the angular speed $\dot{\psi}$. Show that the motion is stable for all $\dot{\psi}$ if $r \geq 2l$, and it is unstable for $\dot{\psi}^2 > 4gl/(4l^2 - r^2)$ if $r < 2l$.

20.6. What do the Euler equations imply for the case of a heavy spherical top as regards constants of the motion?

20.7. Let a top consist of an infinitely thin disk of radius r, mounted midway on a massless rod of length l so that the rod is normal to the plane of the disk. One end of the rod is fixed at a point, the other is free. A constant force $\boldsymbol{P}$ acts for all time through a point on the rim of the disk parallel to the rod. Calculate the reaction at the fixed end.

20.8. Show how the Euler equations for the force-free top may be found from the Lagrange equations.

20.9. Show that for sufficiently large spin velocity, the fast velocity of steady precession is $\dot{\varphi}_1 = \gamma/\cos\theta_0$ and the slow one is $\dot{\varphi}_2 = a/2\gamma$. Can you explain why the slow precession is independent of the nutation angle θ_0?

20.10. The condition of the sleeping top can be satisfied by either of the curves shown. Show that (a) is stable and (b) unstable.

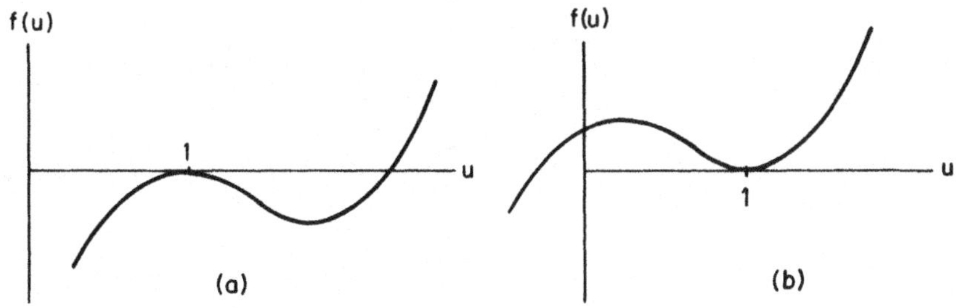

21

Impulsive Motion

21.1. General Remarks

All dynamical theory examined in detail up to this point has been that of strictly Newtonian systems (SN), in which all forces are bounded for every value of time. In this chapter we study Newtonian systems (N) which are not (SN). Therefore, the fundamental dynamical axiom is the axiom of bounded momentum.

It was shown in Section 2.3 that the systems (N) which are not (SN) satisfy Newton's second law "almost always," the exception being the instants t_j at which impulsive forces act, i.e., the t_j for which

$$\lim_{t \to t_j} \int_{t_j}^{t} F^r(x^r, \dot{x}^r, \tau)\, d\tau = P^r, \tag{21.1.1}$$

where $P^r \neq 0$ is a bounded vector. The instants t_j are isolated because of the smoothness properties of particle motion $x^r(t)$.

We shall write

$$\lim_{0<\tau_1 \to 0} f(t_j - \tau_1) = f(t_j - 0) \tag{21.1.2}$$

and refer to $f(t_j - 0)$ as the lower limit of f at t_j, and we write

$$\lim_{0<\tau_2 \to 0} f(t_j + \tau_2) = f(t_j + 0) \tag{21.1.3}$$

and refer to $f(t_j + 0)$ as the upper limit of f at t_j.

In consequence of the properties of particle motion, every particle has one and only one position $x^r(t_j)$ at t_j so that no discontinuous position change occurs at t_j. But, by the axiom of bounded momentum (2.3.1), the velocity changes discontinuously at t_j, and this discontinuity is finite. Let the velocity change be one from $\dot{x}^r(t_j - 0)$ to $\dot{x}^r(t_j + 0)$. Then, for motion in $t > t_j$ we use Newton's law with initial conditions $x^r = x^r(t_j)$, $\dot{x}^r = \dot{x}^r(t_j + 0)$.

Let the impulse (a given vector quantity) be denoted by

$$P^r(t_j) = \big(P_1{}^r(t_j), P_2{}^r(t_j), P_3{}^r(t_j)\big), \tag{21.1.4}$$

and the lower limit of the velocity by

$$\dot{x}^r(t_j - 0) = \big(\dot{x}_1{}^r(t_j - 0), \dot{x}_2{}^r(t_j - 0), \dot{x}_3{}^r(t_j - 0)\big). \tag{21.1.5}$$

The velocity $\dot{x}^r(t_j - 0)$ is known from the motion in t in the interval $[t_0, t_j]$. Then, the quantity

$$\dot{x}^r(t_j + 0) = \big(\dot{x}_1{}^r(t_j + 0), \dot{x}_2{}^r(t_j + 0), \dot{x}_3{}^r(t_j + 0)\big) \tag{21.1.6}$$

may be computed from the axiom of bounded momentum. One finds

$$\begin{aligned} m_r[\dot{x}_1{}^r(+) - \dot{x}_1{}^r(-)] &= P_1{}^r, \\ m_r[\dot{x}_2{}^r(+) - \dot{x}_2{}^r(-)] &= P_2{}^r, \\ m_r[\dot{x}_3{}^r(+) - \dot{x}_3{}^r(-)] &= P_3{}^r, \end{aligned} \tag{21.1.7}$$

where we have utilized the notation

$$\dot{x}_i{}^r(+) = \dot{x}_i{}^r(t_j + 0),$$

and a similar one for $\dot{x}_i{}^r(t_j - 0)$.

All theory of impulsive motion is based on these equations. The theory does not deal with the behavior of the system as time proceeds, but only with the behavior at the instant t_j. Therefore, (21.1.7) are algebraic, not differential, equations. They serve to compute $\dot{x}^r(+)$ from the known quantities $\dot{x}^r(-)$ and P^r.

Example 21.1.1. A particle of unit mass is acted on by constant gravitation. It is released from the origin of the xy coordinate system (y positive up) at $t = 0$ with an initial velocity $\dot{x}^r(0) = v\hat{\imath} + 2v\hat{\jmath}$. At the time $t_0 = v/2g$, it is subjected to an impulse $P^r = P\hat{\jmath}$. Find the trajectory.

From the equations of motion, we have for $0 \leq t \leq t_0(-)$,

$$\begin{aligned} \dot{x} &= v, & \dot{y} &= -gt + 2v, \\ x &= vt, & y &= -\tfrac{1}{2}gt^2 + 2vt. \end{aligned} \tag{a}$$

At the time $t_0 = v/2g$, we obtain

$$x(t_0) = x_0 = v^2/2g, \qquad y(t_0) = \tfrac{7}{8}v^2/g.$$

and

$$\dot{x}(-) = v, \qquad \dot{y}(-) = 3v/2.$$

Applying the formulas (21.1.7), we obtain

$$\dot{x}(+) = v, \qquad \dot{y}(+) = P + 3v/2.$$

Therefore, in $t_0 \leq t < \infty$,

$$x = vt, \qquad y = -\tfrac{1}{2}gt^2 + (P + 2v)t - Pv/2g. \tag{b}$$

Eliminating t from (a), one finds the trajectory in $0 \leq x \leq v^2/2g$ as

$$y = -\frac{1}{2}g\frac{x^2}{v^2} + 2x, \tag{c}$$

and eliminating t from (b), one finds the trajectory in $v^2/2g \leq x < \infty$,

$$y = -\frac{1}{2}g\frac{x^2}{v^2} + \left(\frac{P}{v} + 2\right)x - \frac{Pv}{2g}. \tag{d}$$

21.2. The Fundamental Equation

There exists a fundamental equation of impulsive motion similar to that of the motion of strictly Newtonian systems.

Equation (2.3.1) states that the motion of every particle of a system is governed by

$$m_r[\dot{x}^r(t) - \dot{x}^r(t_0)] = \int_{t_0}^{t} F^r\,d\tau \tag{21.2.1}$$

on any interval $t_0 \leq t < \infty$. In this equation F^r is the resultant of all forces acting on the rth particle at the time τ.

For the entire system, one has (with n the number of particles)

$$\sum_{r=1}^{n}\left\{m_r[\dot{x}^r(t) - \dot{x}^r(t_0)] - \int_{t_0}^{t} F^r\,d\tau\right\} = \sum_{r=1}^{n}\int_{t_0}^{t} F^{r'}\,d\tau, \tag{21.2.2}$$

and in (21.2.2), F^r is the resultant of the *given* forces acting on the rth particle at the time $t = \tau$ while $F^{r'}$ is the resultant of the *constraint* forces acting on the rth particle at $t = \tau$. Then, the *fundamental equation* is

$$\sum_{r=1}^{n}\left\{m_r[\dot{x}^r(t) - \dot{x}^r(t_0)] - \int_{t_0}^{t} F^r\,d\tau\right\}\cdot\delta x^r = \sum_{r=1}^{n}\int_{t_0}^{t} F^{r'}\,d\tau\cdot\delta x^r. \tag{21.2.3}$$

This equation is found from the fundamental equation (9.3.11) by integrating over $[t_0, t]$.

It is clear that, in the *unconstrained* system, the right-hand side of (21.2.3) vanishes because $F^{r'} = 0$ $(r = 1, 2, \ldots, n)$, and in the *constrained* system, it vanishes as well because we may write

$$\int F^{r'}\, d\tau \cdot \delta x^r = \int F^{r'} \cdot \delta x^r\, d\tau, \tag{21.2.4}$$

because δx^r is not a function of τ.

This shows that, in the constrained system, the right-hand side of (21.2.3) is merely the time integral of the virtual work done by the constraint forces, and that work is zero by definition.

We introduce the notation

$$\int_{t_0}^{t} F^r\, d\tau = P^r, \tag{21.2.5}$$

where P^r is called the impulse of the force in the time interval $[t_0, t]$. Hence, we may write (20.2.3) as

$$\sum_{r=1}^{n} \{m_r[\dot{x}^r(t) - \dot{x}^r(t_0)] - P^r\} \cdot \delta x^r = 0, \tag{21.2.6}$$

and this is merely another form of the fundamental equation.

When F^r is continuous or has only (isolated) *finite* discontinuities,

$$\lim_{t\to t_0} P^r = \lim_{t\to t_0} \int_{t_0}^{t} F^r\, d\tau = 0 \qquad (r = 1, 2, \ldots, n).$$

Then, if there are no constraints producing velocity discontinuities, one deduces from (21.2.6) that, in this case,

$$\lim_{\tau\to 0} \dot{x}^r(t_0 + \tau) = \lim_{\tau\to 0} \dot{x}^r(t_0 - \tau) = \dot{x}^r(t_0)$$

so that no information can be gained from (21.2.6) which is not also deducible from the original form of the fundamental equation.

The situation is different if m of the F^r are unbounded at t_0 in such a way that the

$$\lim_{t\to t_0} \int_{t_0}^{t} F^r\, d\tau = P^r \neq 0 \qquad (r = 1, 2, \ldots, m \leq n)$$

are bounded. Then, (21.2.6) becomes in this case

$$\sum_{r=1}^{n} \{m_r[\dot{x}^r(t_0 + 0) - \dot{x}^r(t_0 - 0)] - P^r\} \cdot \delta x^r = 0 \tag{21.2.7}$$

and this is the *fundamental equation of impulsive motion*. We note that in (21.2.7), we may put $m = n$ in the upper limit of the sum because the quantity in the square brackets and P^r vanish for every $r = m + 1, m + 2, \ldots, n$.

The situation is also different if there are constraints that impose velocity discontinuities, and this may occur in the presence, or in the absence, of impulsive forces. Such constraints are called impulsive constraints, and they will be discussed in the next section. The presence (or absence) of such constraints does not invalidate (21.2.7) any more than constraint equations invalidate the fundamental equation for (SN). It merely means that the δx^r are not entirely free but must satisfy certain conditions. However, in the absence of impulsive constraints, the velocity discontinuities are determined solely by (21.2.7).

We have seen (Sections 9.7 and 9.8) that *possible* velocity changes $\Delta \dot{x}^r$ (which need not be small) satisfy the same conditions as the δx^r. Therefore, we may write, in place of (21.2.7),

$$\sum_{r=1}^{n} \{m_r[\dot{x}^r(t_0 + 0) - \dot{x}^r(t_0 - 0)] - P^r\} \cdot \Delta \dot{x}^r = 0. \qquad (21.2.8)$$

As before, it will be convenient at times to write the equations in terms of the N displacement components u_s $(s = 1, 2, \ldots, N)$ instead of the $3n$ variables $x_1{}^r, x_2{}^r, x_3{}^r$ $(r = 1, 2, \ldots, n)$. Then, (21.2.8) becomes

$$\sum_{s=1}^{N} \{m_s[\dot{u}_s(t_0 + 0) - \dot{u}_s(t_0 - 0)] - P_s\} \Delta \dot{u}_s = 0. \qquad (21.2.9)$$

The properties of impulsive motion will be deduced from (21.2.8) or (21.2.9) according to whether it is more convenient to deal with 3-space or with N-space.

21.3. Impulsive Constraints

The problem of impulsive constraints is very complex and cannot be discussed fully here. We shall only discuss some types of impulsive constraints and indicate when and how they arise.

Consider a set of acatastatic constraints

$$\sum_{s=1}^{N} A_{rs}\, du_s + A_r\, dt = 0 \qquad (r = 1, 2, \ldots, L < n). \qquad (21.3.1)$$

Each of them gives rise to a constraint force

$$F_r' = \lambda_r(t)\Big[\sum_{s=1}^{N} A_{rs}\dot{u}_s + A_r\Big], \tag{21.3.2}$$

where $\lambda_r(t)$ is a Lagrange multiplier. We define:

A constraint is said to be impulsive if and only if its constraint force is impulsive.

Therefore, a given constraint is impulsive if and only if

$$I = \lim_{t\to t_0}\int_{t_0}^{t}\Big\{\lambda(t)\Big[\sum_{s=1}^{N} A_s(\tau)\dot{u}_s(\tau) + A(\tau)\Big]\Big\}\,d\tau = P' \neq 0, \tag{21.3.3}$$

where P' is bounded, and where we have suppressed the index r. When we wish to consider more than one constraint we shall restore the index.

We write, for short,

$$\sum_{s=1}^{N} A_s(t)\dot{u}_s(t) + A(t) = a(t) \tag{21.3.4}$$

and we use the notation

$$\int a(t)\,dt = B(t), \qquad \int \lambda(t)\,dt = \Lambda(t). \tag{21.3.5}$$

Then, integrating (21.3.3) by parts in two different ways, we find the two equations

$$I = a(t_0)[\Lambda(t_0+0) - \Lambda(t_0-0)] - \lim_{t\to t_0}\int_{t_0}^{t}\Lambda(\tau)a'(\tau)\,d\tau, \tag{21.3.6}$$

and

$$I = \lambda(t_0)[B(t_0+0) - B(t_0-0)] - \lim_{t\to t_0}\int_{t_0}^{t} B(\tau)\lambda'(\tau)\,d\tau. \tag{21.3.7}$$

In all that follows we shall suppose that the constraint is impulsive, i.e., $I = P^{r'} \neq 0$ is bounded. Then, (21.3.6) and (21.3.7) show that this may be caused by different circumstances.

(*a*) *Case I.* Assume that $\lambda(t)$ is discontinuous at t_0, but $a(t)$ is continuous. For simplicity, we assume $a(t)$ to be smooth as well although that conditions could be relaxed somewhat without affecting the results.

Under these assumptions

$$\lim_{t\to t_0}\int_{t_0}^{t} \Lambda(\tau)a'(\tau)\,d\tau = 0. \tag{21.3.8}$$

Then, it follows from (21.3.6) that

$$\Lambda(t_0+0) - \Lambda(t_0-0) = \lim_{t\to t_0}\int_{t_0}^{t} \lambda(\tau)\,d\tau = Q_1 \neq 0, \tag{21.3.9}$$

where Q_1 is a bounded constant; if it were otherwise, the constraint would not be impulsive, contrary to our initial assumption. But (21.3.9) implies that the Lagrange multiplier λ is unbounded at $t = t_0$. It follows that a constraint may be impulsive because the Lagrange multiplier is unbounded.

It is not difficult to construct examples of this type of impulsive constraint. Let a system be subject to the nonholonomic configuration constraint

$$f(x^1, x^2, \ldots, x^n, t) \leq 0,$$

and let

$$x_0 = (x_0{}^1, x_0{}^2, \ldots, x_0{}^n, t)$$

be the points for which

$$f(x_0{}^1, x_0{}^2, \ldots, x_0{}^n, t) = 0.$$

We now consider a motion such that in $t < t_0$, the configuration satisfies

$$f(x^1, x^2, \ldots, x^n, t) < 0,$$

but, at $t = t_0$, the configuration is

$$x(t_0) = (x_0{}^1, x_0{}^2, \ldots, x_0{}^n),$$

and the lower limit of the velocity

$$\dot{x}(t_0 - 0) = \big(\dot{x}^1(t_0 - 0), \dot{x}^2(t_0 - 0), \ldots, \dot{x}^n(t_0 - 0)\big)$$

does not lie in the tangent plane of $f = 0$. Then there is a discontinuous velocity change at t_0, i.e., the constraint force is impulsive. Since f is continuous and smooth, this impulsive constraint force must be due to a discontinuity in λ.

One can also see from purely physical arguments that λ must have a discontinuity at t_0. In $t < t_0$, the constraint was inactive; the motion might as well have been unconstrained. Now, the constraint force is

$$F' = \lambda \operatorname{grad} f$$

(see Section 9.3), and if that force is zero, $\lambda(t_0 - 0) = 0$ because grad $f \neq 0$. However, at $t = t_0 + 0$, the constraint "goes into action" so that $\lambda(t_0 + 0)$ is, then, not zero. Thus, there must be a discontinuity in λ at t_0.

The simplest case of the above example is that of a particle moving toward a fixed surface, striking it at $t = t_0$ with a velocity which is not in the tangent plane of the surface.

It should be noted that the possible velocity changes are no longer free, but they are still "largely arbitrary" because they have to satisfy only the condition that $\dot{x}^r(t_0 + 0)$ lie in the tangent plane of the surface; the direction in that surface is not prescribed.

(*b*) *Case II.* Next we assume that $a(t)$ is discontinuous at t_0, but λ is continuous and smooth. In that case

$$\lim_{t \to t_0} \int_{t_0}^{t} B(\tau)\lambda'(\tau)\, d\tau = 0, \tag{21.3.10}$$

and it follows from (21.3.7) that

$$B(t_0 + 0) - B(t_0 - 0) = \lim_{t \to t_0} \int_{t_0}^{t} a(\tau)\, d\tau = Q_2 \neq 0, \tag{21.3.11}$$

where Q_2 is a bounded constant; if it were not, the constraint would not be impulsive, contrary to our assumption. It follows that a constraint may be impulsive because $a(t)$ is unbounded at t_0.

Because of the definition of $a(t)$, given in (21.3.4), we now must consider

$$\lim_{t \to t_0} \int_{t_0}^{t} \left[\sum_{s=1}^{N} A_{rs}(\tau)\dot{u}_s(\tau) + A_r(\tau) \right] d\tau \qquad (r = 1, 2, \ldots, L < n). \tag{21.3.12}$$

We use the notation

$$\int A_{rs}(t)\, dt = E_{rs}(t), \qquad \int A_r(t)\, dt = E_r(t) \tag{21.3.13}$$

so that the expression in (21.3.12) is

$$\lim_{t \to t_0} \sum_{s=1}^{N} [E_{rs}(t) - E_{rs}(t_0)]\dot{u}_s + [E_r(t) - E_r(t_0)],$$

or

$$\sum_{s=1}^{N} [E_{rs}(t_0 + 0) - E_{rs}(t_0 - 0)]\dot{u}_s + [E_r(t_0 + 0) - E_r(t_0 - 0)]. \qquad (21.3.14)$$

Now it is not at all clear that all A_{rs} and all A_r are unbounded at t_0. We shall suppose that at some $t = t_j$,

$$E_{rs} \neq 0 \qquad (s = 1, 2, \ldots, L_1),$$
$$E_r \neq 0 \qquad (s = 1, 2, \ldots, L_2),$$

and we denote by L' the larger of L_1 and L_2.

Then the impulsive constraints to be considered are

$$\sum_{s=1}^{N} E_{rs}\dot{u}_s + E_r = 0 \qquad (r = 1, 2, \ldots, L'). \qquad (21.3.15)$$

It is not difficult to construct examples in which such constraints occur. Consider, for the moment, a constraint

$$f_r(u_1, u_2, \ldots, u_N, t) = 0.$$

For such a constraint, one has

$$A_{rs} = \partial f_r/\partial u_s, \qquad A_r = \partial f_r/\partial t.$$

Suppose now that some of the A_{rs} are unbounded, but all A_r are bounded. This implies that the constraint is, in fact, not holonomic, i.e., the surface is not smooth everywhere. Then there exists a subspace of this surface defined by some function

$$g_r(u_{10}, u_{20}, \ldots, u_{N0}, t) = 0,$$

where some of the $\partial g_r/\partial u_s$ do not exist. In other words, the surface has either a ridge or a discontinuity at the points $u_0 = (u_{10}, u_{20}, \ldots, u_{N0})$ for which $g_r = 0$. Now, if there is a motion such that

$$u_s(t_0) = u_{s0} \qquad (s = 1, 2, \ldots, N)$$

and the velocity $\dot{u}(t_0) = (\dot{u}_{10}(t_0), \dot{u}_{20}(t_0), \ldots, \dot{u}_{N0}(t_0))$ is not in the tangent plane of $g_r = 0$, there will be a velocity discontinuity at t_0. This type of impulsive constraint is called *inert*. A simple example is a particle moving at the time $t = t_0$ across a ridge of an otherwise smooth surface.

Suppose next that some of the A_r are unbounded, but all A_{rs} are bounded. Now, by virtue of the definition of f_r, the surface moves or changes in time. If some $A_r = \partial f_r/\partial t$ do not exist (A_r is unbounded) the surface motion has a discontinuity at t_0; the surface executes a "jump" at that instant. Such an impulsive constraint is called *live.*†

One can establish a connection between live constraints and the case when the multiplier is unbounded by imagining a constraint surface that moves without resistance with the representative point in configuration space until $t = t_0$, when it becomes suddenly rigid. In this manner, the case $\lambda(t_0) = \infty$ is regarded as a special case of a live constraint. However, this view seems artificial because it supposes the existence of a constraint in $t < t_0$ which is completely inactive in that interval.

21.4. The Fundamental Equation with Impulsive Constraints

It is now evident that impulsive motion may occur either because of impulsive constraints (i.e., *constraint* forces are impulsive) or because *given* forces are impulsive, or because of a combination of the two.

When given forces are impulsive, and the constraint forces are not, the discontinuous velocities are governed by

$$\sum_{s=1}^{N} \{m_s[\dot{u}_s(t+0) - \dot{u}_s(t-0)] - P_s\} \, \Delta\dot{u}_s = 0, \tag{21.4.1}$$

where the $\Delta\dot{u}_s$ are bounded, but are otherwise completely free.

When there are no impulsive given forces, but impulsive motion takes place, nevertheless, it is due to impulsive constraint forces. In that case, the discontinuous velocity changes are governed by

$$\sum_{s=1}^{N} \{m_s[\dot{u}_s(t+0) - \dot{u}_s(t-0)]\} \, \Delta\dot{u}_s = 0, \tag{21.4.2}$$

and the possible velocity changes $\Delta\dot{u}_s$ are not entirely free. In the case of *inert* constraints, they must satisfy

$$\sum_{s=1}^{N} E_{rs} \, \Delta\dot{u}_s = 0 \qquad (r = 1, 2, \ldots, L). \tag{21.4.3}$$

Thus, while not entirely free, they are largely arbitrary.

† The terms "live" and "inert" are quite descriptive of the physical cause of these impulsive constraints. However, when an impulsive constraint is due to a "jump" of a surface with a "ridge," and the ridge is reached at the instant of the jump, one must conclude that the impulsive constraint is both live and inert.

When the impulsive constraints are *live*, the $E_{rs} = 0$ for all r and s. Then, it follows from (21.3.15) that the possible velocities are not constrained; hence, the $\Delta \dot{u}_s$ are free in that case. However, before and after the impulse the possible velocities are subject to the nonimpulsive constraints

$$\sum_{s=1}^{N} A_{rs}\dot{u}_s + A_r = 0 \qquad (r = 1, 2, \ldots, L)$$

for all $t < t_0$ and $t > t_0$, where t_0 is the instant when the impulsive constraint force acts.

In the general case where impulsive, given, and constraint forces act simultaneously, the actual velocity changes are governed by (21.4.1), while the possible velocity changes are governed by (21.4.3) when the constraints are inert.

21.5. Impulsive Motion Theorems

In this section we derive some general properties of impulsive motion. As may be expected, these deal largely with kinetic energy changes because the effect of impulsive motion is to produce velocity changes. Because impulsive constraints do not occur in this section, it is convenient to deal with the position vectors x^r rather than the u_s, and with the fundamental equation in the form (21.2.8) rather than (21.2.9).

To simplify the notation, we use in this section the convention

$$\begin{aligned} \dot{x}^r(t_0 - 0) &= \dot{x}_0{}^r, \\ \dot{x}^r(t_0 + 0) &= \dot{x}^r. \end{aligned} \tag{21.5.1}$$

Thus, $\dot{x}_0{}^r$ is the velocity "just prior to the impulse," and $\dot{x}^r$ is the velocity "just after the impulse."

We first prove for impulsive motion *Gauss' principle of least constraint*:

The quantity

$$Z = \frac{1}{2} \sum_{r=1}^{n} [m_r(\dot{x}^r - \dot{x}_0{}^r - P^r/m_r)^2] \tag{21.5.2}$$

is a minimum for the actual velocity $\dot{x}^r$ relative to all other possible velocities $\dot{\bar{x}}^r$.

To prove this theorem, we compute the quantity

$$\Delta Z = \bar{Z} - Z, \tag{21.5.3}$$

where

$$\bar{Z} = Z(\dot{\bar{x}}^r) = Z(\dot{x}^r + \Delta\dot{x}^r) \tag{21.5.4}$$

and $\Delta\dot{x}^r$ is a possible velocity change. Then

$$\Delta Z = \frac{1}{2}\sum_{r=1}^{n}\left\{m_r\left(\dot{x}^r + \Delta\dot{x}^r - \dot{x}_0{}^r - \frac{P^r}{m_r}\right)^2\right\} - \left(\dot{x}^r - \dot{x}_0{}^r - \frac{P^r}{m_r}\right)^2\Bigg\}$$

$$= \sum_{r=1}^{n} m_r\left(\dot{x}^r - \dot{x}_0{}^r - \frac{P^r}{m_r}\right)\cdot\Delta\dot{x}^r + \frac{1}{2}\sum_{r=1}^{N} m_r(\Delta\dot{x}^r)^2.$$

But, by (20.2.8), the first sum is zero, or

$$\Delta Z = \frac{1}{2}\sum_{r=1}^{n} m_r(\Delta\dot{x}^r)^2.$$

This quantity is positive for all possible $\Delta\dot{x}^r$ so that $\bar{Z}$ exceeds Z, which was to be shown.

The quantity Z is called the "constraint" of impulsive motion; this is an unfortunate term since our theorem holds whether the system is constrained or not, and whether the impulsive motion is due to given forces or to constraints.

We shall now deduce from this principle an equation which is particularly useful in the solution of problems. The principle states

$$Z = \frac{1}{2}\sum_{r=1}^{n} m_r\left(\dot{x}^r - \dot{x}_0{}^r - \frac{P^r}{m_r}\right)^2 = \text{min}.$$

On expanding this expression, one has

$$Z = T + T_0 + \frac{1}{2}\sum_{r=1}^{n}\frac{(P^r)^2}{m_r} - \sum_{r=1}^{n} m_r\dot{x}^r\dot{x}_0{}^r - \sum_{r=1}^{n}(\dot{x}^r - \dot{x}_0{}^r)P^r = \text{min},$$

where

$$T = \frac{1}{2}\sum_{r=1}^{n} m_r(\dot{x}^r)^2$$

is the kinetic energy after impulse, and

$$T_0 = \frac{1}{2}\sum_{r=1}^{n} m_r(\dot{x}_0{}^r)^2$$

is the kinetic energy prior to the impulse. A necessary condition for Z to be a minimum is that $dZ = 0$, or

$$\frac{\partial T}{\partial \dot{x}^r} - m_r \dot{x}_0{}^r - P^r = 0 \qquad (r = 1, 2, \ldots, n). \tag{21.5.5}$$

This equation is, in general, sufficient to deduce the velocities immediately following the impulse.

Example 21.5.1. Two uniform rods, each of mass M and length l, are smoothly hinged together at one extremity. They lie on a smooth, horizontal table, as shown in Fig. 21.5.1. Originally they are not in a straight line, but at the instant t_1 they are suddenly put into a straight line. Find the motion immediately after the impulse.

Let the extremities of the rods be AB and BC. They are shown immediately after the impulse in Fig. 20.5.1. Let the velocity of the rod end A normal to AB be p, and let q be the velocity of B, also normal to the rod, but in the opposite direction from p; the velocity of AB, i.e., of any point of AB, in the direction of the rod is r. The velocity at C normal to BC is s.

It will be noted that, when $P^r = 0$, the quantity Z is the kinetic energy of the velocity difference immediately before and after the impulse. The kinetic energy of AB is (see Problem 7.1)

$$\begin{aligned} T_{AB} &= \tfrac{1}{6}M(p \cdot p - p \cdot q + q \cdot q) + \tfrac{1}{2}Mr \cdot r \\ &= \tfrac{1}{6}M(p^2 - pq + q^2 + 3r^2) \end{aligned}$$

and that of BC is

$$T_{BC} = \tfrac{1}{6}M(q^2 - qs + s^2 + 3r^2).$$

Hence

$$\begin{aligned} z = \tfrac{1}{6}M[&(p - p_0)^2 + (s - s_0)^2 + 2(q - q_0)^2 + 6(r - r_0)^2 \\ &- (p - p_0)(q - q_0) - (s - s_0)(q - q_0)]. \end{aligned} \tag{a}$$

The constraint requires that the angular velocities of AB and BC be the same, i.e.,

$$s + 2q = 0. \tag{b}$$

Then a necessary condition for Z to be a minimum is that $dZ = 0$ under the constraint (b). One finds

$$\begin{aligned} 2(p - p_0) - (q - q_0) &= 0, \\ 4(q - q_0) - (p - p_0) - (s - s_0) + 2\lambda &= 0, \\ 12(r - r_0) &= 0, \\ 2(s - s_0) - (q - q_0) + \lambda &= 0 \end{aligned} \tag{c}$$

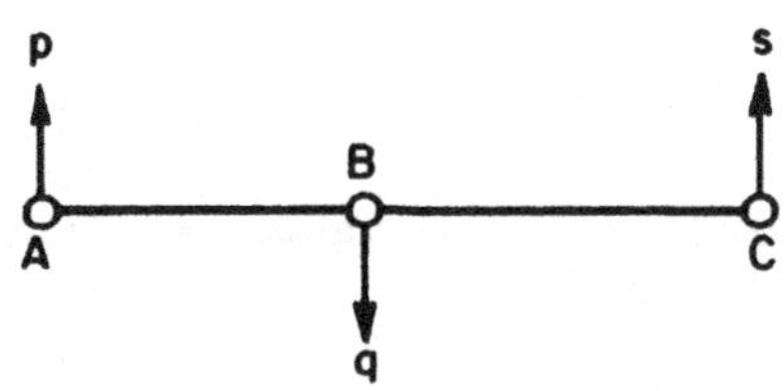

Fig. 21.5.1. Hinged links of Example 21.5.1.

together with (b). Solving (b) and (c) simultaneously, one has

$$p = p_0 - \frac{10}{31} q_0 - \frac{5}{31} s_0,$$

$$q = \frac{11}{31} q_0 - \frac{10}{31} s_0,$$

$$r = r_0,$$

$$s = -\frac{22}{31} q_0 + \frac{20}{31} s_0.$$

This is the required answer since it gives the velocities immediately after the impulse in terms of those immediately preceding the impulse.

Next we state the

Superposition Theorem. Let the velocity $\dot{x}^r$ *after an impulse* P_i^r *be denoted by* $\dot{x}_i^r$. *Then, the velocity after k simultaneous impulses* $P^r{}_i$ $(i = 1, 2, \ldots, k)$ *is given by*

$$\dot{x}^r = \sum_{i=1}^{k} \dot{x}_i^r,$$

and it is the same as that from a single impulse

$$P^r = \sum_{i=1}^{k} P_i^r.$$

The simple proof of this theorem is left as an exercise.

We come now to some additional theorems. The first of these is the

Energy Theorem. The increase in kinetic energy of a system due to the impulses $P^r(t_0)$ $(r = 1, 2, \ldots, n)$ *is equal to*

$$\sum_{r=1}^{n} P^r \cdot (\dot{x}^r + \dot{x}_0^r)/2. \tag{21.5.7}$$

Let us denote the kinetic energy "immediately preceding" the impulses by

$$T_0 = T(t_0 - 0)$$

and that "immediately following" the impulses by

$$T = T(t_0 + 0).$$

Then Theorem (21.5.7) asserts that the increase in kinetic energy equals the scalar product of each impulse with the corresponding mean velocity before and after the impulse.

The proof is simple and is an example of the type of argument used in most proofs to follow in this section. One uses the fundamental equation and chooses for $\Delta\dot{x}^r$ a quantity which belongs to the class of possible velocity changes, and which is so chosen as to result in the particular property to be demonstrated. In Theorem (21.5.7) the impulses are free, hence the $\Delta\dot{x}^r$ are unrestricted. Thus we may select

$$\Delta\dot{x}^r = \dot{x}^r + \dot{x}_0^r.$$

For that choice of $\Delta\dot{x}^r$, the fundamental equation becomes

$$\sum_{r=1}^{n} \{m_r(\dot{x}^r - \dot{x}_0^r) \cdot (\dot{x}^r + \dot{x}_0^r)\} = \sum_{r=1}^{n} P^r \cdot (\dot{x}^r + \dot{x}_0^r),$$

or

$$\sum_{r=1}^{n} m_r(\dot{x}^r)^2 - \sum_{r=1}^{n} m_r(\dot{x}_0^r)^2 = \sum_{r=1}^{n} P^r \cdot (\dot{x}^r + \dot{x}_0^r).$$

The left-hand side is twice the kinetic energy change, so that

$$T - T_0 = \sum_{r=1}^{n} P^r \cdot (\dot{x}^r + \dot{x}_0^r)/2, \tag{21.5.8}$$

as claimed.

For the system initially at rest, one finds the special result

$$T = \frac{1}{2} \sum_{r=1}^{n} P^r \cdot \dot{x}^r. \tag{21.5.9}$$

Next, we state a famous theorem known as

Bertrand's Theorem.† *If a system at rest is subjected to an impulse, the kinetic energy of the subsequent motion is greater than if the system had been constrained and subjected to the same impulse.* (21.5.10)

For the proof, let T be the kinetic energy after the impulse of the unconstrained system, and let T_1 be the corresponding quantity of the constrained system. Bertrand's theorem asserts that $T > T_1$.

† The author of this theorem is the same person who proved the famous theorem on central force motion discussed in Section 18.4. However, this theorem may actually be due to J. C. F. Sturm.

For the *unconstrained* system starting from rest, the fundamental equation is

$$\sum_{r=1}^{n} m_r \dot{x}^r \cdot \Delta\dot{x}^r = \sum_{r=1}^{n} P^r \cdot \Delta\dot{x}^r. \tag{21.5.11}$$

If we denote the velocity acquired by the *constrained* system by $\dot{x}_1{}^r$, the corresponding equation of that system is

$$\sum_{r=1}^{n} m_r \dot{x}_1{}^r \cdot \Delta\dot{x}^r = \sum_{r=1}^{n} P^r \cdot \Delta\dot{x}^r. \tag{21.5.12}$$

We now choose $\Delta\dot{x}^r = \dot{x}_1{}^r$. This is surely an admissible velocity change for both systems. Substituting this velocity change in both (21.5.11) and (21.5.12), and subtracting the latter from the former, we find

$$\sum_{r=1}^{n} m_r \dot{x}_1{}^r(\dot{x}^r - \dot{x}_1{}^r) = 0 \tag{21.5.13}$$

because P^r is the same in both cases. We now substitute in (21.5.13) the identity

$$\dot{x}_1{}^r(\dot{x}^r - \dot{x}_1{}^r) = \tfrac{1}{2}[(\dot{x}^r)^2 - (\dot{x}_1{}^r)^2 - (\dot{x}^r - \dot{x}_1{}^r)^2]$$

and utilize the relations

$$T = \frac{1}{2}\sum_{r=1}^{n} m_r(\dot{x}^r)^2,$$

$$T_1 = \frac{1}{2}\sum_{r=1}^{n} m_r(\dot{x}_1{}^r)^2.$$

We then find

$$T - T_1 = R_1 = \frac{1}{2}\sum_{r=1}^{n} m_r(\dot{x}^r - \dot{x}_1{}^r)^2 > 0, \tag{21.5.14}$$

which was to be shown.

A theorem closely related to that of Bertrand is

Kelvin's Theorem.† *If any points of a connected system are suddenly set into motion with prescribed velocity, the kinetic energy of the resulting motion is a minimum relative to other possible motions of the system when the given points have the same prescribed velocity.* (21.5.15)

† Bertrand's and Kelvin's theorems are stated essentially in the form given in Easthope, C. E., *Three-Dimensional Dynamics*, Academic Press, Inc., New York and London (1958).

Let the system start from rest; then the fundamental equation is again given by

$$\sum_{r=1}^{n} m_r \dot{x}^r \cdot \Delta \dot{x}^r = \sum_{r=1}^{n} P^r \cdot \Delta \dot{x}^r. \tag{21.5.16}$$

Now, let the particles P_r $(r = 1, 2, \ldots, n)$ be those subjected to impulses P^r, and let their velocities be prescribed, in accordance with the theorem; the velocities of the remaining $n - m$ particles are not prescribed. Also, let the velocity vector

$$\dot{x} = (\dot{x}^1, \dot{x}^2, \ldots, \dot{x}^m, \dot{x}^{m+1}, \ldots, \dot{x}^n) \tag{21.5.17}$$

be that of the actual motion after impact, and let

$$\dot{x}_2 = (\dot{x}^1, \dot{x}^2, \ldots, \dot{x}^m, \dot{x}_2^{m+1}, \ldots, \dot{x}_2{}^n) \tag{21.5.18}$$

be another possible velocity for which the first m particles have the same velocity as in $\dot{x}$. We now choose as a possible velocity change

$$\Delta \dot{x}^r = \dot{x}^r - \dot{x}_2{}^r \qquad (r = 1, 2, \ldots, n). \tag{21.5.19}$$

Then, the fundamental equation may be written as

$$\sum_{r=1}^{n} m_r \dot{x}^r \cdot (\dot{x}^r - \dot{x}_2{}^r) = \sum_{r=1}^{m} P^r \cdot (\dot{x}^r - \dot{x}_2{}^r) + \sum_{r=m+1}^{n} P^r \cdot (\dot{x}^r - \dot{x}_2{}^r). \tag{21.5.20}$$

Now, it is easy to see that the sums on the right vanish identically. The first of them vanishes because, for $r = 1, 2, \ldots, m$, one has $\dot{x}^r = \dot{x}_2{}^r$; the second vanishes because, for $r = m + 1, m + 2, \ldots, n$, the impulses $P^r = 0$. Thus, (21.5.20) is

$$\sum_{r=1}^{n} m_r \dot{x}^r \cdot (\dot{x}^r - \dot{x}_2{}^r) = 0. \tag{21.5.21}$$

We now make use of an identity similar to that used in the proof of Bertrand's theorem; it is

$$\dot{x}^r \cdot (\dot{x}^r - \dot{x}_2{}^r) = \tfrac{1}{2}[(\dot{x}^r)^2 - (\dot{x}_2{}^r)^2 + (\dot{x}^r - \dot{x}_2{}^r)^2].$$

It results in

$$T_2 - T_1 = R_2 = \frac{1}{2} \sum_{r=1}^{n} m_r (\dot{x}^r - \dot{x}_2{}^r)^2 > 0, \tag{21.5.22}$$

which was to be shown.

G. I. Taylor answered the question as to how the gain in Kelvin's theorem is related to the loss in Bertrand's theorem by stating:

Taylor's Theorem. The gain R_2 in Kelvin's theorem is greater than the loss R_1 in Bertrand's theorem. (21.5.23)

To prove it, we note that, when we put $\Delta\dot{x}^r = \dot{x}_1{}^r$ in Bertrand's theorem, we found

$$\sum_{r=1}^{n} m_r \dot{x}_1{}^r \cdot (\dot{x}^r - \dot{x}_1{}^r) = 0. \tag{21.5.24}$$

However, $\Delta\dot{x}^r = \dot{x}_2{}^r$ is also a possible velocity change in Bertrand's theorem; thus, one also has

$$\sum_{r=1}^{n} m_r \dot{x}_2{}^r \cdot (\dot{x}^r - \dot{x}_1{}^r) = 0. \tag{21.5.25}$$

Substracting (21.5.24) from (21.5.25) gives

$$\sum_{r=1}^{n} m_r (\dot{x}^r - \dot{x}_1{}^r) \cdot (\dot{x}_2{}^r - \dot{x}_1{}^r) = 0. \tag{21.5.26}$$

This last equation may be written as

$$\sum_{r=1}^{n} m_r (\dot{x}_1{}^r - \dot{x}^r) \cdot [(\dot{x}_1{}^r - \dot{x}^r) - (\dot{x}_2{}^r - \dot{x}^r)] = 0.$$

If we write $\dot{x}_1{}^r - \dot{x}^r = u$, and $\dot{x}_2{}^r - \dot{x}^r = v$, the factor multiplying m_r in (21.5.26) is $\frac{1}{2}[u^2 - v^2 + (u - v)^2]$. Thus, utilizing (21.5.14) and (21.5.22), the equation (21.5.26) may be rewritten as

$$R_2 - R_1 = R_{12},$$

where

$$R_{12} = \frac{1}{2} \sum_{r=1}^{n} m_r (\dot{x}_2{}^r - \dot{x}_1{}^r)^2 > 0,$$

which was to be shown.

We shall show the application of several of these theorems by an example.

Example 21.5.2. (Pars, pp. 243–245). A chain of three uniform equal rods AB, BC, CD, each having total mass M, is smoothly hinged and lies in a straight line on a smooth horizontal table. An impulse P is applied to the point A at one extremity of the chain in the direction normal to the chain, and in the plane of the table. Discuss the motion.

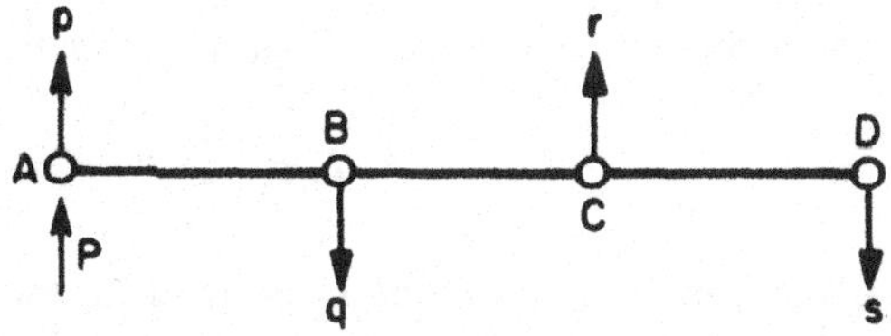

Fig. 21.5.2. Hinged links of Example 21.5.2.

Consider Fig. 21.5.2. To determine the motion immediately following the impulse, we use Gauss' theorem. Using the method of the previous example, we find

$$\begin{aligned} T &= \tfrac{1}{6} M[(p^2 - pq + q^2) + (q^2 - qr + r^2) + (r^2 - rs + s^2)] \\ &= \tfrac{1}{6} M(p^2 - pq + 2q^2 - qr + 2r^2 - rs + s^2). \end{aligned} \tag{a}$$

We now apply (21.5.5). Then, the necessary conditions in this example are

$$\begin{aligned} \frac{1}{6} M\left[(2p - q) - \frac{6P}{M}\right] &= 0, \\ -p + 4q - r &= 0, \\ -q + 4r - s &= 0, \\ -r + 2s &= 0. \end{aligned} \tag{b}$$

Solving these equations simultaneously we find

$$\begin{aligned} p &= \frac{52}{15}\frac{P}{M}, \\ q &= \frac{14}{15}\frac{P}{M}, \\ r &= \frac{4}{15}\frac{P}{M}, \\ s &= \frac{2}{15}\frac{P}{M}. \end{aligned} \tag{c}$$

To calculate the energy created we use (21.5.8), i.e.,

$$T = \frac{1}{2} P \cdot p = \frac{1}{2} \cdot \frac{52}{15}\frac{P^2}{M} = \frac{26}{15}\frac{P^2}{M}. \tag{d}$$

Let us suppose that, in a second experiment, the point D is held fixed, so that $s = 0$. In that case, we find instead of (c) the first three of equations (b) with $s = 0$, and the last is absent. This gives

$$\begin{aligned} p &= \frac{45}{13}\frac{P}{M}, \\ q &= \frac{12}{13}\frac{P}{M}, \\ r &= \frac{3}{13}\frac{P}{M}. \end{aligned} \tag{e}$$

Then, the energy created in this new situation is, instead of (d),

$$T_1 = \frac{1}{2} P \cdot p = \frac{45}{26} \frac{P^2}{M}. \tag{f}$$

If the impulse is the same in both cases, the energy loss is

$$T - T_1 = \left(\frac{26}{15} - \frac{45}{26}\right) \frac{P^2}{M} = \frac{1}{390} \frac{P^2}{M}. \tag{g}$$

If p is denoted by p_0 in the "free" experiment, i.e.,

$$p_0 = \frac{52}{15} \frac{P}{M},$$

we have

$$T - T_1 = \frac{15}{70{,}304} M p_0^2.$$

If $p = p_0$ in both experiments,

$$T_2 - T = \frac{1}{2} \left(\frac{13}{45} - \frac{15}{52}\right) M p_0^2 = \frac{1}{4680} M p_0^2.$$

The latter gain exceeds the previous loss by

$$T_1 + T_2 - 2T = \left(\frac{1}{4680} - \frac{15}{70{,}304}\right) M p_0^2 = \frac{1}{3{,}163{,}680} M p_0^2.$$

As a final theorem, we give a famous theorem which applies to systems in which the impulsive motion is due to impulsive constraints, not to impulsive given forces. This is:

Carnot's Theorem. The energy change due to an impulsive (inert) constraint is always a loss of energy. (21.5.27)

To prove this theorem, we write the fundamental equation for no impulses. It is

$$\sum_{r=1}^{n} m_r (\dot{x}^r - \dot{x}_0{}^r) \cdot \Delta \dot{x}^r = 0, \tag{21.5.28}$$

and we choose for $\Delta \dot{x}^r$ the (certainly possible) value $\dot{x}^r$; thus, we have

$$\sum_{r=1}^{n} m_r \dot{x}^r \cdot (\dot{x}^r - \dot{x}_0{}^r) = 0 \tag{21.5.29}$$

or, in view of the by now familiar identity, (21.5.29) becomes

$$T_0 - T = R_0$$

where

$$R_0 = \frac{1}{2} \sum_{r=1}^{n} m_r(\dot{x}^r - \dot{x}_0{}^r)^2 > 0, \tag{21.5.30}$$

which was to be shown. This loss in energy is equal to the quantity ΔZ, computed in the proof of Gauss' theorem, when $\Delta \dot{x}^r$ is the difference $\dot{x}^r - \dot{x}_0{}^r$.

21.6. Lagrange's Equations for Impulsive Motion

Lagrange's equation for impulsive motion is found by integrating Lagrange's equations

$$\frac{d}{dt}\frac{\partial T}{\partial \dot{q}_s} - \frac{\partial T}{\partial q_s} - Q_s = 0 \qquad (s = 1, 2, \ldots, n),$$

where n is the number of generalized coordinates, over the time interval $t_0 \leq t \leq t_1$ and letting $t_1 \to t_0$ (for simplicity, we omit constraints). Let

$$R_s = \lim_{t_1 \to t_0} \int_{t_0}^{t_1} Q_s \, dt. \tag{21.6.1}$$

If that integral exists and is not zero, there will be velocity discontinuities at t_0, and we find

$$\lim_{t_1 \to t_0} \int_{t_0}^{t_1} \left(\frac{d}{dt}\frac{\partial T}{\partial \dot{q}_s} - \frac{\partial T}{\partial q_s} \right) dt = \Delta\left(\frac{\partial T}{\partial \dot{q}_s} \right), \tag{21.6.2}$$

where $\Delta(\partial T/\partial \dot{q}_s)$ is the discontinuous change in the momentum component $p_s = \partial T/\partial \dot{q}_s$ at the instant t_0. Therefore, Lagrange's equations of impulsive motion are

$$\Delta\left(\frac{\partial T}{\partial \dot{q}_s} \right) = R_s \qquad (s = 1, 2, \ldots, n). \tag{21.6.3}$$

These equations were already derived in (21.5.5), in which

$$\Delta\left(\frac{\partial T}{\partial \dot{q}_s} \right) = \frac{\partial T}{\partial \dot{x}^r} - m_r \dot{x}_0{}^r.$$

Therefore, Example 21.5.1 is, in fact, an application of Lagrange's equations to impulsive motion.

21.7. Problems

21.1. A homogeneous rod of length l and mass m lies on a smooth horizontal table. It is struck a blow in the plane of the table a distance nl from one end ($0 < n < 1$), and in a direction normal to the rod. The blow is such that it would give a velocity v to a particle at rest of mass m. Find the velocities of the ends of the rod immediately after the blow is struck.

21.2. A heavy uniform rod of mass m and length l hangs vertically from a smooth pin a distance nl from one end ($0 < n < 1$). What is the smallest blow that the rod can be struck at the bottom at right angles to the bar which will just make the bar reach the inverted position?

21.3. Two homogeneous rods OA and AB have equal mass m, but OA has length l_1 and AB has length l_2. They are smoothly hinged at A and lie on a smooth horizontal table. The rod OA is smoothly hinged to a vertical pin at O. Initially, the rods form a straight line. An impulse P is applied normal to OA and in the plane of the table to a point C lying between O and A. Calculate the reactions at the hinge pins at O and A and the angular velocities of the two rods immediately after the impulse.

21.4. Two equal heavy, homogeneous rods OA and AB of mass m and length l are smoothly hinged at A, and OB is smoothly hinged at O to a fixed point. The rod AB has fixed to it at B a sandbox of mass M and of negligible dimension. This mechanism hangs at rest when a bullet of mass μ is shot with horizontal velocity v into the sandbox as shown. Find the motion of this mechanism immediately after the impact.

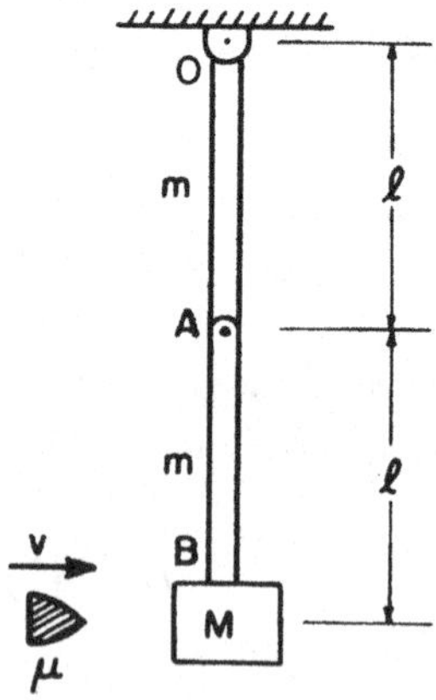

21.5. A box of dimensions $AB = 2a$, $AD = b$, $AA' = 2c$ rests on a smooth horizontal table. It is filled with sand, and the mass of box and sand is M. The density of the material of the box is the same as that of the sand. A bullet of mass m is fired with velocity v into the center of the face $AA'BB'$, as shown. Calculate the kinetic energy imparted to the box by the bullet when the edge DD' is fixed, and compare it to the kinetic energy when DD' is not fixed.

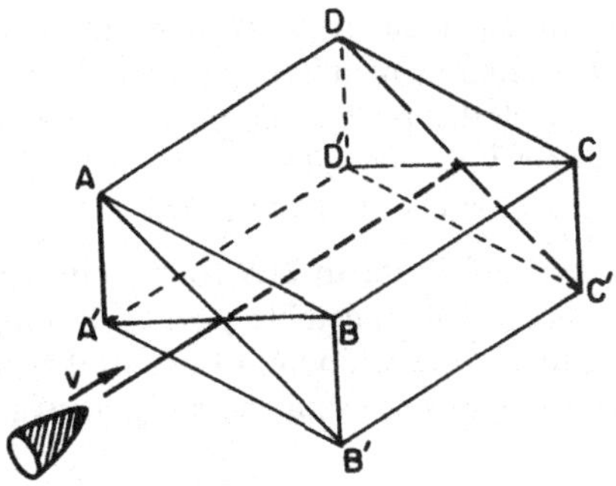

21.6. Four equal, homogeneous rods are smoothly hinged together to form a rhombus which is initially at rest on a smooth horizontal table. Let an impulsive force acting on a hinge point along one of the diagonals give the hinge point a velocity v as shown. Find the angular velocity of the rods immediately after the impact.

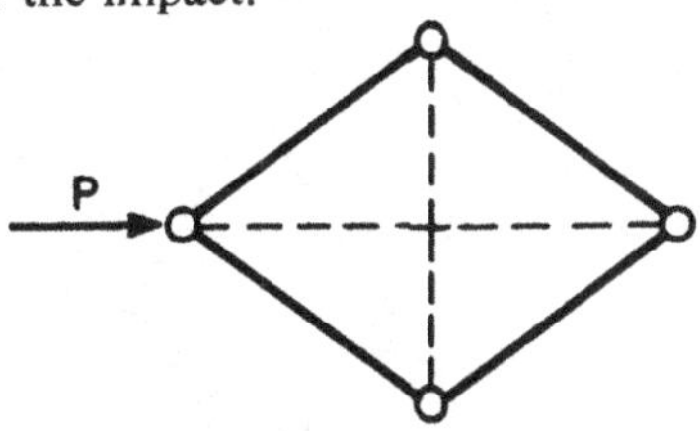

21.7. Two equal homogeneous rods AB and BC of length l and mass m, smoothly hinged at B, lie on a smooth horizontal table so that the angle ABC is a right angle as shown. An impulse in the plane of the table is applied at A so that its line of action intersects BC at a point D between B and C, and $BD = nl$ $(0 < n < 1)$. What is the initial motion of the rods?

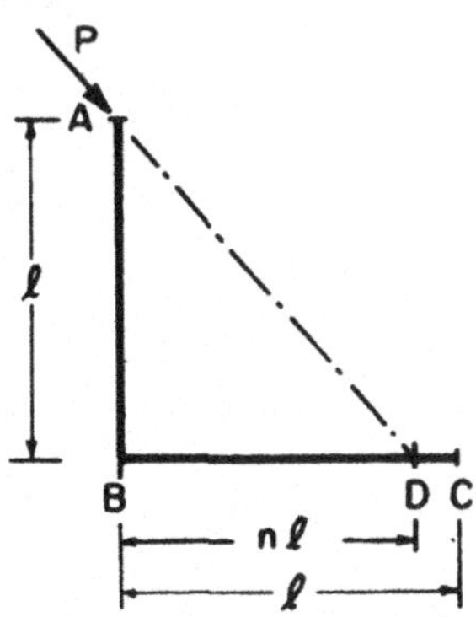

21.8. Three heavy equal homogeneous rods AB, BC, CD are smoothly hinged at B and C. This assembly is suspended at A and D from smooth pins so that the distance between A and D is twice the rod length, and so that all rods are in a vertical plane. An impulse is applied at B in such a way as to give B a velocity v toward D. Find the initial motion.

21.9. Four equal, homogeneous rods are smoothly hinged together at their ends so as to form a square which lies on a smooth, horizontal table. An impulse P in the plane of the table is applied to one of the rods in a direction making an angle $\theta \neq \pi/2$ with one of the rods. Find the kinetic energy of the rods immediately after the impulse.

21.10. Let the four rods of Problem 21.9 rotate with constant angular velocity ω about a line normal to the table passing through the intersection of the diagonals. Suddenly the midpoint of one of the rods is fixed. Show that the angular velocity of this rod remains the same while that of an adjacent rod is $2\omega/5$.

21.11. Let the four rods of Problem 21.9 form a rhombus which rotates in its own plane about one of its corners as shown. Evidently the rhombus will "open up." At the instant when it forms a square, the fixed corner A is set free and the midpoint of CD is fixed. Show that the loss of kinetic energy is $26ml^2\omega^2/5$, where m is the mass and $2l$ is the length of each rod.

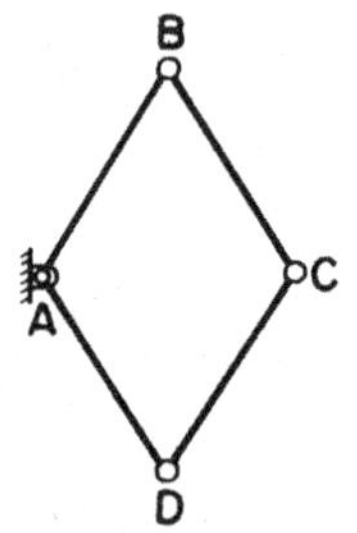

Bibliography

Sources which were used only occasionally in the writing of this book have been given in footnotes. Here we list, in alphabetical order of their authors, frequently used sources and texts which are considered helpful to a study of Lagrangean mechanics.

BYERLY, WILLIAM E., *An Introduction to the Use of Generalized Coordinates in Mechanics and Physics*, Dover Publications, Inc., New York (1965) (originally published 1916).

GOLDSTEIN, HERBERT, *Classical Mechanics*, Addison-Wesley Publishing Co., Reading, Massachusetts (1950).

HAMEL, GEORG, *Theoretische Mechanik*, Springer-Verlag, Berlin (1949).

LAINÉ, E., *Exercises de mécanique*, Librairie Vuibert, Paris (1958).

LANDAU, L. D., and LIFSHITZ, E. M., *Mechanics*, Pergamon Press, Oxford–London–New York–Paris (1960).

LUR'E, A. I., *Analytical Mechanics* (in Russian), Gosudarstvennoe Izdat'elstvo Fiziko-Mat'emat'iskoi Literaturi, Moscow (1961).†

OSGOOD, WILLIAM, F., *Mechanics*, The Macmillan Company, New York (1948) (originally published 1937).

PARS, L. A., *A Treatise on Analytical Dynamics*, John Wiley and Sons, Inc., New York (1968).

SYNGE, JOHN L., "Classical Dynamics" in *Encyclopedia of Physics* **3**/1 (S. Flügge, ed.), Springer, Göttingen–New York (1960).

WEBSTER, ARTHUR, G., *The Dynamics of Particles and of Rigid, Elastic, and Fluid Bodies*, Dover Publications, Inc., New York (1959) (originally published 1904).

WHITTAKER, E. T., *A Treatise on the Analytical Dynamics of Particles and Rigid Bodies*, Dover Publications (1944) (originally published 1904).

† French translation *Mécanique Analytique*, Librairie Universitaire, Louvain, 1968; author erroneously listed as L. Lur'e.

Index

About the author

Reinhardt M. Rosenberg is Professor of Applied Mechanics at the University of California, Berkeley. He received his D.h.c. degree from the University of Besançon, France, his M.S. degree from Purdue, and his undergraduate degree from the University of Pittsburgh. He continues to contribute extensively to the fields of applied mechanics and aeronautical engineering. Professor Rosenberg also has a strong interest in the area of bioengineering.

Made in the USA
Las Vegas, NV
01 January 2026

38283061R00247